T0292619

Studies in Fuzziness and Soft Computing

Volume 339

Series editor

Janusz Kacprzyk, Polish Academy of Sciences, Warsaw, Poland
e-mail: kacprzyk@ibspan.waw.pl

About this Series

The series "Studies in Fuzziness and Soft Computing" contains publications on various topics in the area of soft computing, which include fuzzy sets, rough sets, neural networks, evolutionary computation, probabilistic and evidential reasoning, multi-valued logic, and related fields. The publications within "Studies in Fuzziness and Soft Computing" are primarily monographs and edited volumes. They cover significant recent developments in the field, both of a foundational and applicable character. An important feature of the series is its short publication time and world-wide distribution. This permits a rapid and broad dissemination of research results.

More information about this series at http://www.springer.com/series/2941

Tomasa Calvo Sánchez · Joan Torrens Sastre
Editors

Fuzzy Logic and Information Fusion

To Commemorate the 70th Birthday
of Professor Gaspar Mayor

 Springer

Editors
Tomasa Calvo Sánchez
University of Alcalá
Alcalá de Henares, Madrid
Spain

Joan Torrens Sastre
University of Balearic Islands
Palma, Islas Baleares
Spain

ISSN 1434-9922   ISSN 1860-0808   (electronic)
Studies in Fuzziness and Soft Computing
ISBN 978-3-319-30419-9   ISBN 978-3-319-30421-2   (eBook)
DOI 10.1007/978-3-319-30421-2

Library of Congress Control Number: 2016933220

Printed on acid-free paper

This Springer imprint is published by Springer Nature
The registered company is Springer International Publishing AG Switzerland

To
Gaspar Mayor Forteza

Preface

This book has been written to commemorate the 70th birthday of the Prof. Gaspar Mayor, a full Professor at the University of the Balearic Islands (UIB) and a prolific researcher in the areas of fuzzy sets and information aggregation. The book includes 14 chapters which deal with different topics on fuzzy logic, aggregation functions, multidistances, fuzzy consensus models, fuzzy decision models, and so on. Most of them are related to some publications of Prof. Gaspar Mayor or to his area of interest. Nowadays, there are several monographs and books related to the topics of this book, but in essence they are rather different. The book contains nice and original contributions or reviews, authored by some of the most outstanding researchers in the field, for this reason it will be very useful both for the novel researchers and the seniors. The book also contains some chapters focussed on different fields of application of the mentioned areas. In general, the target audience of this book are computer scientists, knowledge engineers and decision scientists, as well as mathematicians.

The title of the book agrees with the one of the research groups of Prof. Mayor, that is, Fuzzy Logic and Information Fusion (LOBFI), for their initials in Catalan, currently a well-consolidated and recognized research group in the fuzzy community. We think that the book will contribute to make clear the intense work of Prof. Gaspar Mayor and to discover its relation or influence with the works of other researchers.

On the other hand, our purpose is to give the opportunity to the fuzzy community of knowing the more significant contributions of the Prof. Gaspar Mayor and provide these researchers with a small number of new contributions related to the previous one. All mathematical notations are auto-explained in each chapter of the book and also in the corresponding references.

Chapter 1 provides a brief summary of the most representative publications of the Prof. Gaspar Mayor. It collects the most important contributions on different topics: fuzzy sets, aggregation functions, multidistances, including in each case a brief comment on the essence and importance of the corresponding contribution. This chapter has been written by the members of the LOBFI research group to thank him all his support and efforts to put it on a relevant and distinguished position in the international fuzzy community.

The remaining chapters focus on different aims of the soft computing area. Chapter 2 by Esteva, García-Cerdaña, and Godo starts from the well-known result by Prof. G. Mayor about the characterization of divisible (or smooth) discrete t-norms, and presents a summary of results on BL-algebras and finite BL-chains, which allow an equational characterization of any divisible finite t-norm. The authors base their results on an alternative decomposition of a finite divisible t-norm as an ordinal sum of hoops.

Chapter 3 by Fernández-Sánchez, Quesada-Molina, and Úbeda-Flores is devoted to some aspects on associative copulas, presenting a very interesting and complete survey on the topic. The authors focus on the important properties of associative copulas and also on some of their applications on distribution functions and statistics. They finish by posing a small collection of open problems on associative copulas.

In Chap. 4, Boixader and Recasens deal with power stable aggregation functions with respect to powers of t-norms and t-conorms (i.e., satisfying $A(x^{(r)}, y^{(r)}) = (A(x, y))^{(r)}$ where $x^{(r)}$ is the r-th power of $x \in [0, 1]$ with respect to a t-norm or a t-conorm). Special cases when the considered t-norm is continuous Archimedean (strict or not) are characterized. The results on power stable aggregation functions given in this chapter generalize those already known on log-ratio invariant aggregation functions.

Chapter 5, written by Mas, Monreal, Monserrat, Riera, and Torrens, deals with the study and the characterization of the uninorm-based implication functions that satisfy the most used inference schemes in approximate reasoning, namely the Modus Ponens and Modus Tollens. This is done for residual implications derived from uninorms, or RU-implications, specially for the cases when the uninorm is in one of the most usual classes of uninorms, that is, uninorms in $\mathcal{U}_{\min}$, idempotent uninorms and representable uninorms.

Chapter 6, by Bustince, Barrenechea, Pagola, Fernández et al., presents a review of the main concepts related to Atanassovs intuitionistic fuzzy relations (A-IFRs) and their structures, and provides the basic concepts and notions to non-specialist researchers. The authors focus on the construction of A-IFRs, which preserve some specific properties. They achieve this by using Atanassovs operators, which allow to reduce Atanassovs intuitionistic fuzzy sets to fuzzy sets. Moreover, in this chapter some other construction methods are considered.

The next chapter, by Calvo, Fuentes and Fuster-Parra, provides a new lattice L determined by the class of weighting triangles as a base of L-fuzzy subsets. Furthermore, they introduce some extended orders and operators which are obtained by fuzzy binary relations F_Δ associated to a weighting triangle. Moreover, some new expressions have been found for extended ordered weighted averaging operators and extended aggregation functions.

In Chap. 8, the problem of aggregating a family of bounded asymmetric distances is analyzed by Aguiló, Calvo, Fuster-Parra, Martín, Valero, and Suñer. With this purpose, they introduce the notion of bounded asymmetric distance aggregation function and present a full description of such functions. The obtained results are illustrated with examples. Furthermore, the relationship between asymmetric aggregation functions and the bounded ones is discussed.

Chapter 9, by Martínez-Panero, García-Lapresta, and Meneses, provides the formal notion of absolute dispersion measure and its properties are analyzed. This notion is satisfied by some classical dispersion measures used in the statistics, and also by the absolute Gini index, used in the welfare economics for measuring inequality. The notion of the absolute dispersion measure has some common properties with the notion of multidistance introduced and analyzed by Martín and Mayor in several recent papers. The authors of this chapter compare absolute dispersion measures and multidistances and they establish that these two notions are compatible by showing some functions that are simultaneously absolute dispersion measures and multidistances. Also, the connection with the remainders of exponential means is analyzed. Finally, this chapter includes some conjectures for further research and some concluding remarks.

In the next chapter, Pérez, Cabrerizo, Alonso, Chiclana, and Herrera-Viedma describe the group decision-making (GDM) framework and the consensus reaching processes based on moderator. Different kinds of fuzzy consensus models are presented, depending on whether they are based on the reference domain, on the coincidence method, on the generation of recommendation method, or on the guidance measures used to drive the consensus process. This chapter also shows the new challenges in the development of consensus models. Among them, they point out some consensus models based on new technologies as social networks or agent theory.

Chapter 11 is devoted to obtaining the value of the weights in any decision problem, since it is of great importance and can change the course of action for the final decision. The value of these weights is approximate due to the vagueness and ambiguity of the data. The contents of this chapter are based on the analytic hierarchy process (AHP) and its relation with the rioritized aggregation operators. In this work, Caples, Lamata, and Verdegay propose their obtaining starting from a proportionality relationship, and they study the main properties of the prioritized operator with proportionality ratio and linear scale.

Next chapter authored by Fernández and Herrera presents a complete taxonomy for evolutionary fuzzy systems. It focuses the attention toward the imbalanced classification problems, reviewing several significant proposals made in this research area that have been developed for addressing the classification with imbalanced datasets. Finally, it shows a case study from which we will highlight the good behavior of evolutionary fuzzy systems in this particular context.

In Chap. 13, written by Bibiloni, González-Hidalgo, Massanet, Mir, and Ruiz-Aguilera, the starting point is the well-known family of t-norms characterized by Prof. G. Mayor, currently known by his name: Mayor–Torrens t-norms. In this chapter, the authors use this parameterized family of t-norms, together with their residual implications, in fuzzy mathematical morphology and apply them to many different applications such as edge detection, noise reduction, and shape and pattern recognition. The results in the chapter prove that adequate members of this family of t-norms can be selected in each one of these applications to obtain competitive and even better results.

Last chapter by Trillas describes a virtual conversation between two imagined characters, Karl and Carla. They try to debate on how, in fuzzy set algebras, what "is not covered under a linguistic label" should be represented. Along the conversation both the negation and the opposites of a predicate, as well as the relations between them, are analyzed and discussed.

The editors want to dedicate this book to Prof. Gaspar Mayor, and they also wish that this edited volume will contribute to promote new theoretical and practical results. They want to thank each one of the authors for all their contribution to this book, because without their enthusiastic efforts the book would have not been possible. Also, they present their gratitude to Springer Verlag for giving them the opportunity of publishing this book.

Alcalá de Henares   Tomasa Calvo Sánchez
Palma   Joan Torrens Sastre
December 2015

Contents

Contributors

Isabel Aguiló Universitat de les Illes Balears, Palma de Mallorca, Illes Balears, Spain

Sergio Alonso Department of Software Engineering, University of Granada, Granada, Spain

Edurne Barrenechea Departamento of Automática y Computación and Institute of Smart Cities, Universidad Publica de Navarra, Pamplona, Spain

P. Bibiloni Department of Mathematics and Computer Science, University of the Balearic Islands, Palma, Spain

D. Boixader Secció Matemàtiques i Informàtica, ETS Arquitectura del Vallès, Universitat Politècnica de Catalunya, Sant Cugat Del Vallès, Spain

Humberto Bustince Departamento of Automática y Computación and Institute of Smart Cities, Universidad Publica de Navarra, Pamplona, Spain

E. Cables Dpto. Informatica, Universidad de Holguin "Oscar Lucero Moya", Holguin, Cuba

Francisco Javier Cabrerizo Department of Software Engineering and Computer Systems, Distance Learning University of Spain, Madrid, Spain

Tomasa Calvo Sánchez LOBFI Research Group, Department of Mathematics and Computer Science, University of the Balearic Islands, Palma, Spain; Universidad de Alcalá, Alcalá de Henares, Madrid, Spain

Francisco Chiclana Faculty of Technology, Centre for Computational Intelligence, De Montfort University, Leicester, UK

Francesc Esteva IIIA-CSIC, Campus de la UAB, Bellaterra, Spain

Javier Fernandez Departamento of Automática y Computación and Institute of Smart Cities, Universidad Publica de Navarra, Pamplona, Spain

A. Fernández Department of Computer Science, University of Jaén, Jaén, Spain

Juan Fernández-Sánchez Department of Mathematics, University of Almería, Almería, Spain

Ramón Fuentes-González Universidad Pública de Navarra, Pamplona, Spain

Pilar Fuster-Parra Universitat Illes Balears, Palma de Mallorca, Balears, Spain

Àngel García-Cerdaña IIIA-CSIC, Campus de la UAB, Bellaterra, Spain; DTIC, Universitat Pompeu Fabra, Barcelona, Spain

José Luis García-Lapresta PRESAD Research Group, BORDA Research Unit, IMUVa, Departamento de Economía Aplicada, Universidad de Valladolid, Valladolid, Spain

Lluís Godo IIIA-CSIC, Campus de la UAB, Bellaterra, Spain

M. González-Hidalgo Department of Mathematics and Computer Science, University of the Balearic Islands, Palma, Spain

F. Herrera Department of Computer Science and Artificial Intelligence, University of Granada, Granada, Spain

Enrique Herrera-Viedma Department of Computer Science and Artificial Intelligence, University of Granada, Granada, Spain

M.T. Lamata Departamento Ciencias de la Computacion e Inteligencia Artificial, Universidad de Granada, Granada, Spain

Javier Martín Universitat de les Illes Balears, Palma de Mallorca, Illes Balears, Spain

Miguel Martínez-Panero PRESAD Research Group, BORDA Research Unit, IMUVa, Departamento de Economía Aplicada, Universidad de Valladolid, Valladolid, Spain

M. Mas Department of Mathematics and Computer Science, University of the Balearic Islands, Palma de Mallorca, Spain

S. Massanet Department of Mathematics and Computer Science, University of the Balearic Islands, Palma, Spain

Luis Carlos Meneses PRESAD Research Group, IMUVa, Departamento de Economía Aplicada, Universidad de Valladolid, Valladolid, Spain

A. Mir Department of Mathematics and Computer Science, University of the Balearic Islands, Palma, Spain

J. Monreal Department of Mathematics and Computer Science, University of the Balearic Islands, Palma de Mallorca, Spain

M. Monserrat Department of Mathematics and Computer Science, University of the Balearic Islands, Palma de Mallorca, Spain

Javier Montero Department of Statistics and Operations Research I, Faculty of Mathematics, Universidad Complutense de Madrid, Madrid, Spain

Raul Orduna Departamento of Automática y Computación, Universidad Publica de Navarra, Pamplona, Spain

Miguel Pagola Departamento of Automática y Computación and Institute of Smart Cities, Universidad Publica de Navarra, Pamplona, Spain

Ignacio Javier Perez Department of Computer Sciences and Engineering, University of Cadiz, Puerto Real, Spain

José Juan Quesada-Molina Department of Applied Mathematics, University of Granada, Granada, Spain

J. Recasens Secció Matemàtiques i Informàtica, ETS Arquitectura del Vallès, Universitat Politècnica de Catalunya, Sant Cugat Del Vallès, Spain

J.V. Riera Department of Mathematics and Computer Science, University of the Balearic Islands, Palma de Mallorca, Spain

D. Ruiz-Aguilera Department of Mathematics and Computer Science, University of the Balearic Islands, Palma, Spain

Jaume Suñer Universitat de les Illes Balears, Palma de Mallorca, Illes Balears, Spain

Joan Torrens Sastre Department of Mathematics and Computer Science, University of the Balearic Islands, Palma de Mallorca, Spain

Enric Trillas European Centre for Soft Computing Mieres (Asturias), Mieres, Spain

Oscar Valero Universitat de les Illes Balears, Palma de Mallorca, Illes Balears, Spain

J.L. Verdegay Departamento Ciencias de la Computacion e Inteligencia Artificial, Universidad de Granada, Granada, Spain

Manuel Úbeda-Flores Department of Mathematics, University of Almería, Almería, Spain

Chapter 1
Gaspar Mayor: A Prolific Career on Fuzzy Sets and Aggregation Functions

LOBFI Research Group

Abstract The LOBFI Research Group was founded many years ago by Professor Gaspar Mayor at the beginning of his research career at the University of the Balearic Islands. Since then, he has been the leader of the group and has devoted many of his efforts to getting LOBFI to a prestigious and recognized group in the fuzzy community. This chapter has been written by all the current members of the LOBFI group in gratitude to him, and it is devoted to recall his main scientific achievements in the field of fuzzy sets theory and aggregation fusion.

1.1 Gaspar Mayor

Scientific researchers are usually people completely committed and devoted to their work in many aspects, not only in the continuous search of new results and achievements in their fields, but also in the interest of making accessible these advances to as many people as possible. Both aspects are totally present in Professor Gaspar Mayor's career that can not be easily summarized in the length of this chapter. Let us give only some hints about the teaching part and we will focus our attention on his research activity. He received the B.S. degree in Mathematics from the University of Barcelona in 1971 and the Ph.D. degree in Mathematics from the University of the Balearic Islands (UIB) in 1985. He has been working as a teacher and a researcher at the UIB since 1971 and he is currently a full Professor at this university. During this time he has taught many courses in different degrees and Masters. It has been a constant in his career his commitment of training new scientists and new researchers. He founded in particular the LOBFI research group and many of its current members have obtained their Ph.D. degree under his supervision.

LOBFI Research Group (✉)
Department of Mathematics and Computer Science,
University of the Balearic Islands, 07122 Palma, Spain

© Springer International Publishing Switzerland 2016
T. Calvo Sánchez and J. Torrens Sastre (eds.), *Fuzzy Logic and Information Fusion*,
Studies in Fuzziness and Soft Computing 339, DOI 10.1007/978-3-319-30421-2_1

Regarding his research activity he has been working from the very beginnings in fuzzy sets theory, aggregation functions, functional equations and related fields. He was one of the pioneers in the development of aggregation functions and his Ph.D. thesis focused in this direction. He has been and still he is a prolific researcher in the fuzzy community, as it can be seen in the non-exhaustive list of his bibliography that is recalled at the end of this chapter.

It has been exciting for us to have the chance of coincide and jointly work with Professor Gaspar Mayor. Among his works, there are many interesting and quite representative papers that we have had the pleasure of sharing with him. In the sequel we want to highlight some ones that have had a great impact or have represented a special advance in some concrete field.

- *"On a family of t-norms"* (1991). In the study of continuous t-norms an important topic is to check whenever they are completely determined by their values on the diagonal section. A family of t-norms that satisfies this property was characterized in this paper corresponding to [1], through the functional equation:

$$T(x, y) = \max(T(\max(x, y), \max(x, y)) - |x - y|, 0)$$

  for all $x, y \in [0, 1]$. Note that from this equation it is clear that the value of the t-norm in any point (x, y) is known from the values of the t-norm on the diagonal. This functional equation was solved in [1], obtaining as solutions a family of t-norms which is simply given by ordinal sums of only one Łukasiewicz summand on the interval $[0, a]$ for a given $a \in [0, 1]$. This family of t-norms is currently known under the name of Professor Gaspar Mayor (*Mayor-Torrens* t-norms) and it has been studied and developed in many further works (see [2]). In particular, this family of t-norms gives also a parameterized family of copulas from the Minimum or upper Fréchet-Hoeffding bound copula to the Lukasiewicz or lower Fréchet-Hoeffding bound copula (see [3, 4]). Additionally to its high impact, this paper is in fact quite representative of Gaspar's work since many other of his papers also deal with functional equations involving t-norms and related operators, see for instance [5–10].
- *"On a class of operators for expert systems"* (1993). This paper corresponds to [11] and, due to its impact and interest, it was rewritten and completed in several aspects leading to the book chapter in [12]. This is a work that was pioneer in the study of aggregation functions on finite scales or discrete aggregation functions and depicted the starting point of many later papers devoted to this topic, which is the basis of many linguistic approaches for decision making and linguistic preference relations. The smoothness property was firstly introduced in this paper and since then, it has been taken as the counterpart of continuity for operators defined on finite chains since then. In fact it is proved in [12] that smoothness is equivalent in this framework to divisibility or also to the intermediate value theorem. Smooth discrete t-norms and t-conorms were characterized in terms of ordinal sums of Lukasiewicz summands in [1, 12] and the total number of such operators was determined. They have been later used to characterize many new classes of discrete aggregation

functions like uninorms, nullnorms, non-commutative versions of them and so on. Some of these new results are again Professor Gaspar Mayor's contributions (see for instance [13–15]).

- *"t-Operators"* (1999). A field of interest in the study of aggregation functions is the construction or introduction of new classes of this kind of operations. In the associative framework, some families of aggregation functions have appeared generalizing t-norms and t-conorms. The two most important examples of this situation are uninorms and t-operators (or nullnorms). It was in this paper, that corresponds to the Ref. [16], where the second ones were introduced and characterized. In fact, t-operators were defined with the same properties of t-norms and t-conorms, but changing the neutral element by the continuity of the boundary sections. Simultaneously, also the so-called nullnorms were introduced as operations with a zero element with additional conditions, and it was proved in [17] that both kinds of operations coincide. Nullnorms (or t-operators) and their relations with uninorms have been extensively studied since then, specially in relation with many functional equations involving them. Some of these functional equations were solved again by Professor Gaspar Mayor like for instance in [17–19]. Even some generalizations of nullnorms and t-operators, avoiding the commutativity condition, have been introduced obtaining semi-nullnorms and semi-t-operators that, in this non-commutative case, do not further coincide.

- *"On a type of monotonicity for multidimensional operators"* (1999). This work corresponds to [20] and has its origin in the first study on extended aggregation functions presented in the first IFSA Congress (see [21]). In the later contribution the class of extended aggregation functions was introduced based on a special type of orders, which was newly extended in this other contribution considering two partial orderings based on a t-norm and a t-conorm. Moreover, the monotonicity condition for a function f from the set E of multidimensional ordered lists on [0, 1] was established. In this case, when f is a t-norm the monotonicity condition is characterized in terms of a functional equation. Recently, these results and the related to them have been collected in a chapter of a book of Springer-Verlag editorial to celebrate the 75 anniversary of Professor Enric Trillas [22]. This contribution shows the significance of the Professor G. Mayor's work on the aggregation functions field.

- *"Generation of weighting triangles associated with aggregation functions"* (2000). In this work, a way to determine different families of n-dimensional aggregation functions by means of the definition of several types of weighting triangles is presented. Namely, it provides the characterization of some interesting properties of Extended Ordered Weighted Averaging operators, EOWA, and Extended Quasi-linear Weighted Mean, EQLWM, as well as of their reverse functions. Moreover, this work presents different results about generation of these multi-dimensional aggregation functions by means of sequences and fractal structures. Finally, the degree of orness of a weighting triangle associated with an EOWA operator is considered, and also some results on each class of triangles are mentioned. This contribution is referenced as [23]. The construction method of consistent multidimensional aggregation functions was again reconsidered by

Professor G. Mayor in [24], as well as by other authors in posterior works (see [22, 25, 26]). The extended aggregation functions have been widely treated by many authors in the last three decades (see [27–29], among others). The preliminary studies of Professor Mayor on aggregation functions have their origin in his thesis [30], where he introduced the binary aggregation functions A as monotone and symmetric operators and with two boundary conditions: $A(x, 0) = A(1, 0)x$, $A(x, 1) = (1 - A(1, 0))x + A(1, 0)$. The t-norms and t-conorms are examples of this family as well as the linear convex combination of a t-norm and a t-conorm. This work was considered in the representation theorem of aggregation functions defined as linear convex combination of a non-strict Archimedean t-norm and its dual t-conorm [31], which is a generalization of Ling theorem of Archimedean t-norms. These binary aggregation functions were generalized by Födor and Calvo in [32]. In the former work the two mentioned boundary conditions are replaced by the following ones: $A(x, 0) = T(A(1, 0), x)$, $A(x, 1) = S(A(1, 0), x)$, where T is a continuous t-norm and S a continuous t-conorm. This new class of binary aggregation functions is called $T - S$ aggregation functions. This idea was reconsidered in some sense by Pradera et al. [33].

- *"Balanced discrete fuzzy measures"* (2000). Fuzzy measures play a special role in the context of aggregation functions since by means of them it is possible to construct different types of aggregation functions, as they are the Choquet and Sugeno integrals. The main contribution of Professor Mayor in this framework was the introduction of the class of balanced fuzzy measures. These measures satisfy a strong condition of the standard monotonicity: the monotonicity w.r.t the cardinality. Moreover, the balanced additive measures and S-decomposable measures were studied in [34]. This contribution is referenced as [35]. The results of this aim are collected in [34, 35]. They are also related to the framework of aggregation functions because these measures allow us to define specific aggregation functions.

- *"The distributivity condition for uninorms and t-operators"* (2002). Related to the problem of finding additional properties for some classes of aggregation functions, in this paper (that corresponds to the Refs. [17, 19]) the functional equation derived from the distributivity condition involving uninorms and nullnorms is solved. The importance of the distributive property lies not only in the field of fuzzy sets, but also in the pseudo-analysis and pseudo-measures. Since t-norms, t-conorms and uninorms have been the connectives more frequently used in managing pseudo-operations, the study of the distributivity between these classes of aggregation functions has been of great importance. In [17] the distributivity between two uninorms (in $\mathscr{U}_{\min}$ or $\mathscr{U}_{\max}$), a uninorm over a t-operator, a t-operator over a uninorm and between two t-operators is solved. The impact of this paper can be checked from the great quantity of works dealing with distributivity and following the ideas in [17], that have successively appeared since then. Most of these papers extend the results from [17] to more general classes of aggregation operations like different classes of uninorms, semi-uninorms, semi-nullnorms, semi-t-operators, 2-uninorms and so on.

- *"On locally internal monotonic operations"* (2003). This paper corresponds to the Ref. [36] and it has been especially interesting mainly for two reasons. On one hand, locally internal monotonic operations (that is, monotonic operations whose values at any point are always one of their arguments) are a especial kind of idempotent aggregation functions extensively studied along the literature by many authors (in particular, other papers of Professor Gaspar Mayor on this topic are [37, 38]). In this line, a very close relation between associativity and commutativity for this kind of operators is proved, and locally internal monotonic operations with these properties are characterized. On the other hand, the characterization of idempotent uninorms in terms of a decreasing function which is symmetric with respect to the identity map is proved as a consequence of the above mentioned results. This fact was later improved in [39], leading to a lot of works involving this important class of uninorms. Moreover, the characterization of all uninorms with continuous underlying operations has been a goal largely investigated, and it is not totally solved yet. However, there is no doubt that [36] has been a key work and an important step in this direction.
- *"Copula like operations on finite settings"* (2005). Copulas are a special kind of aggregation functions of great importance because of their applications in many fields, specially in statistics and economy. The main reason is the well-known Sklar's theorem that shows how copulas link the joint distribution function of two random variables to their one-dimensional marginal distributions. In this sense, we know that a copula C expresses the dependence among these two random variables, and properties on the copula derive into statistical properties on the random variables. All these results refer to copulas defined on the unit interval $[0, 1]$. In [40], operations with the same properties of copulas but defined on finite chains (known in this framework under the name of discrete copulas) were introduced and similar results were proved, including in [41] a discrete version of the Sklar's theorem. Specifically, a one-to-one correspondence between discrete copulas and permutation matrices is proved in [40], which allows to a complete classification of all possible discrete copulas proving, in particular, that there are exactly $n!$ copulas on a finite chain of $n + 1$ elements. This paper has had a great impact in the field and many papers have later appeared based on it and extending the results to discrete quasi-copulas (see [42–45]).
- *"Additive generators of discrete conjunctive aggregation operations"* (2007). This paper corresponds to [46] and it examines the additive generation of conjunctions and t-norms defined on a totally ordered set $L = \{0, 1, \ldots, n\}$, usually called *discrete conjunctions*. A crucial result states that an additive generator of a discrete conjunction is reduced to a decreasing ordered list of integer numbers $f = (a_0, a_1, \ldots, a_n)$, with $a_0 > a_1 > \cdots > a_n = 0$. Another remarkable result is the description of conjunctions on L that are additively generated in terms of continuous non-strict Archimedean t-norms defined on $[0, n]$. The main goal is to give a procedure for deciding whether a conjunctive aggregation operation is additively generated or not, through the study of the consistency of a system of linear inequalities obtained from the definition of an additive generator. It is shown that the behaviour of additive generators between the discrete case and the $[0, 1]$-case

is completely different, and this fact opens a new research line in this direction. In fact, some works have been derived from [46] (see [47, 48]) where it is proved for instance that all divisible discrete t-norms are additively generated, but not all t-norms are. Also the additive generation of the nesting of t-norms, a similar construction of the ordinal sums, is studied. Moreover, convex and concave generators are investigated, which become smooth and Archimedean conjunctions, respectively, as well as symmetric generators, which generate t-conorms S in such a way that the corresponding S-implication I_S satisfies the Identity Principle, the Ordering Property and the Generalized Modus Ponens.

- *"Aggregation of asymmetric distances in Computer Science"* (2010). A field of great interest in practical problems is the aggregation of information which comes from sources of a different nature. In particular in many situations that arise in applied sciences, as for instance in image processing, decision making, control theory, medical diagnosis, robotics or biology, the incoming data is symbolized via numerical values and thus, the fusion methods must be based on numerical aggregation operators. Besides, a wide class of techniques of aggregation used in the aforesaid fields impose a constraint in order to select the most suitable aggregation operator for the problem to be solved. In general, this constraint consists of considering only those operators that provide an output data which preserves relevant properties of the input data. A typical example of this type of situation is given when one wants to merge a few (generalized) metrics in order to obtain a new one which will provide the numerical values that will be taken into account in order to make a decision. Inspired by the utility of asymmetric distances (a type of general metrics also called quasi-metrics) in several fields of Artificial Intelligence and Computer Science, in [49] was studied the problem of how to combine through a function a collection (not necessarily finite) of asymmetric distances in order to obtain a single one as the output. To this end, the notion of asymmetric distance aggregation function was introduced in the aforementioned reference and a characterization of such functions was obtained. In fact, it was proved that such functions are exactly monotone, sub-additive and they vanish only at the neutral element. Furthermore, the obtained results were applied to asymptotic complexity analysis of algorithms in Computer Science.

- *"Multi-argument distances"* (2011). In this paper [50], a proposal for a formal definition of a multi-argument function distance, multidistance for short, is introduced. In this sense, the conventional definition of a distance between two elements of a set was extended, with the aim of applying it to any finite collection of elements. Some general properties are investigated and significant examples are exhibited. Finally, the concept of ball centred at a list is also introduced, in order to extend the usual balls in metric spaces.

  The main advantage of this approach, concerning the general treatment of the problem of measuring the multidistance of more than two points, is that it is made by means of an axiomatic procedure. It allows to avoid the conventional way of doing it, that is, by means of a function that takes into account the pairwise distances, usually the arithmetic mean.

- *"On fuzzy implications: An axiomatic approach"* (2013). A recent thought of Professor Gaspar Mayor has been the search and proposal of non-functionally expressible fuzzy implications. This topic was first studied by Klement for the intersection, union and complementation of fuzzy sets in [51], but it remained unexplored for fuzzy implications until the publication of the paper corresponding to Ref. [52]. This study represents a landmark in the study of fuzzy implications, mainly focused to the study of fuzzy implication functions, binary functions $I : [0, 1]^2 \rightarrow [0, 1]$ used to model a fuzzy implication $A \rightarrow B$ between two fuzzy sets A, B through $(A \rightarrow B)(x, y) = I(A(x), B(y))$. Note that in this case, the value $(A \rightarrow B)(x, y)$ depends only on the values taken by A and B at the points x and y but in many cases fuzzy implications should depend on the complete information about the involved fuzzy sets and they need not be functionally expressible. Following up on this idea, an axiomatic definition of fuzzy implication as a connective acting on two possibly different systems of fuzzy sets $[0, 1]^X$ and $[0, 1]^Y$ with an output in $[0, 1]^{X \times Y}$ is proposed. Furthermore, the characterization of functionally expressible fuzzy implications (and other fuzzy operations) is given and several examples of fuzzy operations which are not functionally expressible are included.
- *"Convex Linear T-S Functions: A Generalization of Frank's Equation"* (2013). The main purpose of this work is to solve the functional equation $T + \lambda S = Min + \lambda Max$, for $0 \leq \lambda \leq 1$ and a pair $(T; S)$ of a t-norm T and a t-conorm S. This equation also arises in a natural way when we consider operators which are convex linear combinations of a t-norm and a t-conorm, and set out the problem of finding the intersection of the segments determined by the pair (Min; Max) and any other pair $(T; S)$. This equations has just the solution (Min; Max). Moreover, the equation is stated in a more general way by means of an OWA operator. This publication is considered here since it is one of the works of Professor Mayor on functional equations. Namely, it is dedicated to solve a generalization of the well-known Frank's functional equation. This paper corresponds to the Ref. [53]. The functional equations constitute one of the research lines of Professor Mayor in which he has widely contributed.
- *"On the symmetrization of quasi-metrics: an aggregation perspective"* (2013). The problem of how to symmetrize an asymmetric distance has attracted much attention in the literature since the 50s. It is a known fact that every asymmetric distance induces in a natural way a metric. In fact if d is an asymmetric distance on a nonempty set X, then the functions $d_{\max}$ and d_+, defined for all $x, y \in X$ by $d_{\max}(x, y) = \max\{d(x, y), d(y, x)\}$ and $d_+(x, y) = d(x, y) + d(y, x)$, are metrics on X. Motivated by the fact that the preceding symmetrization techniques can be formulated in the context of aggregation theory, the problem of how to symmetrize an asymmetric distance from an aggregation perspective was posed formally in [54]. Thus, in the preceding reference, a complete solution to such a problem was provided by means of the asymmetric distance aggregation framework. To this end, the notion of metric generating function, those functions that allow to generate a metric from an asymmetric distance, was introduced and a full description of such functions was given from an aggregation perspective. The

relevance of the preceding paper was recognized by M.M. Deza who included the description of such method in his celebrated book *"Encyclopedia of Distances"*, Springer-Verlag, Berlin (2013).

- *"On distances derived from t-norms"* (2014). Over the last years there has been a growing interest in the generation of new metrics due to its wide range of applications. Motivated by generalizations of the classical symmetric differences of sets, some authors introduced distances generated from a t-norm and its dual t-conorm. This paper corresponds to [55], where the previous idea is extended to the more general case of any t-norm T and any t-conorm S not necessarily the dual of T. In this work a sufficient condition on a pair $(T; S)$ has been obtained in order to induce a distance $d_{(T,S)}$ and also a necessary condition in terms of the distance $d_{(W,W^*)}$ where W denotes the Łukasiewicz t-norm. The main result is the characterization of a class of t-norms defining distances, that is, in the case that the t-norms have the same zero region as the Łukasiewicz one. This way of generating distances from these pairs has been translated to multi-distances, where the associativity of the t-norm and the t-conorm has all its meaning.

It is clear that the previously commented papers by Professor Gaspar Mayor constitute only a particular choice of some of his papers selected among those that are most representative, important or with highest impact. Of course, there are many other papers that could appear here and some of them are listed in the bibliography. It is possible that some other article was selected by the proper Professor Gaspar Mayor in order to appear in a list of his best papers. If it is the case, let us only say: "Sorry Gaspar for our bad choice".

In any case, we have done this chapter with our best intention and we would like to dedicate it to Professor Gaspar Mayor in his 70th birthday with our gratitude and appreciation.

Acknowledgments This paper has been partially supported by the Spanish grant TIN2013-42795-P.

References

1. G. Mayor, J. Torrens, On a family of t-norms. Fuzzy Sets Syst. **41**, 161–166 (1991)
2. E.P. Klement, R. Mesiar, E. Pap, *Triangular norms* (Kluwer Academic Publishers, Dordrecht, 2000)
3. F. Durante, A. Kolesárová, R. Mesiar, C. Sempi, Copulas with given diagonal sections: novel constructions and applications. Int. J. Uncertain. Fuzziness Knowl. Based Syst. **15**, 397–410 (2007)
4. F. Durante, R. Mesiar, C. Sempi, On a family of copulas constructed from the diagonal section. Soft Comput. **10**, 490–494 (2006)
5. C. Alsina, G. Mayor, M.S. Tomàs, J. Torrens, A characterization of a class of aggregation functions. Fuzzy Sets Syst. **53**, 33–38 (1993)
6. C. Alsina, G. Mayor, M.S. Tomàs, J. Torrens, Subdistributive De Morgan triplets. Serdica **19**, 258–266 (1993)
7. G. Mayor, On the equation $A = (A \cap B) \cup (A \cap B^c)$. Fuzzy Sets Syst. **47**, 265–267 (1992)

8. G. Mayor, On a family of quasi-arithmetic means. Aequationes Mathematicae **48**, 137–142 (1994)
9. G. Mayor, J. Torrens, Duality for a class of binary operations on [0,1]. Fuzzy Sets Syst. **47**, 77–80 (1992)
10. G. Mayor, J. Torrens, De Rham systems and the solution of a class of functional equations. Aequationes Mathematicae **47**, 43–49 (1994)
11. G. Mayor, J. Torrens, On a class of operators for expert systems. Int. J. Intell. Syst. **8**, 771–778 (1993)
12. G. Mayor, J. Torrens, Triangular norms on discrete settings. In *Logical, Algebraic, Analytic, and Probabilistic Aspects of Triangular Norms*, ed. by E.P. Klement, R. Mesiar (Elsevier, 2005), pp. 189–230
13. J. Casasnovas, G. Mayor, Discrete t-norms and operations on extended multisets. Fuzzy Sets Syst. **159**, 1165–1177 (2008)
14. A. Kolesarova, G. Mayor, R. Mesiar, Weighted ordinal means. Inf. Sci. **177**, 3822–3830 (2007)
15. M. Mas, G. Mayor, J. Torrens, t-Operators and uninorms on a finite totally ordered set. Int. J. Intell. Syst. **14**, 909–922 (1999)
16. M. Mas, G. Mayor, J. Torrens, t-Operators. Int. J. Uncertain. Fuzziness Knowl. Based Syst. **7**, 31–50 (1999)
17. M. Mas, G. Mayor, J. Torrens, The distributivity condition for uninorms and t-operators. Fuzzy Sets Syst. **128**, 209–225 (2002)
18. M. Mas, G. Mayor, J. Torrens, The modularity condition for uninorms and t-operators. Fuzzy Sets Syst. **126**, 207–218 (2002)
19. M. Mas, G. Mayor, J. Torrens, Corrigendum to "The distributivity condition for uninorms and t-operators" [Fuzzy Sets and Systems, 128 (2002) 209–225]. Fuzzy Sets Syst. **153**, 297–299 (2005)
20. G. Mayor, T. Calvo, On a type of monotonicity for multidimensional operators. Fuzzy Sets Syst. **104**, 121–123 (1999)
21. G. Mayor, T. Calvo, On extended aggregation functions. In *Proc IFSA'97*, Prague, 1997, pp. 281–285
22. T. Calvo, G. Mayor, J. Suñer, Globally Monotone Extended Aggregation Functions. In A Passion for Fuzzy Sets, Studies in Fuzziness and Soft Computing, ed. by E. Trillas (Springer, Berlin, 2015), pp. 49–66
23. T. Calvo, G. Mayor, J. Torrens, J. Suñer, M. Mas, M. Carbonell, Generation of weighting triangles associated with aggregation functions. Int. J. Uncertain. Fuzziness Knowl. Based Syst. **8**(4), 417–452 (2000)
24. T. Calvo, G. Mayor, R. Mesiar (eds.), *Aggregation Operators. New Trends and Applications* (Physica-Verlag, Heidelberg, 2002)
25. G. Beliakov, S. James, Stability of weighted penalty-based aggregation functions. Fuzzy Sets Syst. **226**, 1–18 (2013)
26. D. Gómez, J. Montero, J. Tinguaro-Rodríguez, K. Rojas, Stability in aggregation operators. Adv. Comput. Intell. Commun. Comput. Inf. Sci. **299**, 317–325 (2012)
27. G. Beliakov, A. Pradera, T. Calvo, *Aggregation Functions: A Guide for Practitioners. Studies in Fuzziness and Soft Computing, 221* (Springer, Heidelberg, 2007)
28. M. Grabisch, J. Marichal, R. Mesiar, E. Pap, *Aggregation Functions*. Encyclopedia of Mathematics and Its Aplications, vol. 127 (Cambridge University Press, NY, 2009)
29. V. Torra, V. Narukawa, *Modeling Decisions. Information Fusion and Aggregation Operators* (Springer, Berlin, 2007)
30. G. Mayor, *Contribució al models matemàtics per a la lògica de la vaguetat*. Tesis doctoral, Universidad de las Islas Baleares (1984)
31. G. Mayor, E. Trillas, On the representation of some aggregation functions. In *Proceedings of the Sixteenth International Symposium on Multiple-Valued Logic*, Blacksburg, Virginia, USA, 1986, pp. 110–114
32. J. Fodor, T. Calvo, Aggregation functions defined by t-norms and t-conorms. In *Aggregation of evidence under fuzziness* (in the series "Studies in Fuzziness"), vol. 12, ed. by B. Bouchon-Meunier (Physica Verlag, 1998), pp. 36–48

33. A. Pradera, E. Trillas, T. Calvo, A general class of triangular norm-based aggregation operators: quasi-linear T-S. Int. J. Approx. Reason. **30**, 57–72 (2002)
34. J. Martín, G. Mayor, T. Calvo, Aggregation operators based on some families of symmetric fuzzy measures. In *Proceedings of IPMU 2004*, Perugia, Italia, 2004, pp. 343–350
35. T. Calvo, J. Martín, G. Mayor, J. Torrens, Balanced discrete fuzzy measures. Uncertain. Fuzziness Knowl. Based Syst. **8–6**, 665–676 (2000)
36. J. Martín, G. Mayor, J. Torrens, On locally internal monotonic operations. Fuzzy Sets Syst. **137**, 27–42 (2003)
37. J. Martín, G. Mayor, J. Suñer, On binary operations with finite external range. Fuzzy Sets Syst. **146**, 19–26 (2004)
38. G. Mayor, J. Martín, Locally internal aggregation functions. Int. J. Uncertain. Fuzziness Knowl. Based Syst. **7**, 235–241 (1999)
39. D. Ruiz-Aguilera, J. Torrens, B. De Baets, J. Fodor, Some Remarks on the Characterization of Idempotent Uninorms, in *Lecture Notes in Artificial Intelligence*, vol. 6178, ed. by E. Hüllermeier, R. Kruse, F. Hoffmann, 2010, pp. 425–434
40. G. Mayor, J. Suñer, J. Torrens, Copula-like operations on finite settings. IEEE Trans. Fuzzy Syst. **13**, 468–477 (2005)
41. G. Mayor, J. Suñer, J. Torrens, Sklar's theorem on finite settings. IEEE Trans. Fuzzy Syst. **15**, 410–416 (2007)
42. I. Aguiló, J. Suñer, J. Torrens, Matrix representation of discrete quasi-copulas. Fuzzy Sets Syst. **159**, 1658–1672 (2008)
43. I. Aguiló, J. Suñer, J. Torrens, Matrix representation of copulas and quasi-copulas defined on non-square grids of the unit square. Fuzzy Sets Syst. **161**, 254–268 (2010)
44. A. Kolesárová, R. Mesiar, J. Mordelová, C. Sempi, Discrete copulas. IEEE Trans. Fuzzy Syst. **14**, 698–705 (2006)
45. A. Kolesárová, J. Mordelová, Quasi-copulas and copulas on a discrete scale. Soft Comput. **10**, 495–501 (2006)
46. G. Mayor, J. Monreal, Additive generators of discrete conjunctive aggregation operations. IEEE Trans. Fuzzy Syst. **15**(6), 1046–1052 (2007)
47. J. Martín, G. Mayor, J. Monreal, The problem of the additive generation of finitely-valued t-conorms. Mathware Soft Comput. **16**, 17–27 (2009)
48. G. Mayor, J. Monreal, On some classes of additive generators. Fuzzy Sets Syst. **264**, 110–120 (2015)
49. G. Mayor, O. Valero, Aggregation of asymmetric distances in computer science. Inf. Sci. **180**, 803–812 (2010)
50. J. Martín, G. Mayor, Multi-argument distances. Fuzzy Sets Syst. **167**, 92–100 (2011)
51. E. Klement, Operations on fuzzy sets-an axiomatic approach. Inf. Sci. **27**(3), 221–232 (1982)
52. S. Massanet, G. Mayor, R. Mesiar, J. Torrens, On fuzzy implications: an axiomatic approach. Int. J. Approx. Reason. **54**, 1471–1482 (2013)
53. T. Calvo, G. Mayor, J. Martín, Convex linear T-S functions: a generalization of Frank's equation. Fuzzy Sets Syst. **226**, 67–77 (2013)
54. J. Martín, G. Mayor, O. Valero, On the symmetrization of quasi-metrics: an aggregation perspective. In *Advances in Intelligence and Soft Computing*, ed. by H. Bustince et al. (Springer, 2013), pp. 319–332
55. I. Aguiló, J. Martin, G. Mayor, J. Suñer, On distances derived from t-norms. Fuzzy Sets Syst. **278**, 40–47 (2015)

Chapter 2
Smooth Finite T-norms and Their Equational Axiomatization

Francesc Esteva, Àngel García-Cerdaña and Lluís Godo

Abstract In this paper, as homage to Professor Gaspar Mayor in his 70 anniversary, we present a summary of results on BL-algebras and related structures that, using the one-to-one correspondence between divisible finite t-norms and finite BL-chains, allows us to provide an equational characterization of any divisible finite t-norm.

2.1 Introduction

In the early 90s, Mayor and Torrens introduced in [15] the notion of *divisible finite t-norms* and proved they can be represented as finite ordinal sums of copies of finite Łukasiewicz and finite Gödel t-norms. Some years later, Hájek introduced in his influential monograph [14] his Basic Fuzzy logic (BL), that has become the reference system in Mathematical fuzzy logic, and showed it was complete with respect to the class of linearly ordered BL-algebras, or BL-chains. BL-chains were characterized by Hájek [13] and Cignoli et al. [7] as ordinal sums of Łukasiewicz, Gödel and Product linearly ordered algebras, but also as ordinal sums of Wajsberg hoops by Aglianò and Montagna [1]. Moreover the variety generated by a finite BL-chain has been proved to be finitely axiomatizable see e.g. Busaniche and Montagna [5].

In this paper, as homage to Professor Gaspar Mayor in his 70 anniversary, we present a summary of these results that, using the one-to-one correspondence between divisible finite t-norms and finite BL-chains, allows us how to provide an equational characterization of any divisible finite t-norm. In more detail, after this short introduction, we first overview in Sect. 2.2 the main results by Mayor and Torrens on finite

F. Esteva · À. García-Cerdaña · L. Godo (✉)
IIIA-CSIC, Campus de la UAB, 08013 Bellaterra, Spain
e-mail: godo@iiia.csic.es

F. Esteva
e-mail: esteva@iiia.csic.es

À. García-Cerdaña
e-mail: angel@iiia.csic.es

À. García-Cerdaña
DTIC, Universitat Pompeu Fabra, 08002 Barcelona, Spain

© Springer International Publishing Switzerland 2016
T. Calvo Sánchez and J. Torrens Sastre (eds.), *Fuzzy Logic and Information Fusion*,
Studies in Fuzziness and Soft Computing 339, DOI 10.1007/978-3-319-30421-2_2

11

t-norms, while in Sect. 2.3 we focus on the relationship between finite t-norms and their residua. Then in the first part of Sect. 2.4 we recall the decomposition of finite t-norms as ordinal sums of Wajsberg hoops, which is used in the second part to show how to derive a set of equations that characterize a given finite divisible t-norm. We end up with some conclusions.

2.2 Mayor and Torrens' Results on T-norms over Finite Chains

In the paper [15] Mayor and Torrens study *directed algebras* over totally ordered finite sets, inspired by the structures on the real unit interval $[0, 1]$ called De Morgan triplets and defined by a t-norm, a strong negation and its dual t-conorm.

Definition 2.2.1 A *directed algebra* is a structure $\langle L, \leq, 0, 1, T, S, N \rangle$, where:

(1) $(L, \leq, 0, 1)$ is a bounded linearly ordered finite set,
(2) T, S are associative and commutative binary operations on L such that $T(1, x) = x$ and $S(0, x) = x$,
(3) N is an order-reversing involution,
(4) for all $x, y \in L$, $N(T(x, y)) = S(N(x), N(y))$,
(5) T and S are divisible, that is, for all $x, y \in L$,
    $x \leq y$ if and only if there exists $z \in L$ such that $x = T(y, z)$, and
    $x \leq y$ if and only if there exists $z \in L$ such that $y = S(x, z)$.

Since L is finite and linearly ordered, N is obviously univocally defined on L, and S is also determined from T and N by duality (item (4) of the definition). Therefore, a directed algebra over a finite chain L is univocally defined by a binary operation T on L satisfying conditions (2) and (5). Moreover, as the authors observe, T satisfies all the conditions of a *continuous* t-norm but over a finite set instead of $[0, 1]$, and, dually, S satisfies all the conditions of a continuous t-conorm in a finite setting. Notice also that the *divisibility condition* in item (5) stipulates that any element $x \in L$ in the interval $[0, y]$ belongs to the image of the unary operation $T(y, \cdot) : L \to L$. In fact, in $[0, 1]$ this condition is equivalent to the continuity for a t-norm (see e.g. [2] for a proof).

Consider the following definition of a *finite* t-norm operation.

Definition 2.2.2 Let C be the chain $a_0 < a_1 < \cdots < a_n$. A *finite t-norm* over C is a binary operation $* : C \times C \to C$ such that:

- the operation $*$ is associative, commutative and non-decreasing in each variable,
- a_0 is an absorbent element, i.e., for all $x \in C$, $x * a_0 = a_0$,
- a_n is a neutral element, i.e., for all $x \in C$, $x * a_n = x$.

Therefore, the operation T in a directed algebra $\langle L, \leq, 0, 1, T, S, N \rangle$ is nothing but a divisible finite t-norm in L. Main examples of divisible finite t-norms on a chain $C = \{a_0 < a_1 < \cdots < a_n\}$ are the $(n + 1)$-valued Łukasiewicz t-norm

$$a_i *_{\text{Ł}} a_j = a_{\max(0,i+j-n)},$$

and the $(n + 1)$-valued minimum t-norm

$$a_i *_{\min} a_j = a_{\min(i,j)}.$$

The notion of ordinal sum of t-norms naturally extends to the finite setting.

Definition 2.2.3 Let C be the chain $a_0 < a_1 < \cdots < a_m < a_{m+1} < \cdots < a_n$ and let $*_1$ be a finite t-norm on the sub-chain $C_1 = \{a_0 < a_1 < \cdots < a_m\}$, and let $*_2$ be a finite t-norm on sub-chain $C_2 = \{a_m < a_{m+1} < \cdots < a_n\}$. Then the ordinal sum of $*_1$ and $*_2$ is the finite t-norm on C defined as follows:

$$x *_{1,2} y = \begin{cases} x *_i y, & \text{if } x, y \in C_i \\ \min(x, y), & \text{otherwise} \end{cases}$$

The main result of Mayor and Torrens's paper [15] is the characterization of *divisible finite t-norms*.

Theorem 2.2.4 [15] *The only divisible finite t-norms over a chain of n elements are the Łukasiewicz n-valued t-norm ($*_{\text{Ł}_n}$), the minimum n-valued t-norm ($\min_n$) and ordinal sums of copies of finite Łukasiewicz and minimum t-norms.*

This is a result that extends to divisible finite t-norms the well-known Mostert and Shields ordinal sum representation theorem of continuous t-norms.[1]

On the other hand, in a previous paper [10], with the goal of avoiding arbitrary numerical representations of linguistically expressed uncertainty, Godo and Sierra considered operators over a linearly ordered, finite set of linguistic terms or labels. In fact, in [10] the authors introduced what they called *r-smooth t-norms* over finite chains $C = \{a_0 < a_1 < \cdots < a_n\}$ to model conjunction operators. These are finite t-norms $* : C \times C \to C$ such that, for any $a_i, a_j, a_k, a_s \in C$,

$$\text{If } a_i * a_j = a_k \text{ and } a_i * a_{j+1} = a_s, \text{ then } s - k \leq r.$$

Here we will be interested in 1-*smooth* t-norms that, for simplicity, will be simply called *smooth* in what follows.

In [16], Mayor and Torrens prove a very interesting fact for our purposes.

Theorem 2.2.5 [16] *A finite t-norm is smooth if and only if it is divisible.*

The basic idea of the proof is that the two properties are equivalent to the fact that, given a finite t-norm $* : C \times C \to C$, for any $x \in C$, the x-row of the table of $*$ has to contain all the elements of the interval $[a_0, x]$. In some sense, these properties correspond to the continuity of a t-norm operation with respect to the order topology

---

[1]Take into account that there are no finite product chains different from the Boolean chain of two elements.

in any infinite complete chain, like $[0, 1]$, where the divisibility is equivalent to the continuity (see [2, 11] for a complete study of this problem). As a consequence we have the following result.

Theorem 2.2.6 *A finite t-norm is smooth if and only if it is a finite ordinal sum of copies of finite Łukasiewicz and minimum t-norms.*

As a direct consequence of this result, Mayor and Torrens further prove the following results.

Proposition 2.2.7

(i) *A smooth (divisible) finite t-norm $*$ is univocally determined by the set I_* of its idempotent elements.*
(ii) *There are as many smooth t-norms over a chain $C = \{a_0 < a_1 < \cdots < a_n\}$ as subsets of the set $C \setminus \{a_0, a_n\}$, i.e. 2^{n-1}.*

The first result (that it is not true for divisible t-norms in general) follows from the fact that the set of idempotent elements univocally determines the structure of the t-norm, i.e. the sequence of Łukasiewicz and Gödel components. In particular, maximal sets of consecutive elements in I_* correspond to Gödel components, the rest determine intervals that are Łukasiewicz components. The second result is an easy consequence of the first one.

2.3   About Smooth (Divisible) Finite T-norms and Their Residua

As usual in logic, in order to define a logical calculus over a finite set of truth-values or linguistic terms, it is necessary to have some form of implication operation defined. In fuzzy logic two main types of implications are usually considered: S-*implications* and R-*implications*.

Definition 2.3.1 Let $\langle C, \leq \rangle$ be a complete (bounded) chain.

- A S-*implication* on C is a binary operation defined as $x \rightarrow_S y = \neg_C x \oplus y$, where $\neg_C$ is an involutive negation on C and $\oplus$ is a t-conorm on C.
- A R-*implication* on C is a binary operation defined as $x \rightarrow_R y = \sup\{z \mid x * z \leq y\}$, where $*$ is a t-norm on C.

Some fuzzy logicians (see e.g. [14]) argue that S-implications are not adequate since, in general, they are not compatible with the (linear) order of the chain of truth values, and hence they advocate the use of R-implications (i.e. residuated implications) as they have a better behaviour in this respect.

Definition 2.3.2 Let $\langle C, \leq \rangle$ be a complete (bounded) chain and let $*$ be a t-norm over C. Then, the *residuum* of $*$ is a binary operation $\rightarrow_*$ on C such that the following property is satisfied for all $x, y, z \in C$:

$$x * y \leq z \text{ if and only if } x \leq y \rightarrow_* z \quad \text{(Residuation condition)}.$$

The residuum of a t-norm does not always exist. Indeed, if $C = [0, 1]$, a t-norm $*$ on C has residuum if and only if the t-norm is left-continuous. This condition makes clear that the residuum of $*$ and the R-implication associated to $*$ are not exactly the same notion, as the R-implication always exists since $[0, 1]$ is complete, but if the residuum exists (i.e. if $*$ is left-continuous) then they do coincide. Indeed an easy computation shows that a t-norm and its associated R-implication satisfy the residuation condition if and only if the supremum in the definition of the R-implication (see Definition 2.3.1) is, in fact, a maximum.

It is easy to check that if $*$ is left-continuous then:

- its residuum $\rightarrow_*$ is univocally defined as

$$x \rightarrow_* y = \max\{z \mid x * z \leq y\};$$

- $x \rightarrow_* y = 1$ if and only if $x \leq y$.

Therefore if a t-norm $*$ has a residuum, we will denote it as $\rightarrow_*$. Nevertheless we will write only $\rightarrow$ if there is no possibility of confusion.

Finally, in [14] it is proved that if a t-norm $*$ has residuum, then the divisibility condition is equivalent to both the continuity of $*$ and to the satisfaction of the following equation:

$$x * (x \rightarrow_* y) = \min(x, y) \quad \text{(Divisibility equation)}.$$

This equivalence is well known but, for the reader's convenience, we will reproduce the proof for the case of divisible finite t-norms. Suppose $*$ is a finite and divisible t-norm. Then, for each pair $x, y \in C$ such that $x \geq y$, there exists z such that $x * z = y$. Then, if $x \geq y$, by definition of the residuum (that clearly exists for any finite t-norm), it must hold that $x * (x \rightarrow y) = y = \min(x, y)$. On the other hand, it is clear that if $x \leq y$, then $x * (x \rightarrow y) = x * 1 = x = \min(x, y)$. Notice the interest of this equivalence for t-norms on $[0, 1]$, since a topological property like continuity can be equivalently expressed by an equation, the divisibility equation.

In the case of C being a finite chain, the residuum of a (finite) t-norm always exists (the supremum is always a maximum) but, as we have already observed (see Theorem 2.2.4), not all finite t-norms are divisible, as the following example shows:

Example 2.3.3 Let $*$ be the t-norm on the finite set $C = \{0, a, b, 1\}$ with $0 < a < b < 1$, defined by $a * b = a * a = 0$ and $b * b = b$, i.e. the nilpotent minimum over a four elements chain. Obviously $*$ is not divisible since $a < b$ and there is no $x \in C$ such that $b * x = a$.

2.4   Axiomatizing Finite Divisible T-norms

In this section we describe how to obtain a finite equational characterization of any finite divisible t-norm, with equations in the language $\langle *, \rightarrow, \wedge, \vee, 0, 1 \rangle$, that is, using symbols not only for the t-norm operation but also for its residuum. Actually, the reader can wonder whether one could do it with equations in the restricted language $\langle *, \wedge, \vee, 0, 1 \rangle$ without the residuum $\rightarrow$. And it turns out that, as shown by Bou in [3], equations in this language cannot distinguish for instance on a chain of four elements the finite t-norm $Ł_2 \oplus Ł_3$ from the t-norm $Ł_3 \oplus Ł_2$. Indeed, Bou shows [3, Lemma 4] that an equation in the restricted language is valid on an ordinal sum of hoops $\mathbf{A} \oplus \mathbf{B}$ if, and only if, it is valid both in $\mathbf{A}$ and in $\mathbf{B}$. Indeed, this proves that the variety generated by an ordinal sum is indistinguishable from the one generated by any permutation of the components in the ordinal sum. Therefore there is no hope to obtain an equational characterization of any (divisible) t-norm different from the minimum t-norm with equations in the restricted language $\langle *, \wedge, \vee, 0, 1 \rangle$.[2]

Hence we are led to consider equations over a language including an operation for the residuum of the t-norm as well. In doing so, we are actually prompted in fact to consider enriched algebraic structures of the kind $\langle A, \wedge, \vee, *, \rightarrow_*, 0, 1 \rangle$, where the lattice reduct $\langle A, \wedge, \vee, 0, 1 \rangle$ is indeed a finite linearly ordered set, $*$ is a finite divisible t-norm on A and $\rightarrow_*$ is its residuum. These structures are examples of linearly ordered BL-algebras, or BL-chains. BL-algebras are bounded, integral, commutative, pre-linear and divisible residuated lattices, and they are the algebraic counterpart of Hájek's BL logic [14], a logic capturing the common 1-tautologies of all the many-valued calculi on $[0, 1]$ defined by a continuous t-norm and its residuum.

Before describing how to get an equational characterization of (the BL-chain defined by) a finite divisible t-norm, mainly based on results from [5], we first recall an alternative ordinal sum decomposition of a finite BL-chain that has advantages for our purposes.

2.4.1   An Alternative Decomposition of a Finite Divisible T-norm as Ordinal Sum of Hoops

First of all we consider an example in order to stress a problem concerning the ordinal sum of (finite) t-norms when the residuated implication is involved. Let $*$ be a divisible finite t-norm over a chain $\mathbf{A}$ that is an ordinal sum of two non-trivial components $*_1$ and $*_2$, i.e. $* = *_1 \oplus *_2$. Suppose now that $x \leq y$ are elements of the first component. Then, clearly, $x \rightarrow_* y = 1$, but 1 is not an element of the first component. This means that, as an ordinal sum of BL-chains $\mathbf{A} = \langle A_1, \wedge, \vee, *_1, \rightarrow_{*_1}, $

---

[2]Note however, that Bou has shown [4] that there is at least one equation in the language $(*, \wedge, \vee, 0, 1)$ that is valid for all finite divisible t-norms but fails in some finite non-divisible t-norm. In particular the exhibited equation in [4] has 9 variables and it fails on a t-norm over a chain of 33 elements.

$a_0, a_{n_1}\rangle \oplus \langle A_2, \wedge, \vee, *_2, \to_{*_2}, a_{n+1}, a_m \rangle$, the first component $\mathbf{A_1}$ is not a subalgebra of the algebra $\mathbf{A}$ defined over the full chain.

As a particular case of a more general result of Aglianò and Montagna in [1], we recall a slightly different notion of ordinal sum for finite linearly-ordered *Wajsberg hoops*. Actually, a *hoop* is an algebra $\mathbf{A} = \langle A, *, \to, 1 \rangle$ such that $\langle A, *, 1 \rangle$ is a commutative monoid and for all $x, y, z \in A$ the following equations hold: $x \to x = 1$, $x * (x \to y) = y * (y \to x)$, $x \to (y \to z) = (x * y) \to z$. A *Wajsberg hoop* is a hoop satisfying the equation: $(x \to y) \to y = (y \to x) \to x$. A *bounded hoop* is an algebra $\mathbf{A} = (A, *, \to, 1, 0)$ such that $\langle A, *, \to, 1 \rangle$ is a hoop and $0 \leq x$ for all $x \in A$. Then it turns out that bounded Wajsberg hoops are termwise equivalent to MV-algebras, or in other words, BL-algebras satisfying the equation $\neg\neg x = x$, where $\neg x = x \to 0$. Particularly relevant examples of finite Wajsberg hoops are the following.

Lemma 2.4.1 *Any linearly ordered finite (bounded) Wajsberg hoop of n elements is isomorphic to the hoop* $\mathbf{Ł}_n = \langle Ł_n, *, \to, 1 \rangle$, *where*

- *the support of* $Ł_n$ *is the set* $\{0, \frac{1}{n-1}, \ldots \frac{n-2}{n-1}, 1\}$,
- $*$ *is the n-valued Łukasiewicz t-norm, i.e.,* $x * y = \max(0, x + y - 1)$,
- $\to$ *is the corresponding residuum, i.e.,* $x \to y = \min(1, 1 - x + y)$.

Therefore, from now on, when speaking about finite linearly ordered Wajsberg hoops, we will directly refer to the hoops $\mathbf{Ł}_n$. Notice that $\mathbf{Ł}_2$ coincide with the two-element Boolean algebra.

Definition 2.4.2 (*Ordinal sums of Wajsberg hoops*)
Let $Ł_{k_i} = \langle Ł_{k_i}, *_i, \to_i, 1 \rangle$ for $1 \leq i \leq m$ be a finite family of finite linearly ordered Wajsberg hoops such that $Ł_{k_i} \cap Ł_{k_j} = \{1\}$ for all $i \neq j$. The ordinal sum (as hoops) of that family is the hoop

$$Ł_{k_1} \oplus Ł_{k_2} \oplus \cdots \oplus Ł_{k_n} = \langle \bigcup_{i=1}^{n} Ł_{k_i}, *, \to, 1 \rangle,$$

where:

- the order is defined by: $x \leq y$ if either both x and y belong to the same component and $x \leq y$, or $y = 1$, or $x \in Ł_{k_i}$ and $y \in Ł_{k_j}$ and $i < j$.
- $x * y = x *_i y$ if $x, y \in Ł_{k_i}$, and $x * y = \min(x, y)$ otherwise.
- $x \to y$ is either $x \to_i y$ if $x, y \in Ł_{k_i}$, or 1 if $x \leq y$, or y if x, y belong to different components and $x > y$.

A main advantage of this kind of decomposition is that the components $\langle Ł_{k_i}, *_i, \to_{*_i}, 1 \rangle$ are substructures (i.e., subhoops) of the whole hoop structure $Ł_{k_1} \oplus Ł_{k_2} \oplus \cdots \oplus Ł_{k_n}$.

From this definition it is easy to prove the following hoop decomposition theorem for finite divisible t-norms.

Fig. 2.1 T-norm ordinal sum versus hoop ordinal sum

$\mathbf{A} = G_3 \oplus \mathbf{L}_5$ as ordinal sum of t-norms

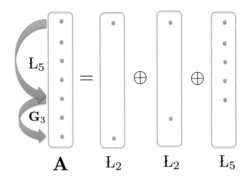

$\mathbf{A} = \mathbf{L}_2 \oplus \mathbf{L}_2 \oplus \mathbf{L}_5$ as ordinal sum of hoops

Theorem 2.4.3 *For any given finite divisible t-norm, its corresponding finite BL-chain is (isomorphic to) an ordinal sum of a finite family of finite linearly ordered Wajsberg hoops.*

Proof Given a finite divisible t-norm we know, by the Mayor and Torrens result, that it is an ordinal sum (as t-norms) of copies of finite minimum t-norm and finite Łukasiewicz t-norms. Take for each minimum component as many $\mathbf{L}_2$ as elements has the component minus 1, and for each finite Łukasiewicz component take the corresponding Wajsberg hoop of the same cardinal. An easy computation shows that the structure $\langle C, *, \rightarrow_*, 1 \rangle$ is in fact an ordinal sum (as hoops) of components of the type $\mathbf{L}_k$ defined before. □

Example 2.4.4 Take the t-norm $*$ defined by $G_3 \oplus \mathbf{L}_5$ as ordinal sum of t-norms over a finite chain of 7 elements C. Then the (hoop) structure $\langle C, *, \rightarrow_*, 1 \rangle$ is the ordinal sum of hoops: $\mathbf{L}_2 \oplus \mathbf{L}_2 \oplus \mathbf{L}_5$ (see Fig. 2.1). As noticed, the components $\langle \mathbf{L}_k, *, \rightarrow_*, 1 \rangle$ are subhoops of $\langle C, *, \rightarrow_*, 1 \rangle$.

2.4.2 Equational Characterization of a Divisible Finite T-norm

As a necessary first step, let us focus on the equational characterization of the finite linearly ordered Wajsberg hoops $\mathbf{L}_n$. In what follows we will denote by x^n the result of the operations $x * \overset{n}{\cdots} * x$, and by $n.x$ the result of the operation $x \oplus \overset{n}{\cdots} \oplus x$, where $\oplus$ is the bounded sum operation (the dual of the Łukasiewicz t-norm), that is definable in each Wajsberg hoop as $x \oplus y := \neg(\neg x * \neg y)$.

Notice that the Wajsberg hoops of the family $Ł_n$, besides satisfying the typical equations of t-norms:

$$x * y = y * x \tag{2.1}$$

$$x * (y * z) = (x * y) * z \tag{2.2}$$

$$1 * x = x \tag{2.3}$$

$$(x \wedge y) * z = (x * z) \wedge (y * z), \tag{2.4}$$

they also satisfy the divisibility equation:

$$x * (x \rightarrow_* y) = \min(x, y), \tag{2.5}$$

the involution equation for the negation:

$$\neg\neg x = x, \tag{2.6}$$

and the $\vee$-definability equation:

$$(x \rightarrow_* y) \rightarrow_* y = \max(x, y). \tag{2.7}$$

Actually, to fully characterize the basic Wajsberg hoops $Ł_n$ we have at hand the axiomatization provided by Grigolia [12] of the $Ł_n$'s as finite MV-algebras (see also [8]). Indeed, $Ł_n$ is equationally characterized as MV-algebra by the (finite) set of equations of axiomatizing the variety of MV-algebras (see e.g. [6]), together with the following equations in one variable:

$$x^n = x^{n-1}, \tag{τ_n}$$

and, if $n \geq 4$:

$$(p \cdot x^{p-1})^n = n \cdot x^p, \tag{$\tau\nu_{np}$}$$

for every $p \in \{2, \ldots, n-2\}$ that does not divide $n-1$.

Since an equation of the kind $t(x) = s(x)$ can be rewritten, using the double implication, as $t(x) \leftrightarrow s(x) = 1$, the above finite set of equations $\{(\tau_n)\} \cup \{(\tau\nu_{np}) : p \in \{2, \ldots, n-2\}$ not dividing $n-1\}$ can be equivalently expressed, using the conjunction, as a single equation on one variable:

$$t_n(x) = 1. \tag{t_n}$$

Therefore, the equations characterizing $Ł_n$ will be those of MV-algebras plus (t_n). For example, for $Ł_3$, the equation (t_3) is:

$$x^3 \leftrightarrow x^2 = 1, \tag{t_3}$$

while the equation (t_4) is:

$$((2x)^4 \leftrightarrow 4x^2) \wedge (x^4 \leftrightarrow x^3) = 1. \tag{t_4}$$

Notice that a set of equations defines a *variety* of algebras, and thus the equations given above actually define the variety of Wajsberg hoops generated by $Ł_n$. This implies that the equations characterizing $Ł_n$ are also satisfied by the subalgebras of $Ł_n$, i.e., by $Ł_k$, where k divides n (but the proper subalgebras satisfy equations that $Ł_n$ does not). Nevertheless, they do provide a univocal characterization in the following sense: given a chain $\langle C, \leq \rangle$ of n elements, then $Ł_n$ is the unique Wajsberg hoop defined by a t-norm on C that satisfies the above equations.

In order to axiomatize any finite ordinal sum of finite Wajsberg hoops we need some preliminary results. In the following, for any natural k we will denote by $t_k^*(x)$ the term obtained from $t_k(x)$ by replacing the constant 0 by x^k.

Lemma 2.4.5 (Cf. [5, Lemma 5.4.1]) *Let $A_1 \oplus \cdots \oplus A_n$ be an ordinal sum of finite linearly ordered Wajsberg hoops and assume A_i is a component with k elements. Then A_i is isomorphic to $Ł_k$ if and only if the equation*

$$t_k^*(x) = 1. \tag{t_k^*}$$

is valid in A_i.

Proof The basic difference between A_i as component of the ordinal sum and $Ł_k$ is the minimum element. The minimum of A_i is not 0 (the minimum of the ordinal sum) but it can be recovered taking x^k for any element $x < 1$ of A_i. Then the result follows.   □

Lemma 2.4.6 *Let C be a finite chain with n elements, and let $*$ be a divisible t-norm defined on C. Then the equation*

$$\bigwedge_{i=1}^{n} ((x_{i+1} \rightarrow x_i) \rightarrow x_i) \leq \bigvee_{i=1}^{n+1} x_i. \tag{λ_n}$$

is valid on the hoop $\langle C, *, \rightarrow, 1 \rangle$ if and only if its decomposition as ordinal sum of hoops $Ł_k$'s has a number of components less or equal than n.

Proof Observe first that if $x_{i+1} \leq x_i$ and they belong to a different component then $((x_{i+1} \rightarrow x_i) \rightarrow x_i) = 1 \rightarrow x_i = x_i$ and thus the inequality holds. Moreover if x_i, x_{i+1} belong to the same component then $((x_{i+1} \rightarrow x_i) \rightarrow x_i) = x_i \vee x_{i+1}$, and thus the inequality holds as well. Thus, in order to check whether the inequality does not hold, we only need to take into account a sequence of $n + 1$ elements x_i such that they are strictly increasing and each x_i belonging to a different component. If the number of components is less or equal than n then such a sequence does not exist, and thus the inequality holds. However, if the number of components is greater than n then an strictly increasing sequence x_i where each element belong to a different component and $x_{n+1} \neq 1$ exists. But for this sequence and for each $i \in \{1, \ldots, n\}$, $((x_{i+1} \to x_i) \to x_i) = 1$ and $\bigvee_{i=1}^{n+1} x_i = x_{n+1} \neq 1$. Thus the inequality does not hold. $\qquad\square$

Lemma 2.4.7 *Let C be a finite chain and let $*$ be a divisible t-norm defined on C such that $\langle C, *, \to, 1 \rangle = Ł_{k_1} \oplus Ł_{k_2} \oplus \cdots \oplus Ł_{k_n}$, i.e. the ordinal sum decomposition has n components. Then the equation*

$$\bigwedge_{i=1}^{n-1} ((x_{i+1} \to x_i) \to x_i) \leq \bigvee_{i=1}^{n} x_i \vee (\bigwedge_{i=1}^{n} t_{k_i}^*(x_i)) \qquad (\varepsilon_n)$$

is valid on the hoop $\langle C, *, \to, 1 \rangle$.

Proof Like in the proof of the previous lemma, the inequality clearly holds in the case that either $x_{i+1} \leq x_i$, and then $(x_{i+1} \to x_i) \to x_i = x_i$, or both x_i, x_{i+1} belong to the same component and then $(x_{i+1} \to x_i) \to x_i = x_i \vee x_{i+1}$. Then, since the number of components is n, a strictly increasing sequence x_i where each element belong to a different component with $x_n \neq 1$ exists. Then, for each x_i the corresponding equation $t_{k_i}^*(x_i)$ defining $Ł_{k_i}$, has to hold. $\qquad\square$

Therefore, the problem is to fix that the number of components is exactly n, but this is not definable directly because a set of equations defines a variety and if a variety contain $Ł_k$ have to contain its subalgebras in particular $Ł_r$ for r divisor of k. In the paper [1], Aglianò and Montagna solve the problem in the following way.

Lemma 2.4.8 *Let C be a finite chain and let $*$ be a divisible t-norm defined on C such that $\langle C, *, \to, 1 \rangle = Ł_{k_1} \oplus Ł_{k_2} \oplus \cdots \oplus Ł_{k_n}$ Then $(C, *, \to, 1, 0)$ is characterized by the equations:*

$$\bigwedge_{i=1}^{n} ((x_{i+1} \to x_i) \to x_i) \leq \bigvee_{i=1}^{n+1} x_i \qquad (\lambda_n)$$

$$t_{k_1}^*(\neg\neg x) = 1 \qquad (t_{k_1}^*)$$

together with the set of equations (ε_r) for $r = 2, \ldots, n$:

$$\bigwedge_{i=1}^{r-1} ((x_{i+1} \to x_i) \to x_i) \to \bigvee_{i=1}^{r} x_i \vee \bigvee_{\sigma_r} (t_{k_{\sigma_r(1)}}^*(x_1) \wedge \cdots \wedge t_{k_{\sigma_r(r)}}^*(x_r)) = 1 \qquad (\varepsilon_r)$$

where for every r, σ_r ranges over increasing sequences of r elements out of n

Proof By the previous lemmas, we know that $*$ is an ordinal sum with less than $n + 1$ components and that if it has n components they have to be the components of $*$. It only remains to prove that the satisfaction of equations (ε_r) implies that $*$ cannot have less than n components. Suppose that $*$ has $r < n$ components. Then we can define a strictly increasing sequence $x_1 < x_2 < \cdots < x_r$ and by equation

(ε_r) we know that there is a sequence σ such that $(t_{k_{\sigma(1)}}(x_1) \wedge \cdots \wedge (t_{k_{\sigma(r)}})(x_r)) = 1$. This implies that the r components are $Ł_{k_{\sigma(1)}}, \ldots, Ł_{k_{\sigma(r)}}$, but the sum of the number of elements of these components is less than n and thus $Ł_{k_{\sigma(1)}} \oplus \cdots \oplus Ł_{k_{\sigma(r)}}$ does not define a t-norm over C. $\square$

To finish the paper we give two examples, the latter being a simpler axiomatic system for the particular case that the decomposition of $*$ as hoops has only two components.

Example 2.4.9 Suppose the decomposition of $\langle C, *, \rightarrow, 1 \rangle$ as ordinal sum is $Ł_s \oplus Ł_t \oplus Ł_r$. Then the following equations determine $*$:

$$((x_4 \rightarrow x_3) \rightarrow x_3) \wedge ((x_3 \rightarrow x_2) \rightarrow x_2) \wedge ((x_2 \rightarrow x_1) \rightarrow x_1) \leq x_1 \vee x_2 \vee x_3 \vee x_4 \tag{λ_3}$$

$$t_s^*(\neg\neg x) = 1 \tag{t_s^*}$$

$$((x_2 \rightarrow x_1) \rightarrow x_1) \rightarrow [x_1 \vee x_2 \vee (t_s^*(x_1) \wedge t_t^*(x_2)) \vee (t_s^*(x_1) \wedge t_t^*(x_2)) \vee (t_t^*(x_1) \wedge t_r^*(x_2))] = 1 \tag{ε_2}$$

$$[((x_3 \rightarrow x_2) \rightarrow x_2) \wedge ((x_2 \rightarrow x_1) \rightarrow x_1)] \rightarrow [x_1 \vee x_2 \vee x_3 \vee (t_s^*(x_1) \wedge t_t^*(x_2) \wedge t_r^*(x_3))] = 1 \tag{ε_3}$$

Example 2.4.10 When the decomposition as hoops of a finite t-norm has only two components, then there is also the following simplified equational characterization with only two equations. Namely, let C be a finite chain of n elements and $*$ be a divisible t-norm over C such that the decomposition of $\langle C, *, \rightarrow, 1 \rangle$ as ordinal sum of hoops is $Ł_s \oplus Ł_t$, i.e., it has only two components. Then, the following simplified pair of equations determine $*$:

$$t_s^*(\neg\neg x) = 1. \tag{2.8}$$

$$t_t^*(\neg\neg x \rightarrow x) = 1. \tag{2.9}$$

The proof is very easy since all the elements of the first component $Ł_s$ are of the form $\neg\neg x$, with $x \in C$, while all the elements of the second component $Ł_t$ are of the form $\neg\neg x \rightarrow x$, with $x \in C$. In other words, $C = \{\neg\neg x \mid x \in C\} \cup \{\neg\neg x \rightarrow x \mid x \in C\}$. To finish the proof, take into account that the chain defined by only the first component would satisfy these equations as well, but it would not be a t-norm over C, since it should coincide with $Ł_s$, and $s < n$.

2.5   Conclusions

The paper has overviewed an approach to characterize divisible t-norms on finite chains by a finite set of equations that use not only the t-norm itself but also its residuum. Thus, in fact, these equations characterize the class (variety) of algebraic structures over finite sets defined by them, namely finite BL-chains.

If we move from finite divisible t-norms to continuous (or divisible) t-norms over $[0, 1]$, then each continuous t-norm defines a *standard* BL-chain, namely the structure $[0, 1]_* = \langle [0, 1], *, \rightarrow_*, 0, 1 \rangle$. In [9] it is proved that there is a finite set of equations (using the t-norm itself but also its residuum) defining the variety $V([0, 1]_*)$ generated by a standard BL-chain $[0, 1]_*$. Nevertheless, only when the t-norm $*$ is a *finite* ordinal sum of copies of Łukasiewicz, product and minimum t-norms, the equations actually characterize the t-norm, since the only standard BL-chain contained in the variety $V([0, 1]_*)$ is $[0, 1]_*$ itself. However when $*$ is an infinite ordinal sum of copies of Łukasiewicz, product and minimum t-norms, there exist an infinite number of continuous t-norms $\circ$ such that $[0, 1]_\circ \in V([0, 1]_*)$.

Dedication

This short note is dedicated to Gaspar Mayor in the occasion of his 70th birthday. We have taken as starting point the research line initiated in his early works about representation of finite divisible t-norms and we have ended with the equational characterization of them. Thanks a lot for your inspiring work and congratulations Gaspar!

Acknowledgments The authors acknowledge support from the MINECO project EdeTRI (TIN2012-39348-C02-01), the LODISCO network TIN2014-56381-REDTLODISCO, and the grants 2014SGR-788 and 2014SGR-118. García-Cerdaña also acknowledges the MICINN project MTM 2011-25747.

References

1. P. Aglianò, F. Montagna, Varieties of BL-algebras I: general properties. J. Pure Appl. Algebra **181**(2–3), 105–129 (2003)
2. D. Boixader, F. Esteva, L. Godo, On the continuity of t-norms. Proc. IFSA **1999**, 476–479 (1999)
3. F. Bou, An exotic MTL-chain. Unpublished Manuscript (2014). Available from the author
4. F. Bou, Introducing an exotic MTL-chain, in *Logic, Algebra and Truth Degrees (LATD 2014): Abstract Booklet*, ed. by M. Baaz, A. Ciabattoni, and S. Hetzl, pp.143–145, 2014
5. M. Busaniche, F. Montagna, Hájek's logic BL and BL-algebras, in *Handbook of Mathematical Fuzzy Logic*, ed. by P. Cintula, P. Hájek, C. Noguera. Studies in Logic, Mathematical Logic and Foundations, vol. 37 (College Publications, London, 2011), pp. 355–447
6. R. Cignoli, I.M.L. D'Ottaviano, D. Mundici, *Algebraic Foundations of Many-Valued Reasoning*. Trends in Logic, vol. 7 (Kluwer, Dordrecht, 1999)
7. R. Cignoli, F. Esteva, L. Godo, A. Torrens, Basic fuzzy logic is the logic of continuous t-norms and their residua. Soft Comput. **4**(2), 106–112 (2000)
8. A. Di Nola, A. Lettieri, Equational characterization of all varieties of MV-algebras. J. Algebra **221**(2), 463–474 (1999)
9. F. Esteva, L. Godo, F. Montagna, Equational characterization of subvarieties of BL-algebras generated by t-norm algebras. Stud. Logica. **76**(2), 161–200 (2004)
10. L. Godo, C. Sierra, A new approach to connective generation in the framework of expert systems using fuzzy logic, in *Proceedings of the 18th IEEE International Symposium on Multiple-Valued Logic*, pp. 144–151 (1988)

11. S. Gottwald, in *A Treatise on Many-Valued Logics*. Studies in Logic and Computation, vol. 9 (Research Studies Press, Baldock, 2001)
12. R. Grigolia, in *Algebraic analysis of Łukasiewicz-Tarski's n-valued logical systems*, ed. by R. Wójcicki, G. Malinowski. Selected Papers on Łukasiewicz Sentential Calculi (Wydawn. Polsk. Akad. Nauk, Wrocław, 1977), pp. 81–92
13. P. Hájek, Basic fuzzy logic and BL-algebras. Soft Comput. **2**(3), 124–128 (1998)
14. P. Hájek, in *Metamathematics of Fuzzy Logic*. Trends in Logic, vol. 4 (Kluwer, Dordrecht, 1998)
15. G. Mayor, J. Torrens, On a class of operators for expert systems. Int. J. Intell. Syst. **8**, 771–778 (1993)
16. G. Mayor, J. Torrens, in *Triangular Norms in Discrete Settings*, ed. by E.P. Klement, R. Messiar. Logical, Algebraic, Analytic, and Probabilistic Aspects of Triangular Norms (Elsevier, Amsterdam, 2005), pp. 189–230

Chapter 3
Associative Copulas: A Survey

Juan Fernández-Sánchez, José Juan Quesada-Molina
and Manuel Úbeda-Flores

Abstract Copulas—functions that join multivariate distribution functions to their one-dimensional margins—are special cases of binary 1-Lipschitz aggregation functions, commonly used in aggregation processes. Here we consider a significant class of copulas: Associative copulas. We explore briefly the subclass of Archimedean copulas, and some of the properties and applications of associative copulas, such as the simultaneous associativity, the Kendall distribution functions, topological aspects, etc. Finally, some open problems are posed.

3.1 Introduction

The class of copulas—a special type of aggregation function—plays an important role in the theory of probabilistic metric spaces (see [1]), in probability theory (see [2, 3]), or in the study of nonparametric measures of dependence for random variables (see [4]), among many other fields. Our aim in this survey is to revisit an important subclass of copulas: Associative copulas. We will review the Archimedean family of copulas and revisit some significant aspects of associative copulas, such as the Kendall distribution functions, the simultaneous associativity, approximation, concavity properties, etc. Finally, some open problems are posed.

J. Fernández-Sánchez · M. Úbeda-Flores (✉)
Research Group of Mathematical Analysis, University of Almería, 04120 Almería, Spain
e-mail: mubeda@ual.es

J. Fernández-Sánchez
e-mail: juanfernandez@ual.es

J.J. Quesada-Molina
Department of Applied Mathematics, University of Granada, 18071 Granada, Spain
e-mail: jquesada@ugr.es

© Springer International Publishing Switzerland 2016
T. Calvo Sánchez and J. Torrens Sastre (eds.), *Fuzzy Logic and Information Fusion*,
Studies in Fuzziness and Soft Computing 339, DOI 10.1007/978-3-319-30421-2_3

3.2   Preliminaries

A (binary) *aggregation function* is a function $A\colon [0, 1]^2 \longrightarrow [0, 1]$ such that (i) $A(0, 0) = 0$ and $A(1, 1) = 1$, and (ii) $A(u_1, v_1) \leq A(u_2, v_2)$ for any $u_1, u_2, v_1, v_2 \in [0, 1]$ such that $u_1 \leq u_2$ and $v_1 \leq v_2$. Aggregation functions play an important role in many applications of fuzzy set theory and fuzzy logic (see [5–7]), among many other fields, including probability and statistics.

An aggregation function A is called a *conjunctor* if it has 0 as absorbing element, i.e. $A(t, 0) = A(0, t) = 0$ for any $t \in [0, 1]$.

An aggregation function A with neutral element 1—i.e., $A(t, 1) = A(1, t) = t$ for any $t \in [0, 1]$—is called a *triangular norm* (briefly, *t-norm*) if it is commutative and associative, i.e. it satisfies $A(u, v) = A(v, u)$ and $A(A(u, v), w) = A(u, A(v, w))$ for all $u, v, w \in [0, 1]$.

Copulas—bivariate probability distribution functions with uniform margins on $[0, 1]$—are special types of conjunctors. In fact, every associative copula is a continuous *t*-norm, and every *t*-norm which satisfies the 1-Lipschitz property is a copula (see [8, 9]). They are used in aggregation processes because they ensure that the aggregation is stable in the sense that small error inputs correspond to small error outputs (see [10]). For a complete review on copulas, we refer to the monographs of [11, 12] (see also [13–16] for some applications).

Specifically, a function $C\colon [0, 1]^2 \longrightarrow [0, 1]$ is a *copula* if the following conditions hold:

(C1) $C(t, 0) = C(0, t) = 0$ and $C(t, 1) = C(1, t) = t$ for every $t \in [0, 1]$;
(C2) C is *2-increasing*, i.e. $C(u_1, v_1) + C(u_2, v_2) \geq C(u_1, v_2) + C(u_2, v_1)$ for every $u_1, u_2, v_1, v_2 \in [0, 1]$ such that $u_1 \leq u_2$ and $v_1 \leq v_2$.

The importance of copulas in probability and statistics comes from *Sklar's theorem* (see [3]), which shows that the joint distribution function H of a pair of random variables and the corresponding marginal distribution functions F and G are linked by a copula C in the following manner:

$$H(x, y) = C(F(x), G(y)), \quad \forall (x, y) \in [-\infty, \infty]^2. \tag{3.1}$$

If F and G are continuous, then the copula is unique; otherwise, the copula is uniquely determined on Range $F \times$ Range G (see [17] for more details).

For any copula C we have the following inequalities:

$$\max(u + v - 1, 0) = W(u, v) \leq C(u, v) \leq M(u, v) = \min(u, v)$$

for all $(u, v) \in [0, 1]^2$, where M and W are themselves (associative) copulas. We note that W is also known as Łukasiewicz *t*-norm in fuzzy logic. Another example of associative copula is the copula for independent random variables, i.e. $\Pi(u, v) = uv$ for all (u, v) in $[0, 1]^2$.

We recall that if a copula C is associative then it is commutative—or exchangeable —(see [8, 18]); however, the converse is not true in general. For example, the copula $(1 - \lambda)W + \lambda M$, for every $\lambda \in (0, 1)$, is commutative but not associative.

Given a copula C, we denote by δ_C the *diagonal section* of C, which is defined by $\delta_C(t) = C(t, t)$ for every $t \in [0, 1]$.

Finally, given a real function f, the left-hand and right-hand derivatives of f at a point $x \in \mathbb{R}$ are denoted, respectively, by $f'(a^-)$ and $f'(a^+)$, whereas the limits exist.

3.3 Some Results on Associative Copulas

In the next subsections we review some—but not all–of the most outstanding facts, properties and results involving associative copulas. We tackle both theoretical results and related applications.

3.3.1 Archimedean Copulas

In this subsection, we review an important class of associative copulas known as Archimedean copulas. They became popular since they model the dependence structure between risk factors, and are used in many applications, such as finance, insurance, or reliability (see, for example, [13, 19]) due to their simple forms and nice properties.

Let $\varphi \colon [0, 1] \longrightarrow [0, \infty]$ be a continuous strictly decreasing function such that $\varphi(1) = 0$, and let $\varphi^{[-1]}$ be the *pseudo-inverse* of φ, i.e., $\varphi^{[-1]}(x) = \varphi^{-1}(\min(\varphi(0), x))$ for $x \in [0, \infty]$. Consider the function given by

$$C_\varphi(u, v) = \varphi^{[-1]}(\varphi(u) + \varphi(v)), \quad (u, v) \in [0, 1]^2. \tag{3.2}$$

The following characterization can be found in [1, 12, 20].

Theorem 3.1 *The function C_φ given by (3.2) is a copula if and only if φ is convex.*

Copulas given by (3.2) are called *Archimedean*—the name is due to a property of associative operations (see [12, 18])—, and the function φ is called the *generator* of C_φ.

When $\varphi(0) = \infty$, the Archimedean copula C is said to be *strict*, and when $\varphi(0) < \infty$, it is said to be *non-strict*. When C is strict, $C(u, v) > 0$ for all $(u, v) \in (0, 1]^2$. Π and W are Archimedean copulas with respective generators $\varphi_1(t) = -\ln t$, with $\varphi_1(0) = \infty$, and $\varphi_2(t) = 1 - t$ for all $t \in [0, 1]$. Moreover, Π is strict. Other important examples of Archimedean copulas are listed in Table 3.1, and for further examples we refer to [12].

It is interesting to note that if C is an Archimedean copula with diagonal section δ_C and $\delta'_C(1^-) = 2$, then C is uniquely determined by its diagonal section (see [21]).

Table 3.1 Some parametric families of Archimedean copulas

Name	$\varphi_\theta(t)$	θ	Strict
Clayton	$\dfrac{1}{\theta}(t^{-\theta} - 1)$	$[-1, \infty)\backslash\{0\}$	$\theta > 0$
Ali-Mikhail-Haq	$\ln\dfrac{1 - \theta(1 - t)}{t}$	$[-1, 1)$	Yes
Gumbel	$(-\ln t)^\theta$	$[1, \infty)$	Yes
Frank	$-\ln\dfrac{e^{-\theta t} - 1}{e^{-\theta} - 1}$	$\mathbb{R}\backslash\{0\}$	Yes

Archimedean copulas are associative and satisfy $\delta_C(u) < u$ for all $u \in (0, 1)$; in fact, we have the following characterization of Archimedean copulas (see [18] for more details).

Theorem 3.2 *Let C be a copula. Then, C is an Archimedean copula if and only if C is associative and $\delta_C(u) < u$ for all $u \in (0, 1)$.*

Furthermore, a sufficient condition for an associative copula to be Archimedean is given in the next result.

Theorem 3.3 *If a copula C is associative and is differentiable at every point (t, t), with $t \in (0, 1)$, then it is Archimedean.*

Finally, we stress that statistical inference procedures for Archimedean copulas—for example, choosing a particular copula to fit a data set—can be found in [22–24].

3.3.2 Ordinal Sums

The ordinal sum construction was firstly introduced in the context of the theory of semigroups; and later, it has been used in the theory of t-norms and copulas.

Let $\{J_i\} = \{[a_i, b_i]\}$, $a_i < b_i$, be a collection of subintervals of $[0, 1]$. Let $\{C_i\}$ be a collection of copulas with the same indexing as $\{J_i\}$ such that, for all indices i for which $J_i = \{a_i\}$ (i.e., $a_i = b_i$), $C_i(u, v) = M(u, v)$ for every (u, v) in $[0, 1]^2$. Then, the *ordinal sum* of the collection $\{C_i\}$ with respect to $\{J_i\}$ is the copula C defined by

$$C(u, v) = \begin{cases} a_i + (b_i - a_i) \cdot C_i\left(\dfrac{u - a_i}{b_i - a_i}, \dfrac{v - a_i}{b_i - a_i}\right), & (u, v) \in J_i^2, \\ M(u, v), & \text{otherwise.} \end{cases}$$

So basically, in an ordinal sum it is redefined the value that a copula assumes on a rectangle scaling copies of the copulas C_i over the squares J_i^2.

A characterization of those copulas that have an ordinal sum representation is the following (see [12]): If C is a copula, then C is an ordinal sum if and only if there exists $t \in (0, 1)$ such that $\delta_C(t) = t$.

In the theory of t-norms (copulas), ordinal sums are used in order to provide both a general method of construction and a model of representing continuous t-norms (copulas).

We have the following characterization of associative copulas in terms of ordinal sums (see [8]).

Theorem 3.4 *A copula is associative if and only if it is an ordinal sum of a collection of copulas $\{C_i\}$ such that, for each i, C_i is Archimedean.*

We want to stress that, according to the definition of an ordinal sum and Theorem 3.4, a copula C is associative if and only if C is Archimedean or $C = M$ or C is an ordinal sum of Archimedean copulas.

3.3.3 Kendall Distribution Functions

Let (X, Y) be a continuous random vector with distribution function H. Then the distribution function of the random variable $H(X, Y)$—which we denote by K—is called *Kendall distribution function*. Kendall distribution functions arise in the study of stochastic orderings of random vectors. Kendall's name is due to the fact that the population version of the measure of association known as *Kendall's tau* (see [12] for a detailed study of this measure) can be expressed as

$$\tau(X, Y) = 3 - 4 \int_0^1 K(t)\, dt$$

(see [22, 25]). In [26] it is proved that every distribution function K satisfying $\lim_{t \to 0^-} K(t) = 0$ and $K(t) \geq t$, with $t \in [0, 1]$, is the Kendall distribution function of some pair of random variables; and each equivalence class of the relation "to have the same Kendall distribution function as" defined on the set of copulas contains a unique Bertino copula (see [27, 28]).

A relationship between Kendall distribution functions and associative copulas is given in the following result (see [29]).

Theorem 3.5 *Any Kendall distribution function is the Kendall distribution function of some associative copula.*

The main interest of Theorem 3.5 is that it permits to show that each equivalence class of the relation "to have the same Kendall distribution function as" contains a unique associative copula.

For an Archimedean copula C with generator φ, the Kendall distribution function K_C can be given in terms of φ in the following manner (see [30]):

$$K_C(t) = t - \frac{\varphi(t)}{\varphi'(t^+)}.$$

Kendall distribution functions can be used to define new classes of orderings (see [31, 32]), in the description of the dependence structure between non-overlapping random vectors (see [33]), and in applications in the field of risk and models of interacting defaults (see [34]).

3.3.4 Topological Aspects of Associative Copulas

It is known that if C is an associative copula then $([0, 1], C)$ is a topological semigroup with neutral element 1 and annihilator 0 (see [35, 36]). In this subsection we review a topological approach of associative copulas. For that, we need some definitions and notation, which we take from [37].

Given a metric space (Ω, d), a subset of Ω is called a *nowhere dense* set in (Ω, d) if its closure has empty interior. A subset of Ω is called of *first category* in (Ω, d) if it can be expressed as (or covered by) a countable union of nowhere dense sets. Let d_∞ denote the standard uniform metric.

Let $\mathscr{C}$, $\mathscr{C}_c$, and $\mathscr{C}_a$ denote the set of all copulas, commutative copulas, and associative copulas, respectively.

The following result is shown in [38].

Theorem 3.6 *Each Cauchy sequence—with the d_∞ metric—of associative copulas converges uniformly to some associative copula.*

Therefore, the (non-convex) set of associative copulas is a compact subset in the set of all functions from $[0, 1]^2$ onto $[0, 1]$ with the uniform norm.

As a consequence of several results given in [38, 39], each associative copula can be approximated by an ordinal sum of finitely many strict Archimedean copulas.

Theorem 3.7 *The set $\mathscr{C}_a$ is the closure of both the set of all strict Archimedean copulas and the set of all non-strict Archimedean copulas with the uniform distance.*

This means, in particular, that each associative copula can be approximated with arbitrary precision by strict as well as by non-strict Archimedean copulas.

The following result, which can be found in [40], solves an open problem posed in [41] (see also [8]).

Theorem 3.8 *The set of all associative copulas $\mathscr{C}_a$ is nowhere dense in $(\mathscr{C}, d_\infty)$, as well as in $(\mathscr{C}_c, d_\infty)$. In particular, $\mathscr{C}_a$ is of first category in $(\mathscr{C}, d_\infty)$.*

Furthermore, as it is shown in [40], Theorem 3.8 also holds for the stronger metric D_1 introduced in [42] (see also [43]) and related to Markov operators.

3.3.5 Constructing Associative Copulas

A method for constructing associative copulas is presented in this subsection. It is based on a binary operation and a convex function on an interval, and which extends Archimedean copulas. Specifically, let $\oplus$ be a continuous associative operation in $[0, a]$, $a \in [0, \infty]$, such that $t \oplus 0 = 0 \oplus t = t$ and $t \oplus a = a \oplus t = a$ for all $t \in [0, a]$. A function $\psi : [0, a] \longrightarrow \mathbb{R}$ is called $\oplus$-*convex* if

$$\psi(r \oplus t) - \psi(r) \leq \psi(s \oplus t) - \psi(s)$$

for every $r \leq s$ and any t. Then the following result holds (see [44]).

Theorem 3.9 *Let* $\phi : [0, 1] \longrightarrow [0, a]$ *be a strictly decreasing continuous surjection. Then the function*
$$C(u, v) = \phi^{-1}(\phi(u) \oplus \phi(v))$$

is an associative copula if and only if ϕ^{-1} *is* $\oplus$-*convex.*

On the other hand, constructions of associative copulas uniformly close can be found in [45], where, among other results, we have the following:

Theorem 3.10 *Given an* $\varepsilon \in (0, 1)$, *let* f *and* g *be two convex and strictly decreasing functions from* $[0, 1]$ *onto* $[0, \infty]$ *such that*

$$\left| f^{-1}(x) - g^{-1}(x) \right| \leq \frac{\varepsilon}{3}$$

for all $x \in [0, \infty)$. *Then the operations* $C_f(u, v) = f^{-1}(f(u) + f(v))$ *and* $C_g(u, v) = g^{-1}(g(u) + g(v))$ *are two associative copulas such that*

$$\sup_{(u,v) \in [0,1]^2} \left| C_f(u, v) - C_g(u, v) \right| \leq \varepsilon.$$

3.3.6 Simultaneous Associativity

In the course of the study of the associativity of certain binary operations defined on the space of probability distributions, the question of finding the solution of the functional equation $T(x, y) + S(x, y) = x + y$ for all $x, y \in [0, 1]$, where T is a t-norm and S is a t-conorm (see [1]), was answered in [46], where, in terms of copulas, the problem can be reformulated as the simultaneous associativity of a copula C and

$$\tilde{C}(u, v) = u + v - C(u, v), \quad (u, v) \in [0, 1]^2,$$

called the *dual copula* of C, and the following fundamental result is proved:

Theorem 3.11 *The only copulas C having the property that both C and $\tilde{C}$ are simultaneously associative are M, Π, W, a member of the Frank family of copulas or an ordinal sum of members of this family.*

If (X, Y) is a pair of continuous random variables with associated copula C, and f and g are two decreasing functions, let C^*, C^{**}, and $\hat{C}$ be the respective copulas associated with the random vectors $(f(X), Y)$, $(X, g(Y))$, and $(f(X), g(Y))$, and which are given by

$$C^*(u, v) = v - C(1 - u, v)$$
$$C^{**}(u, v) = u - C(u, 1 - v)$$
$$\hat{C}(u, v) = u + v - 1 + C(1 - u, 1 - v)$$

for every $(u, v) \in [0, 1]^2$. Simultaneous associativity between the copulas described above can be given, as the following result shows (see [8, 47]).

Theorem 3.12 *Let C be a copula.*

(a) *The copulas C and $\hat{C}$ are simultaneously associative if and only if C is M, Π, W, a member of the Frank family of copulas or an ordinal sum of members of this family.*
(b) *The copulas C and C^* (or C and C^{**}) are simultaneously associative if and only if C is M, Π, W, or a member of the Frank family of copulas.*
(c) *If C is associative, then $C = C^*$ (or $C = C^{**}$) if and only if $C = Π$.*

3.3.7 Inequalities Involving Copulas

Many studies have been devoted to search for concavity properties of (bivariate) distribution functions and, as a consequence, of copulas. Some examples are the results given in [48–52]. First, we recall now some preliminary concepts:

(i) A copula C is *quasi-concave* if, for all $u_1, u_2, v_1, v_2, \lambda \in [0, 1]$,

$$\min(C(u_1, v_1), C(u_2, v_2)) \leq C(\lambda u_1 + (1 - \lambda)u_2, \lambda v_1 + (1 - \lambda)v_2)$$

(ii) A copula C is *Schur-concave* if, for all $u, v, \lambda \in [0, 1]$,

$$C(u, v) \leq C(\lambda u + (1 - \lambda)v, \lambda v + (1 - \lambda)u).$$

The following result concerning associative copulas and quasi-concavity can be found in [8].

Theorem 3.13 *Every associative copula is quasi-concave.*

Since associative copulas are commutative, it follows at once that every associative copula is Schur-concave. But associativity in Theorem 3.13 is not necessary: for example, the non-associative copula $(\Pi + M)/2$ has convex level curves. Furthermore, every convex combination of associative copulas belongs to the class of Schur-concave and commutative copulas (see [49, 53]).

Another type of inequality involving copulas, and posed as an open problem in [54], is the following: Which copulas C satisfy the inequality

$$C(\max(u - a, 0), \min(u + a, 1)) \leq \delta_C(u) \tag{3.3}$$

for all $u \in [0, 1]$ and for all $a \in (0, 1/2)$? By using the fact that every associative copula is Schur-concave, a partial answer is given in [55]:

Theorem 3.14 *If C is an associative copula, then C satisfies (3.3).*

3.3.8 Associative Discrete Copulas

When studying a model for pairs of discrete random variables, one can obtain a bivariate probability function by taking the derivative of Radon-Nikodym of $H(x, y)$ in Eq. (3.1) regarding the accounting measure. Thus, the joint probability function of a pair of discrete random variables (X_1, X_2) can be represented in terms of the discretized version of the copula and the marginal distribution functions. Specifically, if $n \geq 1$ and $L = \{0, 1, \ldots, n\}$, a *discrete copula* C on L is a binary operation on L satisfying the following conditions: (i) $C(i, 0) = C(0, i) = 0$ and $C(i, n) = C(n, i) = i$ for all $i \in L$, and (ii) $C(i, j) + C(i', j') \geq C(i, j') + C(i', j)$ whenever $i \leq i', j \leq j'$ (see [56]). For the statement and proof of Sklar's theorem in a finite setting, see [57].

The following result summarizes several facts concerning the structure of associative discrete copulas (for more details, see [56–58]), where $C_{\text{Ł}}$ denotes the *Łukasiewicz discrete copula*, which is given by $C_{\text{Ł}}(x, y) = \max(0, x + y - n)$ for every $(x, y) \in L^2$.

Theorem 3.15 *Let C be a discrete copula. The following properties hold:*

(a) *C is associative if and only if it is an ordinal sum of $C_{\text{Ł}}$ copulas.*
(b) *The only associative discrete copula with 0 and n as the only idempotent elements is $C_{\text{Ł}}$.*
(c) *If C is associative then it is commutative (see also [59]).*
(d) *The class of associative discrete copulas coincides with the class of divisible t-norms (see also [60, 61]).*
(e) *Any discrete copula is a product of associative discrete copulas.*
(f) *There are 2^{n-1} associative discrete copulas on L.*

3.3.9 Measuring and Testing Non-associativity

Inspired by the idea of a "measure of non-exchangeability" for a given copula C, defined simultaneously in [62, 63] (see also [64]) by

$$\zeta(C) = 3 \cdot \sup_{(u,v)\in[0,1]^2} |C(u, v) - C(v, u)|, \tag{3.4}$$

the following "measure of non-associativity" is studied in [65]:

$$\omega(C) = 3 \cdot \sup_{(u,v,w)\in[0,1]^3} |C(C(u, v), w) - C(u, C(v, w))| \tag{3.5}$$

(observe that $\omega(C) = 0$ when C is associative). The value of this measure lies in the interval $[0, 1]$, and the upper bound is attained. Denoting by $\mathscr{C}_m$ the set of *maximally non-associative* copulas, i.e. the set of copulas C for which $\omega(C) = 1$, we have the following result (see [65]).

Theorem 3.16 *Let C be a copula. Then $C \in \mathscr{C}_m$ if and only if either $C(1/3, 2/3) = 1/3$ and $C(2/3, 1/3) = 0$, or $C(1/3, 2/3) = 0$ and $C(2/3, 1/3) = 1/3$.*

A test of associativity of a copula is proposed in [66] in order to check the validity of the Archimedean zero assumption. Specifically, a two-parameter family of statistical tests is constructed in the following manner:

$$\mathscr{T}_n^{(1)}(u, v) := \sqrt{n}\,[C_n(u, C_n(v, v)) - C_n\,(C_n(u, v), v)]\,, \quad u, v \in (0, 1),$$

where C_n denotes the *empirical copula* (see [12]). However, the pointwise approach is not consistent in some cases (e.g. there exist associative copulas which are not Archimedean). In [67] it is introduced the trivariate process

$$\mathscr{T}_n^{(2)}(u, v, w) := \sqrt{n}\,[C_n(u, C_n(v, w)) - C_n\,(C_n(u, v), w)]\,, \quad u, v, w \in (0, 1),$$

and where is proved its weak convergence in the metric d_∞.

For other nonnegative statistics which measures the associativity of a given sample, see [68, 69].

3.3.10 Associative N-dimensional copulas

The concept of (bivariate) copula can be extended to n dimensions, where $n > 2$ is a natural number. An *n-copula* is an n-dimensional distribution function from $[0, 1]^n$ onto $[0, 1]$ whose one-dimensional margins are uniform. Many of the definitions and theorems for bivariate copulas have analogous multivariate versions (see [1, 12]).

The main problem in the theory of copulas is to determine which sets of (possible different dimensions) copulas are margins of a higher-dimensional copula.

The associativity of n-copulas in the sense of Post (see [70]) is studied in [71]—solving an open problem posed in [72]. Specifically, if $n \geq 2$ is a natural number and S is a nonempty set, an n-ary operation $f : S^n \longrightarrow S$ is *associative* on S if, for any $1 \leq i < j \leq n$, the equality

$$f\left(x_1^{i-1}, f\left(x_i^{n+i-1}\right), x_{n+i}^{2n-1}\right) = f\left(x_1^{j-1}, f\left(x_j^{n+j-1}\right), x_{n+j}^{2n-1}\right)$$

holds for all $x_1, \ldots, x_{2n-1} \in S$, where x_p^q denotes the vector $(x_p, \ldots, x_q)$.

Characterizations of associative n-copulas are given in the next result [71] (see also [73]).

Theorem 3.17 *Let $n \geq 2$ be a natural number, and let $C : [0, 1]^n \longrightarrow [0, 1]$ be a function.*

(a) *C is an associative n-copula if and only if there is a system $\{I_k\}_{k \in \mathcal{K}}$ of pairwise disjoint open subintervals of $(0, 1)$ and a system $\{C_k\}_{k \in \mathcal{K}}$ of associative n-copulas satisfying the inequality $C_k(u, \ldots, u) < u$ for all $u \in (0, 1)$ and $k \in \mathcal{K}$ such that C is an ordinal sum of $\{C_k\}_{k \in \mathcal{K}}$ with respect to $\{I_k\}_{k \in \mathcal{K}}$.*

(b) *C is an associative n-copula satisfying the inequality $C(u, \ldots, u) < u$ for all $u \in (0, 1)$ if and only if there is a generator φ whose pseudo-inverse $\varphi^{[-1]}$ is an $(n - 2)$-times differentiable function with derivatives alternating the sign, such that $(-1)^n \frac{d^{n-2}\varphi^{[-1]}}{dx^{n-2}}$ is a convex function, and*

$$C(u_1, \ldots, u_n) = \varphi^{[-1]}\left(\sum_{i=1}^{n} \varphi(u_i)\right).$$

Finally, we note that the associativity property enables us to often (not always) extend Archimedean copulas to higher dimensions—for a characterization of Archimedean n-copulas, see [74].

3.4 Some Open Problems

We wish to conclude this brief survey concerning associative copulas with some questions and open problems. The first two are from [8, 41]:

1. What is the significance of associativity in a copula? It is a useful—as we have exposed in several results along this work—algebraic property, of course, but is there some sort of probabilistic interpretation? In the study of t-norms associativity is a natural requirement because they are an analog to the logic operation of "and", and one requires associativity to generalize the triangle inequality (see [1]). Is there a similar motivation for copulas? May we speak about a "set-theoretic"

interpretation of associative copulas? In this vein, results obtained in [75] concerning a geometric interpretation of associativity in terms of the level sets of *t*-norms could help.

2. Can $\tilde{\Pi}$ be expressed as a convex combination of associative copulas?
3. [72] Denote by $\tilde{C}_a$ the closure (with respect to the topology of uniform convergence) of the set of all convex combinations of associative copulas. Does $\tilde{C}_a$ coincide with the class of all Schur-concave and commutative copulas?
4. [72] A binary operation B on the unit interval is called *n-ultramodular*, $n \geq 2$, if

$$\sum_{J \subset \{1,\ldots,k\}} (-1)^{(k+|J|)} B\left(x + \sum_{i \in J} u_i, y + \sum_{i \in J} v_i\right) \geq 0$$

holds for every $k \in \{2, \ldots, n\}$ whenever the arguments of B do not run out of $[0, 1]$ and $(x, y), (u_1, v_1), \ldots, (u_k, v_k) \in [0, 1]^2$. It can be checked that Π is *n*-ultramodular for any natural n (see [76, 77] for more details). Moreover, each associative *n*-ultramodular copula is a trivial sum of Archimedean copulas. The question is: For a fixed $n \geq 2$, characterize all associative copulas which are *n*-ultramodular.
5. [40] A set $A \subset \Omega$ of a metric space (Ω, d) is of second category if it is not of first category. The set of all Archimedean copulas $\mathscr{C}_{Ar}$ is of second category in $(\mathscr{C}_a, d_\infty)$ (see [40]). Is $\mathscr{C}_{Ar}$ of second category in $(\mathscr{C}_a, D_1)$?
6. A pair of continuous random variables (X, Y) is said to have a *Schur-constant* joint survival function if

$$P(X > x, Y > y) = S(x + y)$$

for an appropriate function S (see [78, 79]). In [80] it is shown that there is a one-to-one correspondence between Schur-constant models (also for the multivariate case) and Archimedean copulas. Is it possible to extend this result in the case of associative copulas?
7. Different studies of the measure of non-exchangeability given by (3.4) can be found in [48, 81–86]; and a set of axioms for measures of non-exchangeability is introduced in [87]. Is it possible to do similar studies for measures of non-associativity in the sense of Eq. (3.5)?

Acknowledgments The authors wish to thank two anonymous referees for helpful comments. The authors also acknowledge the support by the Ministerio de Economía y Competitividad (Spain) under research project MTM2014-60594-P.

References

1. B. Schweizer, A. Sklar, *Probabilistic Metric Spaces* (Dover Publications, New York, 2005)
2. M. Fréchet, Sur les tableaux de corrélation dont les marges sont données. Ann. Univ. Lyon Sect. A **9**, 53–77 (1951)
3. A. Sklar, Fonctions de répartition à n dimensions et leurs marges. Publ. Inst. Statist. Univ. Paris **8**, 229–231 (1959)
4. B. Schweizer, E.F. Wolff, On nonparametric measures of dependence for random variables. Ann. Statist. **9**, 879–885 (1981)
5. G. Beliakov, A. Pradera, T. Calvo, *Aggregation Functions: A Guide for Practitioners*. Studies in Fuzziness and Soft Computing, vol. 221 (Springer, Berlin, 2007)
6. T. Calvo, A. Kolesárová, M. Komorníkova, R. Mesiar, in *Aggregation Operators: Properties, Classes and Construction Methods*, eds. by T. Calvo, G. Mayor, R. Mesiar. Aggregation Operators, New Trends and Applications (Physica-Verlag, Heidelberg, 2002), pp. 3–104
7. E.P. Klement, R. Mesiar (eds.), *Logical, Algebraic, Analytic, and Probabilistic Aspects of Triangular Norms* (Elsevier, Amsterdam, 2005)
8. C. Alsina, M.J. Frank, B. Schweizer, *Associative Functions: Triangular Norms and Copulas* (World Scientific, Singapore, 2006)
9. R. Moynihan, On τ_T-semigroups of probability distribution functions. II. Aequationes Math. **17**, 19–40 (1978)
10. M. Grabisch, J.L. Marichal, R. Mesiar, E. Pap, Aggregation Functions, in *Encyclopedia of Mathematics and its Applications*, vol 127 (Cambridge University Press, Cambridge, 2009)
11. F. Durante, C. Sempi, *Principles of Copula Theory* (Chapman & Hall, London, 2015)
12. R.B. Nelsen, *An Introduction to Copulas*, 2nd edn. (Springer, New York, 2006)
13. U. Cherubini, E. Luciano, W. Vecchiato, *Copula Methods in Finance*. Wiley Finance Series (Wiley, Chichester, 2004)
14. P. Jaworski, F. Durante, W. Härdle (eds.). *Copulae in Mathematical and Quantitative Finance*. Lecture Notes in Statistics–Proceedings (Springer, Berlin–Heidelberg, 2013)
15. P. Jaworski, F. Durante, W. Härdle, T. Rychlik (eds.). *Copula Theory and its Applications*. Lecture Notes in Statistics–Proceedings (Springer, Berlin–Heidelberg, 2010)
16. G. Salvadori, C. De Michele, N.T. Kottegoda, R. Rosso, *Extremes in Nature, An Approach Using Copulas* (Springer, Dordrecht, 2007)
17. E. de Amo, M. Díaz Carrillo, J. Fernández-Sánchez, Characterization of all copulas associated with non-continuous random variables. Fuzzy Sets Syst. **191**, 103–112 (2012)
18. C.H. Ling, Representation of associative functions. Publ. Math. Debrecen **12**, 189–212 (1965)
19. A.J. McNeil, R. Frey, P. Embrechts, *Quantitative Risk Management: Concepts, Techniques, and Tools* (Princeton University Press, Princeton, 2005)
20. C. Genest, J. MacKay, Copules archimédiennes et familles de lois bidimensionnelles dont les marges sont données. Canad. J. Statist. **14**, 145–159 (1986)
21. M.J. Frank, Diagonals of copulas and Schröeder's equation. Aequationes Math. **51**, 150 (1996)
22. C. Genest, L.-P. Rivest, Statistical inference procedures for bivariate Archimedean copulas. J. Amer. Statist. Assoc. **55**, 698–707 (1993)
23. V. Schmitz, Revealing the dependence structure between $X_{(1)}$ and $X_{(n)}$. J. Statist. Plann. Infer. **123**, 41–47 (2004)
24. W. Wang, M.T. Wells, Model selection and semiparametric inference for bivariate failure-time data. J. Amer. Statist. Assoc. **95**, 62–76 (2000)
25. C. Genest, L.-P. Rivest, On the multivariate probability integral transformation. Statist. Probab. Lett. **53**, 391–399 (2001)
26. R.B. Nelsen, J.J. Quesada-Molina, J.A. Rodríguez-Lallena, M. Úbeda-Flores, Kendall distribution functions. Statist. Probab. Lett. **65**, 263–268 (2003)
27. S. Bertino, Sulla dissomiglianza tra mutabili cicliche. Metron **35**, 53–88 (1977)
28. G.A. Fredricks, R.B. Nelsen, The Bertino family of copulas, in *Distributions with Given Marginals and Statistical Modelling*, ed. by C. Cuadras, J. Fortiana, J.A. Rodríguez-Lallena (Kluwer Academic Publishers, Dordrecht, 2002), pp. 81–91

29. R.B. Nelsen, J.J. Quesada-Molina, J.A. Rodríguez-Lallena, M. Úbeda-Flores, Kendall distribution functions and associative copulas. Fuzzy Sets Syst. **160**, 52–57 (2009)
30. P. Barbe, C. Genest, K. Ghoudi, B. Rémillard, On Kendall's process. J. Multivar. Anal. **58**, 197–229 (1996)
31. R.L. Fountain, J.R. Herman, D.L. Rustvold, An application of Kendall distributions and alternative dependence measures: SPX vs. VIX. Insur.: Math. Econom. **42**, 469–472 (2008)
32. R.B. Nelsen, J.J. Quesada-Molina, J.A. Rodríguez-Lallena, M. Úbeda-Flores, Distribution functions of copulas: a class of bivariate probability integral transforms. Statist. Probab. Lett. **54**, 277–282 (2001)
33. U. dos Anjos, N. Kolev, An application of Kendall distributions. Rev. Bus. Econ. Res. **1**, 95–102 (2005)
34. G. Nappo, F. Spizzichino, Kendall distributions and level sets in bivariate exchangeable survival models. Inform. Sci. **179**, 2878–2890 (2009)
35. P.S. Mostert, A.L. Shields, On the structure of semigroups on a compact manifold with boundary. Ann. Math. **65**, 117–143 (1957)
36. B. Schweizer, A. Sklar, Associative functions and statistical triangle inequalities. Publ. Math. Debrecen **8**, 169–186 (1961)
37. J.C. Oxtoby, *Measure and Category: A Survey of the Analogies between Topological and Measure Spaces*, 2nd edn. (Springer, New York, 1980)
38. E.P. Klement, R. Mesiar, E. Pap, Uniform approximation of associative copulas by strict and non-strict copulas. Illinois J. Math. **45**, 1393–1400 (2001)
39. E.P. Klement, R. Mesiar, E. Pap, *Triangular Norms* (Kluwer Academic Publishers, Dordrecht, 2000)
40. F. Durante, J. Fernández-Sánchez, W. Trutschnig, Baire category results for exchangeable copulas. Fuzzy Sets Syst. (2015). doi:10.1016/j.fss.2015.04.010
41. C. Alsina, M.J. Frank, B. Schweizer, Problems on associative functions. Aequationes Math. **66**, 128–140 (2003)
42. W. Trutschnig, On a strong metric on the space of copulas and its induced dependence measure. J. Math. Anal. Appl. **384**, 690–705 (2011)
43. J. Fernández-Sánchez, W. Trutschnig, Conditioning-based metrics on the space of multivariate copulas and their interrelation with uniform and levelwise convergence and Iterated Function Systems. J. Theor. Probab. (2015). doi:10.1007/s10959-014-0541-4
44. P. Mikusiński, M.D. Taylor, A remark on associative copulas. Comment. Math. Univ. Carolinae **40**, 789–793 (1999)
45. C. Alsina, R. Ger, On associative copulas uniformly close. Internat. J. Math. Math. Sci. **11**, 439–448 (1988)
46. M.J. Frank, On the simultaneous associativity of $F(x, y)$ and $x + y − F(x, y)$. Aequationes Math. **19**, 194–226 (1979)
47. E.P. Klement, R. Mesiar, E. Pap, Invariant copulas. Kybernetika **38**, 275–285 (2002)
48. E. Alvoni, P.L. Papini, Quasi-concave copulas, asymmetry and transformations. Comment. Math. Univ. Carolin. **48**, 311–319 (2007)
49. F. Durante, C. Sempi, Copulæ and Schur-concavity. Internat. Math. J. **3**, 893–905 (2003)
50. W. Fenchel, *Convex Cones, Sets and Functions*, Lecture Notes (Priceton University, Princeton, 1953)
51. B. de Finetti, Sulle stratificazioni convesse. Ann. Mat. Pura Appl. **30**, 173–183 (1949)
52. A. Marshall, I. Olkin, *Inequalities: Theory of Majorization and its Applications* (Academic Press, New York, 1979)
53. C. Alsina, in *On Schur-concave t-norms and Triangle Functions*, ed. by W. Walter. General Inequalities, vol. 4 (Birkhäuser Verlag, Basel, 1984), pp. 241–248
54. E.P. Klement, R. Mesiar, E, Pap, Problems on triangular norms and related operators. Fuzzy Sets Syst. **145**, 471–479 (2004)
55. F. Durante, Solution of an open problem for associative copulas. Fuzzy Sets Syst. **152**, 411–415 (2005)

56. G. Mayor, J. Suñer, J. Torrens, Copula-like operations on finite settings. IEEE Trans. Fuzzy Syst. **13**, 468–477 (2005)
57. G. Mayor, J. Suñer, J. Torrens, Sklar's theorem in finite settings. IEEE Trans. Fuzzy Syst. **15**, 410–416 (2007)
58. A. Kolesárová, J. Mordelová, Quasi-copulas and copulas on a discrete scale. Soft Comput. **10**, 495–501 (2006)
59. J. Fodor, Smooth associative operations on finite ordinal scales. IEEE Trans. Fuzzy Syst. **8**, 791–795 (2000)
60. G. Mayor, J. Torrens, On a class of operators for expert systems. Int. J. Intell. Syst. **8**, 771–778 (1993)
61. G. Mayor, J. Torrens, Triangular norms on discrete settings, in *Logical, Algebraic, Analytic, and Probabilistic Aspects of Triangular Norms*, ed. by E.P. Klement, R. Mesiar (Elsevier, Amsterdam, 2005), pp. 189–230
62. E.P. Klement, R. Mesiar, How non-symmetric can a copula be? Comment. Math. Univ. Carolin. **47**, 141–148 (2006)
63. R.B. Nelsen, Extremes of nonexchangeability. Stat. Papers **48**, 329–336 (2007)
64. K.F. Siburg, P.A. Stoimenov, Symmetry of functions and exchangeability of random variables. Stat. Papers **52**, 1–15 (2011)
65. F.H. Ferreira, Medidas de assimetria bivariada e dependência local. PhD Thesis, Universidade de São Paulo (2008)
66. P. Jaworski, Testing Archimedeanity, in *Combining Soft Computing and Statistical Methods in Data Analysis*, ed. by C. Borglet, G. González-Rodríguez, W. Trutschnig, M.A. Lubiano, M.A. Gil, P. Grzegorzewski, O. Hryniewicz (Springer, Berlin, 2010), pp. 353–360
67. A. Bücher, H. Dette, S. Volgushev, A test for Archimedeanity in bivariate copula models. J. Multivariate Anal. **110**, 121–132 (2012)
68. A. Erdely, J.M. González-Barrios, R.B. Nelsen, Symmetries of random discrete copulas. Kybernetika **44**, 846–863 (2008)
69. J.M. González-Barrios, Statistical aspects of associativity for copulas. Kybernetika **46**, 149–177 (2010)
70. E.L. Post, Polyadic groups. Trans. Amer. Math. Soc. **48**, 208–350 (1940)
71. A. Stupňanová, A. Kolesárová, Associative n-dimensional copulas. Kybernetika **47**, 93–99 (2011)
72. R. Mesiar, P. Sarkoci, Open problems posed at the tenth international conference on fuzzy set theory and applications (FSTA 2010, Liptovský Ján, Slovakia). Kybernetika **46**, 585–599 (2010)
73. R. Mesiar, C. Sempi, Ordinal sums and idempotents of copulas. Aequationes Math. **79**, 39–52 (2010)
74. A.J. McNeil, J. Nešlehová, Multivariate Archimedean copulas, d-monotone functions and l_1-norm symmetric distributions. Ann. Stat. **37**, 3059–3097 (2009)
75. M. Petrík, P. Sarkoci, Associativity of triangular norms characterized by the geometry of their level sets. Fuzzy Sets Syst. **202**, 100–109 (2012)
76. E.P. Klement, M. Manzi, R. Mesiar, Ultramodularity and copulas. Rocky Mt. J. Math. **44**, 189–201 (2014)
77. M. Marinacci, L. Montrucchio, Ultramodular functions. Math. Oper. Res. **30**, 311–332 (2005)
78. R.E. Barlow, M.B. Mendel, in *Similarity as a Probabilistic Characteristic of Aging*, eds. by R.E. Barlow, C.A. Clarotti, F. Spizzichino. Reliability and Decision Making (Chapman & Hall, London, 1993)
79. F. Spizzichino, *Subjective Probability Models for Lifetimes* (Chapman & Hall/CRC, Boca Raton, 2001)
80. R.B. Nelsen, Some properties of Schur-constant survival models and their copulas. Braz. J. Probab. Stat. **19**, 179–190 (2005)
81. G. Beliakov, B. De Baets, H. De Meyer, R.B. Nelsen, M. Úbeda-Flores, Best-possible bounds on the set of copulas with given degree of non-exchangeability. J. Math. Anal. Appl. **417**, 451–468 (2014)

82. B. De Baets, H. De Meyer, R. Mesiar, Asymmetric semilinear copulas. Kybernetika **43**, 221–233 (2007)
83. F. Durante, R. Mesiar, L^∞-measure of non-exchangeability for bivariate extreme value and Archimax copulas. J. Math. Anal. Appl. **369**, 610–615 (2010)
84. F. Durante, P.L. Papini, A weakening of Schur-concavity for copulas. Fuzzy Sets Syst. **158**, 1378–1383 (2007)
85. F. Durante, P.L. Papini, Componentwise concave copulas and their asymmetry. Kybernetika **45**, 1003–1011 (2009)
86. F. Durante, P.L. Papini, Non-exchangeability of negatively dependent random variables. Metrika **71**, 139–149 (2010)
87. F. Durante, E.P. Klement, C. Sempi, M. Úbeda-Flores, Measures of non-exchangeability for bivariate random vectors. Stat. Papers **51**, 687–699 (2010)

Chapter 4
Powers with Respect to t-Norms
and t-Conorms and Aggregation Functions

D. Boixader and J. Recasens

Abstract Aggregation functions A stable with respect to powers of t-norms and t-conorms (i.e.: satisfying $A(x^{(r)}, y^{(r)}) = (A(x, y))^{(r)}$) where $x^{(r)}$ is the r-th power of $x \in [0, 1]$ with respect to a t-norm or t-conorm) are characterized. This result generalizes the characterization of power stable aggregation functions in [5].

4.1 Introduction

In many situations the preservation of a scale is paramount. In [5] aggregation functions preserving log-ratio scales, i.e. power stable aggregation functions were studied. More explicitly, the aggregation functions $A : [0, 1]^2 \to [0, 1]$ satisfying $A(x^r, y^r) = (A(x, y))^r$ for any constant $r \in]0, \infty[$ and all $x, y \subset [0, 1]$ were characterized.

The power of a number $x \in [0, 1]$ can be considered as the power with respect to the Product t-norm. So it comes the natural generalization of the results in [5] by characterizing aggregation functions preserving powers with respect to a given t-norm or t-conorm [6]. This paper studies this generalization for continuous Archimedean t-norms and t-conorms. Not surprisingly, the results for strict t-norms are similar to the ones in [5], while we have different characterizations for non-strict ones.

The next section recalls the definitions and some properties of the powers of a number of the unit interval with respect to t-norms and t-conorms. Section 4.3 contains the main results of the paper; namely the characterization of stable aggregation

D. Boixader · J. Recasens (✉)
Secció Matemàtiques i Informàtica, ETS Arquitectura del Vallès,
Universitat Politècnica de Catalunya, Pere Serra 1-15, 08190 Sant Cugat Del Vallès, Spain
e-mail: j.recasens@upc.edu

D. Boixader
e-mail: dionis.boixader@upc.edu

© Springer International Publishing Switzerland 2016   41
T. Calvo Sánchez and J. Torrens Sastre (eds.), *Fuzzy Logic and Information Fusion*,
Studies in Fuzziness and Soft Computing 339, DOI 10.1007/978-3-319-30421-2_4

functions with respect to powers of continuous Archimedean t-norms and t-conorms and Sect. 4.4 provides some examples of such aggregation functions.

4.2 Powers with Respect to t-Norms and t-Conorms

In this section, the definition and some properties of powers with respect to t-norms and t-conorms will be recalled. The readers are referred to [4, 6] for further details.

Definition 4.2.1 Given a continuous t-norm T, $T(\overbrace{x, x, ...x}^{n\ times})$-the n-th power of x-will be denoted by $x_T^{(n)}$ or simply by $x^{(n)}$ if the t-norm is clear.

The n-th root $x_T^{(\frac{1}{n})}$ of x with respect to T is defined by

$$x_T^{(\frac{1}{n})} = \sup\{z \in [0, 1] \mid z_T^{(n)} \le x\}$$

and for $m, n \in \mathbb{N}$, $x_T^{(\frac{m}{n})} = \left(x_T^{(\frac{1}{n})}\right)_T^{(m)}$.

Lemma 4.2.2 If $k, m, n \in \mathbb{N}$, $k, n \ne 0$ then $x_T^{(\frac{km}{kn})} = x_T^{(\frac{m}{n})}$.

The powers $x_T^{(\frac{m}{n})}$ can be extended to irrational exponents in a straightforward way.

Definition 4.2.3 If $r \in \mathbb{R}^+$ is a positive real number, let $\{a_n\}_{n\in\mathbb{N}}$ be a sequence of rational numbers with $\lim_{n\to\infty} a_n = r$. For any $x \in [0, 1]$, the power $x_T^{(r)}$ is

$$x_T^{(r)} = \lim_{n\to\infty} x_T^{(a_n)}.$$

Continuity assures the existence of the last limit and independence of the sequence $\{a_n\}_{n\in\mathbb{N}}$.

Proposition 4.2.4 Let T be a continuous Archimedean t-norm with additive generator t, $x \in [0, 1]$ and $r \in \mathbb{R}^+$. Then

$$x_T^{(r)} = t^{[-1]}(rt(x)).$$

Lemma 4.2.5 If $x_T^{(r)} \ne 0$, then $x_T^{(r)} = t^{-1}(rt(x))$.

The powers $x_S^{(r)}$ with respect to t-conorms can be defined in the same way and the following proposition is analogous to Proposition 4.2.4.

Proposition 4.2.6 Let S be a continuous Archimedean t-conorm with additive generator s, $x \in [0, 1]$ and $r \in \mathbb{R}^+$. Then

$$x_S^{(r)} = s^{[-1]}(rs(x)).$$

Similarly to Lemma 4.2.5, we obtain the following result for t-conorms.

Lemma 4.2.7 *Let S be a continuous Archimedean t-conorm with additive generator s, $x \in [0, 1]$ and $r \in \mathbb{R}^+$. If $x_S^{(r)} \neq 1$, then $x_S^{(r)} = s^{-1}(rs(x))$.*

4.3 Aggregation Functions Power Stable with Respect to t-Norms and t-Conorms

Definition 4.3.1 An aggregation function $A : [0, 1]^2 \to [0, 1]$ is called power stable with respect to a t-norm T if and only if for all $r \in \mathbb{R}^+$ and all $x, y \in [0, 1]$

$$A(x^{(r)}, y^{(r)}) = (A(x, y))^{(r)}$$

whenever $x^{(r)} \neq 0$, $y^{(r)} \neq 0$, $A(x, y) \neq 0$ and $(A(x, y))^{(r)} \neq 0$.

Note: Power stability with respect to the Product t-norm coincides with the regular notion of power stability [5].

Power stability with respect to a t-conorm can be defined in a similar way.

Definition 4.3.2 An aggregation function $A : [0, 1]^2 \to [0, 1]$ is called power stable with respect to a t-conorm S if and only if for all $r \in \mathbb{R}^+$ and all $x, y \in [0, 1]$ it holds

$$A(x^{(r)}, y^{(r)}) = (A(x, y))^{(r)}.$$

whenever $x^{(r)} \neq 1$, $y^{(r)} \neq 1$ $A(x, y) \neq 1$ and $(A(x, y))^{(r)} \neq 1$.

Proposition 4.3.3 *Every continuous Archimedean t-norm T is stable with respect to itself.*

Proof If $x^{(r)} \neq 0$, $y^{(r)} \neq 0$, $T(x, y) \neq 0$ and $(T(x, y))^{(r)} \neq 0$, then

$$\begin{aligned}
(T(x, y))^{(r)} &= t^{-1}(rt(T(x, y))) \\
&= t^{-1}(t(t^{-1}(r(t(x) + t(y))))) \\
&= t^{[-1]}(r(t(x) + t(y))) \\
&= t^{[-1]}(rt(x) + rt(y)) \\
&= t^{[-1]}(t(t^{-1}(rt(x))) + t(t^{-1}(rt(y)))) \\
&= T(x^{(r)}, y^{(r)}).
\end{aligned}$$

$\square$

In a similar way,

Proposition 4.3.4 *Every continuous Archimedean t-conorm S is stable with respect to itself.*

Proposition 4.3.5 *A non-constant function $F :]0, 1[^2 \to [0, 1]$ is power stable with respect to a non-strict continuous Archimedean t-norm T with additive generator t normalized to $t(0) = 1$ if and only if there is a non-zero function $f :]0, \infty[\to \mathbb{R}$ such that for all $x, y \in]0, 1[$*

$$F(x, y) = t^{[-1]}(f(\frac{t(y)}{t(x)}) \max(t(x), t(y))).$$

Proof Consider $x, y \in]0, 1[$ with $x^{(r)} \neq 0$, $y^{(r)} \neq 0$, $F(x, y) \neq 0$ and $(F(x, y))^{(r)} \neq 0$.

$\Leftarrow)$

$$
\begin{aligned}
F(x^{(r)}, y^{(r)}) &= t^{[-1]}(f(\frac{t(y^{(r)})}{t(x^{(r)})}) \max(t(x^{(r)}), t(y^{(r)}))) \\
&= t^{[-1]}(f(\frac{t(t^{-1}(rt(y)))}{t(t^{-1}(rt(x)))}) \max(t(t^{-1}(rt(x))), t(t^{-1}(rt(y))))) \\
&= t^{[-1]}(f(\frac{rt(y)}{rt(x)}) \max(rt(x), rt(y))) \\
&= t^{[-1]}(f(\frac{t(y)}{t(x)}) \max(rt(x), rt(y))).
\end{aligned}
$$

If $F(x, y) \neq 0$, then $F(x, y) = t^{-1}(f(\frac{t(y)}{t(x)}) \max(t(x), t(y)))$.

$$
\begin{aligned}
(F(x, y))^{(r)} &= t^{[-1]}(rt(F(x, y))) \\
&= t^{[-1]}(rt(t^{-1}(f(\frac{t(y)}{t(x)}) \max(rt(x), rt(y))))) \\
&= t^{[-1]}(rf(\frac{t(y)}{t(x)}) \max(rt(x), rt(y))).
\end{aligned}
$$

$\Rightarrow)$

Define $f(z) = t(F(t^{-1}(\min(1, \frac{1}{z})), t^{-1}(\min(1, z))))$. Putting $z = \frac{t(y)}{t(x)}$,

$$
\begin{aligned}
F(x, y) &= F(t^{-1}(\min(1, \frac{t(x)}{t(y)}))^{(\max(t(x), t(y)))}, t^{-1}(\min(1, \frac{t(y)}{t(x)}))^{(\max(t(x), t(y)))}) \\
&= (F(t^{-1}(\min(1, \frac{t(x)}{t(y)})), t^{-1}(\min(1, \frac{t(y)}{t(x)}))))^{(\max(t(x), t(y)))} \\
&= t^{[-1]}(\max(t(x), t(y)) t(F(t^{-1}(\min(1, \frac{t(x)}{t(y)})), t^{-1}(\min(1, \frac{t(y)}{t(x)}))))) \\
&= t^{[-1]}(\max(t(x), t(y)) f(\frac{t(y)}{t(x)})).
\end{aligned}
$$

$\square$

Proposition 4.3.6 *A non-constant function* $F :]0, 1[^2 \rightarrow \mathbb{R}$ *is power stable with respect to a strict continuous Archimedean t-norm* T *with additive generator* t *if and only if there is a non-zero function* $f :]0, \infty[\rightarrow \mathbb{R}$ *such that for all* $x, y \in]0, 1[$

$$F(x, y) = t^{-1}(f(\frac{t(y)}{t(x)})t(x)).$$

Proof

$\Leftarrow)$

$$F(x^{(r)}, y^{(r)}) = t^{[-1]}(f(\frac{t(y^{(r)})}{t(x^{(r)})})t(x^{(r)}))$$

$$= t^{[-1]}(f(\frac{t(t^{-1}(rt(y)))}{t(t^{-1}(rt(x)))})t(t^{-1}(rt(x))))$$

$$= t^{[-1]}(f(\frac{rt(y)}{rt(x)})rt(x))$$

$$= t^{[-1]}(f(\frac{t(y)}{t(x)})rt(x)).$$

$$(F(x, y))^{(r)} = t^{[-1]}(rt(F(x, y)))$$

$$= t^{[-1]}(rt(t^{[-1]}(f(\frac{t(y)}{t(x)})t(x)))) = t^{[-1]}(rf(\frac{t(y)}{t(x)})t(x)).$$

$\Rightarrow)$

Define $f(z) = t(F(t^{-1}(1), t^{-1}(z)))$. Putting $z = \frac{t(y)}{t(x)}$,

$$F(x, y) = F(t^{-1}(1)^{(t(x))}, t^{-1}(z)^{(t(x))}) = (F(t^{-1}(1), t^{-1}(z)))^{(t(x))}$$

$$= t^{[-1]}(t(x)t(F(t^{-1}(1), t^{-1}(z)))) = t^{[-1]}(t(x)f(\frac{t(y)}{t(x)})).$$

$\square$

In a similar way,

Proposition 4.3.7 *A non-constant function* $F :]0, 1[^2 \rightarrow [0, 1]$ *is power stable with respect to a non-strict continuous Archimedean t-conorm* S *with additive generator* s *normalized to* $s(0) = 1$ *if and only if there is a non-zero function* $f :]0, \infty[\rightarrow \mathbb{R}$ *such that for all* $x, y \in]0, 1[$

$$F(x, y) = s^{[-1]}(f(\frac{s(y)}{s(x)}) \max(s(x), s(y))).$$

Proposition 4.3.8 *A non-constant function $F :]0, 1[^2 \to \mathbb{R}$ is power stable with respect to a strict continuous Archimedean t-conorm S with additive generator s if and only if there is a non-zero function $f :]0, \infty[\to \mathbb{R}$ such that for all $x, y \in]0, 1[$*

$$F(x, y) = s^{-1}(f(\frac{s(y)}{s(x)})s(x)).$$

4.4 Examples

The characterizations of the previous section allow us to find interesting power stable aggregation functions.

Example 4.4.1 T, S non-strict:

- If $f(z) = \min(1, \frac{1}{z}) + \min(1, z)$, then $F(x, y)$ is the continuous Archimedean t-norm with additive generator t.
- For the Łukasiewicz t-conorm $(t(x) = x)$, stability means homogeneity: $F(rx, ry) = rF(x, y)$ when $rx < 1$ and $ry < 1$. Then F is homogeneous (stable) if and only if $F(x, y) = \max(x, y)f(\frac{y}{x})$.
- If $f(z) = \frac{\min(1, \frac{1}{z}) + \min(1, z)}{2}$, then $F(x, y) = t^{-1}(\frac{t(x) + t(y)}{2})$ is the quasi-arithmetic mean generated by t (see [1–3] for properties of quasi-arithmetic means).
- In particular, for the Łukasiewicz t-norm and t-conorm, we obtain the arithmetic mean $F(x, y) = \frac{x+y}{2}$ for $f(z) = \frac{\min(1, \frac{1}{z}) + \min(1, z)}{2}$.

Example 4.4.2 T strict:

- If we consider the Product t-norm $(t(x) = -\log x)$, then we recover the classical concept of power stability and $F(x, y) = x^{f(\frac{\log y}{\log x})}$ [5].
- If $f(z) = 1 + z$, then $F(x, y)$ is the continuous Archimedean t-norm with additive generator t.
- If $f(z) = \frac{1+z}{2}$, then $F(x, y) = t^{-1}(\frac{t(x) + t(y)}{2})$ is the quasi-arithmetic mean generated by t.
- In particular, for the Product t-norm, if $f(z) = \frac{1+z}{2}$, then $F(x, y) = \sqrt{xy}$ is the geometric mean.

References

1. J. Aczél, *Lectures on Functional Equations and Their Applications* (Academic Press, NY-London, 1966)
2. G. Beliakov, A. Pradera, T. Calvo, *Aggregation Functions: A Guide for Practitioners*. Studies in Fuzziness and Soft Computing, vol. 221 (Springer, 2007)

3. T. Calvo, A. Kolesárova, M. Komorníková, R. Mesiar, Aggregation operators: properties, classes and construction methods, in *Aggregation Operators: New Trends and Applications*, Studies in Fuzziness and Soft Computing, ed. by R. Mesiar, T. Calvo, G. Mayor (Springer, 2002), pp. 3–104
4. E.P. Klement, R. Mesiar, E. Pap, *Triangular Norms* (Kluwer Academic Publishers, Dordrecht, 2000)
5. A. Kolesárová, R. Mesiar, T. Rückschlossová, Power stable aggregation functions. Fuzzy Sets Syst. **240**, 39–50 (2014)
6. E.A. Walker, C.L. Walker, Powers of t-norms. Fuzzy Sets Syst. **129**, 1–18 (2002)

Chapter 5
Modus Tollens on Fuzzy Implication Functions Derived from Uninorms

M. Mas, J. Monreal, M. Monserrat, J.V. Riera and Joan Torrens Sastre

Abstract The most used inference schemes in approximate reasoning are the so-called Modus Ponens for forward inferences, and Modus Tollens for backward inferences. In this way, finding new fuzzy implication functions satisfying these two properties has become an important topic for researchers. In the framework of fuzzy logic, they can be written as two inequalities involving fuzzy implication functions. In this paper, the property of Modus Tollens with respect to a continuous t-norm and a continuous fuzzy negation is studied for residual implication functions derived from uninorms, that is, for RU-implications. The corresponding inequality is solved in the cases of an RU-implication derived from a uninorm U in the class of $\mathcal{U}_{\min}$, from an idempotent uninorm or from a representable uninorm.

5.1 Introduction

Fuzzy implication functions are one of the most important logical connectives in fuzzy logic and approximate reasoning. They are logical operations used in modelling all fuzzy conditionals and also in the inference process. Moreover, they are also useful in many application fields not only derived from the proper approximate reasoning, but also in other aspects as fuzzy subsethood measures, fuzzy relational equations, fuzzy mathematical morphology, and computing with words among others. For this

M. Mas · J. Monreal · M. Monserrat · J.V. Riera · J. Torrens Sastre (✉)
Department of Mathematics and Computer Science, University of the Balearic
Islands, 07122 Palma de Mallorca, Spain
e-mail: jts224@uib.es

M. Mas
e-mail: mmg448@uib.es

J. Monreal
e-mail: jaumemonreal@gmail.com

M. Monserrat
e-mail: mma112@uib.es

J.V. Riera
e-mail: jvicente.riera@uib.es

© Springer International Publishing Switzerland 2016   49
T. Calvo Sánchez and J. Torrens Sastre (eds.), *Fuzzy Logic and Information Fusion*,
Studies in Fuzziness and Soft Computing 339, DOI 10.1007/978-3-319-30421-2_5

reason, investigations on fuzzy implication functions have been extensively developed in last decades even from the pure theoretical point of view, as it can be seen in the survey [20] and in the books [3, 4], entirely devoted to this kind of logical operations.

One of the main topics in this theoretical study consists on the investigation of additional properties of implication functions, properties that usually come from the concrete applications where implications functions are going to be applied. The study of each one of these additional properties usually leads to solve a functional equation (or inequality) involving fuzzy implication functions (see for instance [6] or Chap. 7 in [4] and the references therein).

Two of these additional properties, that in this case come from approximate reasoning, are known as the (generalized) *Modus Ponens* and *Modus Tollens*. In fact, forward and backward inference schemes in approximate reasoning are usually based on these properties, that are carried out through the well known *Compositional Rule of Inference* (CRI) of Zadeh, based on the sup $-T$ composition, where T is a t-norm (see for instance, Sect. 8.3 in [4]). Thus, if I is a fuzzy implication function, T is a t-norm and N is a fuzzy negation, the Modus Ponens and the Modus Tollens properties for I with respect to T (and N) becomes the functional inequalities:

$$T(x, I(x, y)) \leq y \quad \text{for all } x, y \in [0, 1],$$

$$T(N(y), I(x, y)) \leq N(x) \quad \text{for all } x, y \in [0, 1],$$

respectively.

Both properties have been extensively studied in the literature by some authors (namely [2, 4, 18, 25–28]), mainly for R, (S, N) and QL-implications derived from t-norms, t-conorms and strong fuzzy negations. However, some generalizations of these classes of implications have been introduced, by substituting the t-norm and the t-conorm by more general aggregation functions (for more details see [4] and also [21] with the references therein). One of these generalizations is based on uninorms obtaining the so-called RU-implications [8], (U, N)-implications [5], and even QL and D-implications derived from conjunctive and disjunctive uninorms [17].

Whereas the Modus Ponens has been recently studied for RU-implications (see [15]), this is not the case of the Modus Tollens property and this is the idea of the present paper. Specifically, we want to study the Modus Tollens with respect to any continuous t-norm T and continuous (strict or strong) fuzzy negation N, for the case of RU-implications. We will prove that there are a lot of them that satisfy the Modus Tollens with respect to any t-norm T and fuzzy negation N, and we will characterize the special case when the t-norm T and the negation N are continuous. We will do it for RU-implications derived from three well known classes of uninorms: Uninorms in $\mathcal{U}_{min}$, idempotent uninorms and representable uninorms.

5.2 Preliminaries

We will suppose the reader to be familiar with the theory of t-norms, t-conorms and fuzzy negations (all necessary results and notations can be found in [11]). We also suppose that some basic facts on uninorms are known (see for instance [10]) as well as their most usual classes, that is, uninorms in $\mathcal{U}_{\min}$ and $\mathcal{U}_{\max}$ [10], representable uninorms [10] and idempotent uninorms [7, 14, 24].

   We recall here only some facts on implications and uninorms in order to stablish the necessary notation that we will use along the paper.

Definition 5.2.1 [4] A binary operator $I : [0, 1] \times [0, 1] \rightarrow [0, 1]$ is said to be a fuzzy *implication function*, or an *implication*, if it satisfies:

(I1) $I(x, z) \geq I(y, z)$ when $x \leq y$, for all $z \in [0, 1]$.
(I2) $I(x, y) \leq I(x, z)$ when $y \leq z$, for all $x \in [0, 1]$.
(I3) $I(0, 0) = I(1, 1) = 1$ and $I(1, 0) = 0$.

Note that, from the definition, it follows that $I(0, x) = 1$ and $I(x, 1) = 1$ for all $x \in [0, 1]$ whereas the symmetrical values $I(x, 0)$ and $I(1, x)$ are not derived from the definition.

Proposition 5.2.2 [4] *Let I be a fuzzy implication function. The function $N_I :$ $[0, 1] \rightarrow [0, 1]$ given by $N_I(x) = I(x, 0)$ for all $x \in [0, 1]$ is always a fuzzy negation, which is called the* natural negation *of I.*

Definition 5.2.3 [10] A *uninorm* is a two-place function $U : [0, 1]^2 \longrightarrow [0, 1]$ which is associative, commutative, increasing in each place and such that there exists some element $e \in [0, 1]$, called *neutral element*, such that $U(e, x) = x$ for all $x \in [0, 1]$.

   Evidently, a uninorm with neutral element $e = 1$ is a t-norm and a uninorm with neutral element $e = 0$ is a t-conorm. For any other value $e \in]0, 1[$ the operation works as a t-norm in the $[0, e]^2$ square, as a t-conorm in $[e, 1]^2$ and its values are between minimum and maximum in the set of points $A(e)$ given by

$$A(e) = [0, e[\times]e, 1] \cup]e, 1] \times [0, e[.$$

   We will usually denote a uninorm with neutral element e and underlying t-norm and t-conorm, T and S, by $U \equiv \langle T, e, S \rangle$. For any uninorm it is satisfied that $U(0, 1) \in \{0, 1\}$ and a uninorm U is called *conjunctive* if $U(1, 0) = 0$ and *disjunctive* when $U(1, 0) = 1$. On the other hand, let us recall the most usual classes of uninorms in the literature that will be used along the paper.

Theorem 5.2.4 [10] *Let $U : [0, 1]^2 \rightarrow [0, 1]$ be a uninorm with neutral element $e \in]0, 1[$.*

(a) *If $U(0, 1) = 0$, then the section $x \mapsto U(x, 1)$ is continuous except in $x = e$ if and only if U is given by*

$$U(x, y) = \begin{cases} eT_U \left(\frac{x}{e}, \frac{y}{e} \right) & \text{if } (x, y) \in [0, e]^2, \\ e + (1 - e)S_U \left(\frac{x-e}{1-e}, \frac{y-e}{1-e} \right) & \text{if } (x, y) \in [e, 1]^2, \\ \min(x, y) & \text{if } (x, y) \in A(e), \end{cases}$$

where T_U is a t-norm, and S_U is a t-conorm.

(b) *If $U(0, 1) = 1$, then the section $x \mapsto U(x, 0)$ is continuous except in $x = e$ if and only if U is given by the same structure as above, changing minimum by maximum in $A(e)$.*

The set of uninorms as in case (a) will be denoted by $\mathscr{U}_{\min}$ and the set of uninorms as in case (b) by $\mathscr{U}_{\max}$. We will denote a uninorm in $\mathscr{U}_{\min}$ with underlying t-norm T_U, underlying t-conorm S_U and neutral element $e \in]0, 1[$ as $U \equiv \langle T_U, e, S_U \rangle_{\min}$ and in a similar way, a uninorm in $\mathscr{U}_{\max}$ as $U \equiv \langle T_U, e, S_U \rangle_{\max}$.

Idempotent uninorms were characterized first in [7] for those with a lateral continuity and in [14] for the general case. An improvement of this last result was done in [24] as follows.

Theorem 5.2.5 [24] *U is an idempotent uninorm with neutral element $e \in [0, 1]$ if and only if there exists a decreasing function $g : [0, 1] \rightarrow [0, 1]$, symmetric with respect to the identity function, with $g(e) = e$, such that*

$$U(x, y) = \begin{cases} \min(x, y) & \text{if } y < g(x) \text{ or} \\ & (y = g(x) \text{ and } x < g^2(x)), \\ \max(x, y) & \text{if } y > g(x) \text{ or} \\ & (y = g(x) \text{ and } x > g^2(x)), \\ x \text{ or } y & \text{if } y = g(x) \text{ and } x = g^2(x), \end{cases}$$

being commutative in the points (x, y) such that $y = g(x)$ with $x = g^2(x)$.

Any idempotent uninorm U with neutral element e and associated function g, will be denoted by $U \equiv \langle g, e \rangle_{\text{ide}}$ and the class of idempotent uninorms will be denoted by $\mathscr{U}_{\text{ide}}$. Obviously, for any of these uninorms the underlying t-norm T is the minimum and the underlying t-conorm S is the maximum.

Definition 5.2.6 [10] Let $e \in]0, 1[$. A binary operation $U : [0, 1]^2 \rightarrow [0, 1]$ is a representable uninorm if and only if there exists a strictly increasing function $h : [0, 1] \rightarrow [-\infty, +\infty]$ with $h(0) = -\infty$, $h(e) = 0$ and $h(1) = +\infty$ such that

$$U(x, y) = h^{-1}(h(x) + h(y))$$

for all $(x, y) \in [0, 1]^2 \setminus \{(0, 1), (1, 0)\}$ and $U(0, 1) = U(1, 0) \in \{0, 1\}$. The function h is usually called an *additive generator* of U.

Any representable uninorm U with neutral element e and additive generator h, will be denoted by $U \equiv \langle h, e \rangle_{\text{rep}}$ and the class of representable uninorms will be denoted by $\mathscr{U}_{\text{rep}}$. For any of these uninorms the underlying t-norm T is always strict and the underlying t-conorm S is strict as well. Moreover, for any representable uninorm $U \equiv \langle h, e \rangle_{\text{rep}}$ the function

$$N_h(x) = h^{-1}(-h(x)) \quad \text{for all} \quad x \in [0, 1]$$

is always a strong negation and U is auto-dual with respect to N_h, except for the points $(1, 0)$, $(0, 1)$.

On the other hand, different classes of implications derived from uninorms have been studied. We recall here RU-implications.

Definition 5.2.7 [8] Let U be a uninorm. The *residual operation* derived from U is the binary operation given by

$$I_U(x, y) = \sup\{z \in [0, 1] \mid U(x, z) \le y\} \text{ for all } x, y \in [0, 1].$$

Proposition 5.2.8 [8] *Let U be a uninorm and I_U its residual operation. Then I_U is an implication if and only if the following condition holds*

$$U(x, 0) = 0 \quad \text{for all} \quad x < 1. \tag{5.1}$$

In this case I_U is called an RU-implication.

This includes all conjunctive uninorms but also many disjunctive ones, for instance in the classes of representable uninorms (see [8]) and idempotent uninorms (see [22]).

Some properties of RU-implications have been studied involving the main classes of uninorms, specially uninorms in $\mathscr{U}_{\text{min}}$, idempotent uninorms and representable uninorms (for more details see [1, 4, 8, 19, 22, 23]). The Modus Ponens property was also studied for RU-implications in [15], but not the Modus Tollens property, that we will discuss in next section.

5.3 RU-Implications and the Modus Tollens

In this section, fixed a (continuous) t-norm T and a (continuous, strict, strong) fuzzy negation N, we will investigate which RU-implications satisfy the Modus Tollens (MT for short) property with respect to T and N, in a similar way that it was done for the Modus Ponens (MP for short) with respect to T, in [15]. Fisrt of all, let us recall the definition of the Modus Tollens in the framework of fuzzy logic.

Definition 5.3.1 Let I be an implication function, T a t-norm and N a fuzzy negation. It is said that I satisfies the *Modus Tollens property* with respect to T and N if

$$T(N(y), I(x, y)) \leq N(x) \quad \text{for all } x, y \in [0, 1]. \tag{5.2}$$

A general result about MT was already done in [25] that will be used along the paper.

Proposition 5.3.2 [25] *Let T be a continuous t-norm, I_T its residual implication and N a fuzzy negation. Then an implication function I satisfies MT with respect to T and N if and only if $I(x, y) \leq I_T(N(y), N(x))$ for all $x, y \in [0, 1]$.*

Remark 5.3.3 Note that, as in the MP-property, when $x \leq y$ we have $N(y) \leq N(x)$ and Eq. (5.2) trivially holds in these cases. Thus, the MT-property needs to be checked only in points (x, y) where $y < x$.

In both properties, MP and MT, the t-norm T is usually considered a continuous t-norm and we will develop our study of the MT-property mainly for this case. On the other hand, there are many differences between the MP and the MT properties, and the first one lies in the negation N. Whereas in the MP-property no negation appears, the role of N in the MT-case is quite important, as it can be seen in the following two examples.

Example 5.3.4 Let us consider N be the least fuzzy negation, that is, $N(x) = N_{lt}(x) = 0$ for all $x > 0$. In this case,

- For all $y > 0$ we have $N_{lt}(y) = 0$ and Eq. (5.2) holds.
- For $y = 0$ we obtain $T(1, I(x, 0)) = I(x, 0) = N_I(x)$.

Thus, a fuzzy implication I satisfies the MT with respect to a fixed t-norm T and the least negation N_{lt} if, and only if, the natural negation of I is the proper N_{lt}. Note that there are many implication functions with natural negation N_{lt} like for instance, R-implications derived from the minimum or from strict t-norms, (S, N), QL and D-implications derived from any t-conorm, t-norm and the least negation, f-generated Yager implications with $f(0) = \infty$, and so on. Among RU-implications, those derived from uninorms in $\mathcal{U}_{\min}$ (with underlying t-norm T_U being the minimum or a strict t-norm) or those derived from idempotent uninorms also satisfy this property.

Example 5.3.5 Let us suppose now that N is the greatest fuzzy negation, that is, $N(x) = N_{gt}(x) = 1$ for all $x < 1$. In this case,

- For all $x < 1$ we have $N_{gt}(x) = 1$ and Eq. (5.2) holds.
- For $x = 1$ we obtain $T(N_{gt}(y), I(1, y)) = I(1, y)$.

Thus, a fuzzy implication I satisfies the MT with respect to a fixed t-norm T and the greatest negation N_{gt} if, and only if, $I(1, y) = 0$ for all $y < 1$. Note that RU-implications derived from representable uninorms or those derived from idempotent uninorms with $g(1) = 0$ satisfy this property.

However, the fuzzy negation N is usually taken a continuous negation (in many cases, only strict or strong negations are considered) and so, from now on, we will suppose N continuous and consequently with a fixed point, usually denoted by $s \in]0, 1[$. Of course, the uninorm used to derive the RU-implication also plays a fundamental role in our framework. Thus we will divide our study in some sections depending on the class where the used uninorm lies. First, let us recall some general facts.

All uninorms in the considered families satisfy that they are locally internal on the boundary, that is, $U(1, y) \in \{y, 1\}$ for all $y \in [0, 1]$ (in fact, this is always the case when the underlying operations of U are continuous, see [16]). Consequently, in all these cases it must be some $\alpha \in [0, 1]$ such that

$$U(1, y) = \begin{cases} y & \text{if } y < \alpha \\ \alpha \text{ or } 1 & \text{if } y = \alpha \\ 1 & \text{if } y > \alpha. \end{cases}$$

Taking this fact in mind, we can give the following general result.

Proposition 5.3.6 *Let T be a t-norm, N a continuous fuzzy negation, and U a uninorm such that $U(1, y) = y$ for all $y < \alpha$ and $U(1, y) = 1$ for all $y > \alpha$, with $\alpha \in]0, 1[$. If I_U satisfies the MT-property with respect to T and N, then*

(i) *$T(N(y), y) = 0$ for all $y \leq \alpha$.*
(ii) *If T is a continuous t-norm then T must be nilpotent with normalized additive generator $t : [0, 1] \to [0, 1]$ and associated negation N_T, which is given by $N_T(x) = t^{-1}(1 - t(x))$, such that $N(y) \leq N_T(y)$ for all $y \leq \alpha$.*

Proof To prove (i) just taking $x = 1$ in Eq. (5.2) we obtain $T(N(y), I_U(1, y)) = 0$. Since

$$I_U(1, y) = \sup\{z \in [0, 1] \mid U(1, z) \leq y\} = y \quad \text{for all} \quad y \leq \alpha,$$

we have $T(N(y), y) = 0$ for all $y \leq \alpha$.

Part (ii) is a direct consequence of the classification theorem of continuous t-norms, see Theorem 5.11 in [11]. ∎

5.3.1 Case When U Is a Uninorm in $\mathscr{U}_{\min}$

In this section we will deal with RU-implications derived from uninorms in $\mathscr{U}_{\min}$, that is, uninorms $U \equiv \langle T_U, e, S_U \rangle_{\min}$ with neutral element $e \in]0, 1[$. Recall that for this kind of uninorms, RU-implications have the following structure.

Proposition 5.3.7 (Theorem 5.4.7 in [4]) *Let $U \equiv \langle T_U, e, S_U \rangle_{\min}$ a uninorm in $\mathscr{U}_{\min}$ and I_U its residual implication. Then*

$$I_U(x, y) = \begin{cases} 1 & \text{if } x \leq y < e \\ e I_{T_U}(\frac{x}{e}, \frac{y}{e}) & \text{if } y < x \leq e \\ e + (1-e)I_{S_U}(\frac{x-e}{1-e}, \frac{y-e}{1-e}) & \text{if } e \leq x \leq y \\ e & \text{if } e \leq y < x \\ y & \text{otherwise.} \end{cases}$$

Note that for any uninorm $U \equiv \langle T_U, e, S_U \rangle_{\min}$ it is always $U(1, y) = y$ for all $y \leq e$ and thus Proposition 5.3.6 holds for $\alpha = e$, the neutral element of U. Moreover, for this kind of RU-implications we have the following result.

Proposition 5.3.8 *Let T be a continuous t-norm, N a continuous fuzzy negation with and $U \equiv \langle T_U, e, S_U \rangle_{\min}$ a uninorm in $\mathscr{U}_{\min}$. If $T(N(y), y) = 0$ for all $y \leq e$ then I_U satisfies the MT-property with respect to T and N for all $y < x$ such that $x \geq e$.*

Proof We only need to check condition (5.2) for all $y < x$ and $x \geq e$ and we will do it by distinguishing two cases:

(i) When $y < e \leq x$. In this case we have $I_U(x, y) = y$ and then condition (5.2) holds because
$$T(N(y), I_U(x, y)) = T(N(y), y) = 0.$$

(ii) When $e \leq y < x$. In this case we have $I_U(x, y) = e$ and $N(y) \leq N(e)$. Thus,
$$T(N(y), I_U(x, y)) = T(N(y), e) \leq T(N(e), e) = 0 \leq N(x). \qquad \blacksquare$$

The previous proposition proves that for a uninorm U in $\mathscr{U}_{\min}$ the MT-property only can fail for values $y < x < e$. In particular, the underlying t-conorm S_U of a uninorm U in $\mathscr{U}_{\min}$ is not relevant in order I_U to satisfy MT with respect to T and N. Only the underlying t-norm T_U is relevant and inequality (5.2) only depends on the neutral element e and the underlying t-norm T_U. In this line, we can give the following results.

Theorem 5.3.9 *Let T be a t-norm, and N be a continuous fuzzy negation with fixed point s such that $T(N(y), y) = 0$ for all $y \leq e$. For any uninorm $U \equiv \langle T_U, e, S_U \rangle_{\min}$ in $\mathscr{U}_{\min}$, the following items hold.*

(i) *If $e \leq s$ then I_U always satisfies the MT-property with respect to T and N.*
(ii) *If $T_U = \min$ then I_U always satisfies the MT-property with respect to T and N.*
(iii) *If $s < e$ then I_U satisfies the MT-property with respect to T and N if and only if*
$$e I_{T_U}\left(\frac{x}{e}, \frac{y}{e}\right) \leq I_T(N(y), N(x)) \quad \text{for all } \; y < x < e.$$

Proof Let us prove the MT-property for all x, y such that $y < x < e$ in each case.

(i) If $e \leq s$, for all $y < x < e$ we have $N(x) \geq s$ and $I_U(x, y) = eI_{T_U}(\frac{x}{e}, \frac{y}{e}) \leq e$. Thus,

$$T(N(y), I_U(x, y)) \leq T(N(y), e) \leq e \leq s \leq N(x).$$

(ii) If $T_U = \min$ then $I_U(x, y) = y$ for all $y < x \leq e$ and the result follows.
(iii) This simply corresponds to apply the general result in Proposition 5.3.2 to the case $y < x < e$.

∎

Example 5.3.10 Take for instance the Łukasiewicz t-norm T_L and the classical negation $N(x) = N_c(x) = 1 - x$ for all $x \in [0, 1]$. Then the following RU-implications satisfy MT with respect to T_L and N_c:

- I_U derived from any uninorm in $\mathcal{U}_{\min}$ with neutral element $e \leq 1/2$, just applying part (i) of Theorem 5.3.9.
- I_U derived from any uninorm in $\mathcal{U}_{\min}$ with neutral element $e > 1/2$ such that $T_U(x, y) = \min(x, y)$ for all $x, y \in [\frac{1}{2e}, 1]$. In this case we have

$$eI_{T_U}\left(\frac{x}{e}, \frac{y}{e}\right) = y \quad \text{for all} \quad x > 1/2 \text{ and } y < x \leq e,$$

whereas

$$eI_{T_U}\left(\frac{x}{e}, \frac{y}{e}\right) \leq 1/2 \quad \text{for all} \quad y < x \leq \frac{1}{2}.$$

In both cases, the result is less than or equal to $I_T(N(y), N(x)) = 1 + y - x$, and so I_U satisfies MT in this case just applying part (iii) of Theorem 5.3.9.

5.3.2 Case When U Is an Idempotent Uninorm

In this section we will deal with RU-implications derived from idempotent uninorms that is, from uninorms $U \equiv \langle g, e \rangle_{\text{ide}}$ with neutral element $e \in {]}0, 1[$ and such that $g(0) = 1$. Let us recall in this case the expression of the residual implication derived from U.

Proposition 5.3.11 (Theorem 5.4.14 in [4]) *Let $U \equiv \langle g, e \rangle_{\text{ide}}$ be an idempotent uninorm with neutral element $e \in {]}0, 1[$ and such that $g(0) = 1$. Then I_U is given by*

$$I_U(x, y) = \begin{cases} \max(g(x), y) & \text{if } x \leq y, \\ \min(g(x), y) & \text{if } x > y. \end{cases}$$

Recall that to get an RU-implication from an idempotent uninorm $U \equiv \langle g, e \rangle_{\text{ide}}$ it must be $g(0) = 1$. On the other hand, note that if $g(1) = \alpha > 0$ then $U(1, y) = y$ for all $y < \alpha$ and Proposition 5.3.6 applies, whereas if $g(1) = 0$ then $U(1, y) = 1$

for all $y > 0$. Moreover, for this kind of uninorms we have the following general result.

Theorem 5.3.12 *Let $U \equiv \langle g, e \rangle_{\text{ide}}$ be an idempotent uninorm with neutral element $e \in \ ]0, 1[$, N a continuous fuzzy negation and T any t-norm. Then I_U satisfies MT with respect to T and N if and only if*

$$\min(T(N(y), y), T(N(y), g(x))) \leq N(x) \quad \text{for all} \quad y < x. \quad (5.3)$$

Proof This is a direct consequence of the monotonicity of T and the expression of I_U given in the previous proposition. ∎

Example 5.3.13 From the previous theorem we can get some general cases of RU-implications derived from idempotent uninorms that satisfy MT as follows.

(i) Let T be a nilpotent t-norm with associated negation N_T and N a continuous fuzzy negation such that $N \leq N_T$. Then I_U satisfies MT with respect to T and N for any idempotent uninorm $U \equiv \langle g, e \rangle_{\text{ide}}$. This is because we have $T(N(y), y) = 0$ for all $y \in [0, 1]$ in this case and so Eq. (5.3) holds.
(ii) Let $U \equiv \langle g, e \rangle_{\text{ide}}$ be any idempotent uninorm and N a continuous fuzzy negation such that $g \leq N$ (in particular, it must be $g(1) = 0$). Then I_U satisfies MT with respect to any t-norm T and N. In this case Eq. (5.3) holds because $g(x) \leq N(x)$ for all $x \in [0, 1]$. As a particular case take N a strong negation and U an idempotent uninorm with associated function $g = N$. Then the corresponding RU-implication, I_U, satisfies MT with respect to any t-norm T and N. This kind of RU-implications are specially interesting because of their properties (see [22]).

In the case when N is a strict negation we can give a characterization of those RU-implications that satisfy MT with respect to any t-norm and the strict negation N as follows.

Proposition 5.3.14 *Let $U \equiv \langle g, e \rangle_{\text{ide}}$ be an idempotent uninorm with neutral element $e \in \ ]0, 1[$ and N a strict fuzzy negation. Then I_U satisfies MT with respect to any t-norm T and N if and only if $g(x) \leq N(x)$ for all $x \geq e$.*

Proof It is enough to prove the result for the minimum t-norm. Suppose first that I_U satisfies MT with respect to the minimum t-norm and N and let us prove that $g(x) \leq N(x)$ for all $x \geq e$ in two steps:

- For all $x > e$ it is $g(x) \leq e < x$ and so we can take y such that $e < y < x$. By the MT-property for this value y we have

$$\min(N(y), \min(g(x), y)) \leq N(x),$$

but, $N(y) > N(x)$ because N is strict and so the previous inequality leads to $g(x) \leq N(x)$.

- For $x = e$ we obtain by MT that

$$\min(N(y), I_U(e, y)) = \min(N(y), y) \leq N(e) \quad \text{for all} \quad y < e.$$

Since for all $y < e$ it is $N(y) > N(e)$ the previous inequality becomes $y \leq N(e)$ for all $y < e$. That is, $e \leq N(e)$.

Conversely, if $g(x) \leq N(x)$ for all $x \geq e$ we can prove MT again in two steps:

- For all $y < x < e$, we have $N(x), N(y) \geq N(e) \geq e$ and so

$$\min(N(y), I_U(x, y)) = \min(N(y), \min(g(x), y)) = y < e \leq N(x).$$

- For $y < x$ and $e \leq x$ we have $g(x) \leq N(x)$ and then MT follows trivially. ■

5.3.3   Case When U Is a Representable Uninorm

In this section we will deal with RU-implications derived from representable uni-norms, that is, from uninorms $U \equiv \langle h, e \rangle_{\text{rep}}$ with neutral element $e \in]0, 1[$. Let us recall in this case the expression of the residual implication derived from U.

Proposition 5.3.15 (Theorem 5.4.10 in [4]) *Let* $U \equiv \langle h, e \rangle_{\text{rep}}$ *be a representative uninorm with neutral element* $e \in]0, 1[$. *Then* I_U *is given by*

$$I_U(x, y) = \begin{cases} 1 & \text{if } (x, y) \in \{(0, 0), (1, 1)\}, \\ h^{-1}(h(y) - h(x)) & \text{otherwise.} \end{cases}$$

Note that for this kind of uninorms it is $U(1, y) = 1$ for all $y > 0$ and so Proposition 5.3.6 does not apply. Consequently, the t-norm T considered for the MT-property can be any one in these cases. Thus, reducing to continuous t-norms, we will distinguish three possible cases, that is, when $T = \min$, when T is a continuous Archimedean t-norm or when T is an ordinal sum of continuous Archimedean t-norms.

Proposition 5.3.16 *Let* $U \equiv \langle h, e \rangle_{\text{rep}}$ *be a representable uninorm with neutral element* $e \in]0, 1[$, *N a continuous fuzzy negation and* $T = \min$. *Then* I_U *never satisfy MT with respect to T and N.*

Proof By continuity of N it must be some interval $[a, b] \subseteq [0, 1]$ such that N is strictly decreasing in $[a, b]$ with $N(x) < e$ for all $x \in [a, b]$. Then, for all $y < x$ with $x \in]a, b[$ it must be $N(y) > N(x)$ and so applying MT we obtain

$$\min(N(y), h^{-1}(h(y) - h(x))) \leq N(x),$$

which implies that $h^{-1}(h(y) - h(x)) \leq N(x)$. By continuity of h we will get $h^{-1}(0) = e \leq N(x)$ leading to a contradiction. ■

However, if we consider non-continuous fuzzy negations N, we can obtain RU-implications from representable uninorms that satisfy MT with respect to $T = \min$ and N, as the following example shows.

Example 5.3.17 Take for instance any representable uninorm $U \equiv \langle h, e \rangle_{\mathrm{rep}}$ with neutral element $e \in \,]0, 1[$ and the fuzzy negation N given by

$$N(x) = \begin{cases} (e-1)x + 1 & \text{if } x < 1 \\ 0 & \text{if } x = 1 \end{cases}$$

Since $I_U(x, y) = h^{-1}(h(y) - h(x)) \leq e$ for all $y < x$, it follows that I_U always satisfies MT with respect to the minimum t-norm (and consequently with respect to any t-norm) and N.

Proposition 5.3.18 *Let $U \equiv \langle h, e \rangle_{\mathrm{rep}}$ be a representable uninorm with neutral element $e \in \,]0, 1[$, N a continuous fuzzy negation and T a continuous Archimedean t-norm with additive generator $t : [0, 1] \to [0, +\infty]$. Then*

(i) I_U satisfies MT with respect to T and N if and only if

$$h^{-1}(h(y) - h(x)) \leq t^{-1}(t(N(x)) - t(N(y))) \quad \text{for all } y \leq x \qquad (5.4)$$

(ii) If T is nilpotent and $N = N_T$, I_U satisfies MT with respect to T and N if and only if φ is sub-additive where $\varphi : [0, 1] \to [-\infty, +\infty]$ is given by $\varphi(x) = h(t^{-1}(x))$ for all $x \in [0, 1]$.

Proof Part (i) is a direct consequence of Proposition 5.3.6 and the structure of the implications I_U and I_T respectively.

To prove part (ii), note that when $N = N_T$ we have $t(N(x)) = 1 - t(x)$ for all $x \in [0, 1]$ and Eq. (5.4) can be written as

$$h^{-1}(h(y) - h(x)) \leq t^{-1}(t(y) - t(x)) \quad \text{for all } y \leq x.$$

Now, taking $t(x) = a$, $t(y) = b$ and $\varphi h \circ t^{-1}$ we obtain that I_U satisfies MT with respect to T and N_T if and only if

$$\varphi(b) - \varphi(a) \leq \varphi(b - a) \quad \text{for all } a \leq b,$$

and the result follows. ∎

Example 5.3.19 Let us consider the Łukasiewicz t-norm T_{L} and its associate negation $N_c(x) = 1 - x$. Let U be the representable uninorm given by

$$U(x, y) = \begin{cases} 0 & \text{if } (x, y) \in \{(0, 1), (1, 0)\} \\ \frac{xy}{xy + (1-x)(1-y)} & \text{otherwise,} \end{cases}$$

which is a well known representable uninorm with neutral element $e = \frac{1}{2}$ and additive generator $h(x) = \ln\left(\frac{x}{1-x}\right)$. In this case, we have $\varphi(x) = h(1-x) = \ln\left(\frac{1-x}{x}\right)$, which is clearly sub-additive. By applying the previous proposition we obtain that I_U satisfies MT with respect to $T_{\mathbf{L}}$ and N_c.

Finally, with respect to the case when T is an ordinal sum, we can give the following partial result.

Proposition 5.3.20 *Let $U \equiv \langle h, e \rangle_{\mathrm{rep}}$ be a representable uninorm with neutral element $e \in]0, 1[$, N a continuous fuzzy negation and T a continuous ordinal sum t-norm.*

(i) *If I_U satisfies MT with respect to T and N, there exists $r \geq e$ such that T is Archimedean on the interval $[0, r]$, that is, $T = \langle (0, r, T_0), (r, 1, T') \rangle$, with T_0 Archimedean and T' continuous.*
(ii) *If $T = \langle (0, e, T_0), \langle e, 1, T' \rangle \rangle$ and N has fixed point e, then I_U satisfies MT with respect to T and N if and only if*

$$h^{-1}(h(y) - h(x)) \leq e\, t^{-1}\left(t\left(\frac{N(x)}{e}\right) - t\left(\frac{N(y)}{e}\right)\right)$$

for all $e < y < x$, where t is an additive generator of T_0.

Proof To prove part (i), suppose that T has an idempotent element $r < e$. Consider $a = \max\{z \mid N(z) = r\}$. Then N will be strictly decreasing in an interval $[a, b]$ with $a < b \leq 1$, and so we have $N(x) < N(y) < r$ for all $a < y < x < b$. On the other hand, $I_U(x, y) = h^{-1}(h(y) - h(x))$ approaches to e when y approaches to x and so there are some $y < x$ such that $r < I_U(x, y) < e$. For these values we obtain a contradiction with the MT-property because

$$T(N(y), I_U(x, y)) = \min(N(y), I_U(x, y)) = N(y) > N(x).$$

To prove (ii), let us discuss some cases separately.

- For all $y < x \leq e$, MT always holds because

$$T(N(y), h^{-1}(h(y) - h(x))) = h^{-1}(h(y) - h(x)) \leq e \leq N(x).$$

- For $e < y < x$ we have $T(N(y), h^{-1}(h(y) - h(x))) \leq N(x)$ if and only if

$$e\, t^{-1}\left(t\left(\frac{N(y)}{e}\right) + t\left(\frac{h^{-1}(h(y) - h(x))}{e}\right)\right) \leq N(x)$$

which is equivalent to the inequality given in the statement for $e < y < x$. Note that by continuity this inequality also holds for $y = e$, which derives into $N_h(x) \leq N(x)$ for all $x > e$.

- For $y \leq e < x$, we have

$$T(N(y), h^{-1}(h(y) - h(x))) = h^{-1}(h(y) - h(x)) \leq N(x)$$

for all $y \leq e < x$. This occurs if and only if $N_h(x) \leq N(x)$ for all $x > e$, but this always holds by the previous step. ∎

Example 5.3.21 Let us consider again U the representable uninorm given in Example 5.3.19 with neutral element $e = \frac{1}{2}$ and additive generator $h(x) = \ln\left(\frac{x}{1-x}\right)$. Consider also $N = N_c$ and let T be the ordinal sum t-norm $T = (\langle 0, e, T_L \rangle, \langle e, 1, T' \rangle)$, where T' is any continuous t-norm. Recall that in this case $N_h = N_c$ and the additive generator of the Łukasiewicz t-norm T_L is $t(x) = 1 - x$. Thus, a simple calculation shows that condition in part (ii) of Preposition 5.3.20 holds and so I_U satisfies MT with respect to T and N_c.

5.4 Conclusions and Future Work

In this paper the property of Modus Tollens (MT) with respect to a continuous t-norm T and a continuous fuzzy negation N has been studied for residual implication functions derived from uninorms, also known as RU-implications. Along this study, a lot of new solutions of the MT-property have been pointed out and the corresponding inequality has been completely characterized in many cases when the considered uninorm is either in the class of $\mathscr{U}_{\min}$, an idempotent uninorm or a representable uninorm.

As a future work, this study could be enlarged by considering other well known classes of uninorms like uninorms continuous in the open unit square, compensatory uninorms or uninorms having continuous underlying operators, that has been recently characterized in some cases (see [9, 12, 13]). Moreover, the Modus Tollens with respect to T and N could be also studied in a more general setting by changing the t-norm T by a conjunctive uninorm U_c. Additionally, the same study could be also done in the discrete case, that is, when t-norms, uninorms and implications are defined on a finite chain.

Acknowledgments The authors want to dedicate this work to Professor Gaspar Mayor in his 70th birthday. We have had the pleasure of jointly work with him in many aspects of aggregation functions and fuzzy implications. He has been and he is, not only a very good teacher and colleague for us, but also a friend. This paper has been partially supported by the Spanish grant TIN2013-42795-P.

References

1. I. Aguiló, J. Suñer, J. Torrens, A characterization of residual implications derived from left-continuous uninorms. Inf. Sci. **180**(20), 3992–4005 (2010)
2. C. Alsina, E. Trillas, When (S, N)-implications are (T, T_1)-conditional functions? Fuzzy Sets Syst. **134**, 305–310 (2003)
3. M. Baczyński, G. Beliakov, H. Bustince-Sola, A. Pradera (eds.), *Advances in Fuzzy Implication Functions. Studies in Fuzziness and Soft Computing*, vol. 300 (Springer, Berlin, 2013)
4. M. Baczyński, B. Jayaram, *Fuzzy Implications. Studies in Fuzziness and Soft Computing*, vol. 231 (Springer, Berlin, 2008)
5. M. Baczyński, B. Jayaram, (U, N)-implications and their characterizations. Fuzzy Sets Syst. **160**, 2049–2062 (2009)
6. H. Bustince, P. Burillo, F. Soria, Automorphisms, negations and implication operators. Fuzzy Sets Syst. **134**, 209–229 (2003)
7. B. De Baets, Idempotent uninorms. Eur. J. Oper. Res. **118**, 631–642 (1999)
8. B. De Baets, J.C. Fodor, Residual operators of uninorms. Soft Comput. **3**, 89–100 (1999)
9. J. Fodor, B. De Baets, A single-point characterization of representable uninorms. Fuzzy Sets Syst. **202**, 89–99 (2012)
10. J.C. Fodor, R.R. Yager, A. Rybalov, Structure of uninorms. Int. J. Uncertain. Fuzziness Knowl. Based Syst. **5**, 411–427 (1997)
11. E.P. Klement, R. Mesiar, E. Pap, *Triangular Norms* (Kluwer Academic Publishers, Dordrecht, 2000)
12. G. Li, H.W. Liu, J. Fodor, Single-point characterization of uninorms with nilpotent underlying t-norm and t-conorm. Int. J. Uncertain. Fuzziness Knowl. Based Syst. **22**, 591–604 (2014)
13. G. Li, H.W. Liu, Distributivity and conditional distributivity of a uninorm with continuous underlying operators over a continuous t-conorm. Fuzzy Sets Syst. **287**, 154–171 (2016)
14. J. Martín, G. Mayor, J. Torrens, On locally internal monotonic operators. Fuzzy Sets Syst. **137**, 27–42 (2003)
15. M. Mas, M. Monserrat, D. Ruiz-Aguilera, J. Torrens, Residual implications derived from uninorms satisfying Modus Ponens, in *Proceedins of IFSA-EUSFLAT-2015 conference*, Gijón (Spain), 2015
16. M. Mas, M. Monserrat, D. Ruiz-Aguilera, J. Torrens, Migrative uninorms and nullnorms over t-norms and t-conorms. Fuzzy Sets Syst. **261**, 20–32 (2015)
17. M. Mas, M. Monserrat, J. Torrens, Two types of implications derived from uninorms. Fuzzy Sets Syst. **158**, 2612–2626 (2007)
18. M. Mas, M. Monserrat, J. Torrens, Modus Ponens and Modus Tollens in discrete implications. Int. J. Approx. Reason. **49**, 422–435 (2008)
19. M. Mas, M. Monserrat, J. Torrens, A characterization of (U, N), RU, QL and D-implications derived from uninorms satisfying the law of importation. Fuzzy Sets Syst. **161**, 1369–1387 (2010)
20. M. Mas, M. Monserrat, J. Torrens, E. Trillas, A survey on fuzzy implication functions. IEEE Trans. Fuzzy Syst. **15**(6), 1107–1121 (2007)
21. S. Massanet, J. Torrens, An overview of construction methods of fuzzy implications, in [3], pp. 1–30 (2013)
22. D. Ruiz, J. Torrens, Residual implications and co-implications from idempotent uninorms. Kybernetika **40**, 21–38 (2004)
23. D. Ruiz-Aguilera, J. Torrens, S- and R-implications from uninorms continuous in $]0, 1[^2$ and their distributivity over uninorms. Fuzzy Sets Syst. **160**, 832–852 (2009)
24. D. Ruiz-Aguilera, J. Torrens, B. De Baets, J. Fodor, Some remarks on the characterization of idempotent uninorms, in *Computational Intelligence for Knowledge-Based Systems Design*, vol. 6178, Lecture Notes in Computer Science, ed. by E. Hüllermeier, R. Kruse, F. Hoffmann (Berlin, 2010), pp. 425–434
25. E. Trillas, C. Alsina, A. Pradera, On MPT-implication functions for fuzzy logic. Revista de la Real Academia de Ciencias. Serie A. Matemáticas (RACSAM) **98**(1), 259–271 (2004)

26. E. Trillas, C. Alsina, E. Renedo, A. Pradera, On contra-symmetry and MPT-conditionality in fuzzy logic. Int. J. Intell. Syst. **20**, 313–326 (2005)
27. E. Trillas, C. Campo, S. Cubillo, When QM-operators are implication functions and conditional fuzzy relations. Int. J. Intell. Syst. **15**, 647–655 (2000)
28. E. Trillas, L. Valverde, On Modus Ponens in fuzzy logic, in *Proceedings of the 15th International Symposium on Multiple-Valued Logic*, Kingston, Canada, 1985, pp. 294–301

Chapter 6
A Survey of Atanassov's Intuitionistic Fuzzy Relations

**Humberto Bustince, Edurne Barrenechea, Miguel Pagola,
Javier Fernandez, Raul Orduna and Javier Montero**

Abstract In this chapter we review several properties of Atanassov's intuitionistic fuzzy relations, recalling the main concepts related to Atanassov's intuitionistic fuzzy relations and the main properties that can be demanded to such conepts. We also consider the use of Atanassov's operators over such relations.

6.1 Introduction

Relations are a very appropriate tool for describing correspondences between objects [21]. The use of fuzzy relations (FRs) originated from the observation that real-life objects can be related to each other to a certain degree, rather than fully related or fully non-related. However, in real-life situations, a person may assume that a certain object A is in relation R with another object B to a certain degree, but it is possible that he/she is not so sure about it. Moreover, even the construction of such degree may

H. Bustince (✉) · E. Barrenechea · M. Pagola · J. Fernandez
Departamento of Automática y Computación and Institute of Smart Cities,
Universidad Publica de Navarra, 31006 Pamplona, Spain
e-mail: bustince@unavarra.es

E. Barrenechea
e-mail: edurne.barrenechea@unavarra.es

M. Pagola
e-mail: miguel.pagola@unavarra.es

J. Fernandez
e-mail: fcojavier.fernandez@unavarra.es

R. Orduna
Departamento of Automática y Computación,
Universidad Publica de Navarra, 31006 Pamplona, Spain
e-mail: raul.orduna@unavarra.es

J. Montero
Department of Statistics and Operations Research I, Faculty of Mathematics,
Universidad Complutense de Madrid, Plaza de Ciencias 3, 28040 Madrid, Spain
e-mail: jamonter@ucm.es

© Springer International Publishing Switzerland 2016
T. Calvo Sánchez and J. Torrens Sastre (eds.), *Fuzzy Logic and Information Fusion*,
Studies in Fuzziness and Soft Computing 339, DOI 10.1007/978-3-319-30421-2_6

be complicate, due to lack of knowledge or uncertainty. A possible solution is to use Atanassov's intuitionistic fuzzy sets, defined by Atanassov in 1983 [5]. Atanassov's intuitionistic fuzzy sets give the possibility to model hesitation and uncertainty by using an additional degree, called non-membership degree [1, 2, 4, 6]. These sets have become nowadays very useful for many different applications [7, 28], including image processing [27] or decision making [8, 20, 30].

From a theoretical point of view, fuzzy relations and Atanassov's intuitionistic fuzzy relations have attracted a lot of interest in last years, see [22–25, 29] for a non-exhaustive list.

In this chapter, we present a survey of Atanassov's intuitionistic fuzzy relations (A-IFRs). We describe the most important concepts and results of this theory, intended to provide the basic concepts and notions to non-specialist researchers and we focus on the way these sets may be built in order to satisfy predetermined properties. We mainly follow the developments in [10–19].

This chapter is organized in the following way. We start by recalling the definition of Atanassov's intuitionistic fuzzy relations and their structures in Sect. 6.2. In Sect. 6.3 we consider the construction of A-IFRs which satisfy some predetermined properties. Section 6.4 is devoted to conclusions and we finish with some references.

6.2 Definitions and Structures with A-IFS

Let X, Y and Z be ordinary finite non-empty sets [1]. An Atanassov's intuitionistic fuzzy set (A-IFS) A on X is given by

$$A = \{\langle x, \mu_A(x), \nu_A(x) \rangle | x \in X\},$$

where $\mu_A(x) : X \to [0, 1]$, $\nu_A(x) : X \to [0, 1]$, with the condition $0 \leq \mu_A(x) + \nu_A(x) \leq 1$ for all $x \in X$.

The numbers $\mu_A(x)$ and $\nu_A(x)$ denote, respectively, the degree of membership and the degree of nonmembership of the element x in the set A. In this chapter we denote by A-IFS(X) the set of all the Atanassov's intuitionistic fuzzy sets in X. When $\nu_A(x) = 1 - \mu_A(x)$ for every x in X, the set A is a fuzzy set. We denote by $FS(X)$ the class of all the fuzzy sets in X. We also call $\pi_A(x) = 1 - \mu_A(x) - \nu_A(x)$ the Atanassov's intuitionistic index of the element x in the set A.

The following expressions are defined in [1, 2, 10] for every $A, B \in$ A-IFS(X).

1. $A \leq B$ if and only if $\mu_A(x) \leq \mu_B(x)$ and $\nu_A(x) \geq \nu_B(x)$ for all $x \in X$.
2. $A \geq B$ if and only if $\mu_A(x) \geq \mu_B(x)$ and $\nu_A(x) \leq \nu_B(x)$ for all $x \in X$.
3. $A = B$ if and only if $A \leq B$ and $B \leq A$.
4. $A_c = \{\langle x, \nu_A(x), \mu_A(x) \rangle | x \in X\}$.

In 1986, Atanassov proposed different ways of reducing an Atanassov's intuitionistic fuzzy set to a fuzzy set. To do so, he defined the following operator.

Definition 6.1 Let $C \in$ A-IFS(X). For $p \in [0, 1]$, the Atanassov's operator is defined by

$$D_p(C) = \{(x, \mu_C(x) + p \cdot \pi_C(x), 1 - \mu_C(x) - p \cdot \pi_C(x)) | x \in X\}$$

Note that $D_p(C) \in FS(X)$. A study of the properties of this operator is made in [3, 16].

Let C be an Atanassov's intuitionistic fuzzy set and let D_p be the operator given in Definition 6.1. We denote the family of all fuzzy sets associated to C through the operator D_p by $\{D_p(C)\}_{p \in [0,1]}$. It is clear that $\{D_p(C)\}_{p \in [0,1]}$ is a totally ordered family of fuzzy sets (with respect to Zadeh's order). It is worth pointing out that for any Atanassov's intuitionistic fuzzy set C, $\mu_{(D_0(C))}(x) = \mu_C(x)$, $\mu_{(D_1(C))_c}(x) = \nu_C(x)$ for every $x \in X$.

An Atanassov's intuitionistic fuzzy relation (A-IFR) E over $X \times Y$ is an Atanassov's intuitionistic fuzzy subset of $X \times Y$: that is,

$$E = \{\langle (x, y), \mu_E(x, y), \nu_E(x, y) \rangle | x \in X, y \in Y\},$$

where $\mu_E : X \times Y \to [0, 1]$, $\nu_E : X \times Y \to [0, 1]$ satisfy the condition $0 \le \mu_E(x, y) + \nu_E(x, y) \le 1$ for every $(x, y) \in X \times Y$.

We denote by A-IFR($X \times Y$) the set of all the Atanassov's intuitionistic fuzzy subsets in $X \times Y$. The complementary relation of an A-IFR E is given by:

$$E_c = \{\langle (x, y), \nu_E(x, y), \mu_E(x, y) \rangle | (x, y) \in X \times Y\}.$$

The most important properties of the Atanassov's intuitionistic fuzzy relations are studied in [9, 16].

Now we consider the definition of the composition of A-IFRs.

Definition 6.2 Let $\alpha, \beta, \lambda, \rho$ be t-norms or t-conorms not necessarily dual to each other, $E \in$ A-IFR($X \times Y$) and $L \in$ A-IFR($Y \times Z$). The composed relation $L \overset{\alpha,\beta}{\underset{\lambda,\rho}{\circ}} E \in$ A-IFR($X \times Z$) is defined by

$$L \overset{\alpha,\beta}{\underset{\lambda,\rho}{\circ}} E = \{((x, z), \mu_{L \overset{\alpha,\beta}{\underset{\lambda,\rho}{\circ}} E}(x, z), \nu_{L \overset{\alpha,\beta}{\underset{\lambda,\rho}{\circ}} E}(x, z)) | x \in X, z \in Z)\}$$

where

$$\mu_{L \overset{\alpha,\beta}{\underset{\lambda,\rho}{\circ}} E}(x, z) = \alpha_y\{\beta[\mu_E(x, y), \mu_L(y, z)]\}$$

and

$$\nu_{L \overset{\alpha,\beta}{\underset{\lambda,\rho}{\circ}} E}(x, z) = \lambda_y\{\rho[\nu_E(x, y), \nu_L(y, z)]\}$$

whenever

$$0 \le \mu_{L \overset{\alpha,\beta}{\underset{\lambda,\rho}{\circ}} E}(x, z) + \nu_{L \overset{\alpha,\beta}{\underset{\lambda,\rho}{\circ}} E}(x, z) \le 1 \ \forall(x, z) \in X \times Z.$$

Notice that the operators α, β are applied to the membership functions and the operators λ and ρ are applied to the non-membership functions. For a further analysis, see [16].

We recall now the main properties of A-IFRs. A complete study of these properties can be found in [16].

Definition 6.3 Let $\triangle \in$ A-IFR$(X \times X)$.

1. $\triangle \in$ A-IFR$(X \times X)$ is called identity relation if:

$$\mu_\triangle(x, y) = \begin{cases} 1 \text{ if } x = y \\ 0 \text{ if } x \neq y \end{cases}, \nu_\triangle(x, y) = \begin{cases} 0 \text{ if } x = y \\ 1 \text{ if } x \neq y \end{cases}, \forall(x, y) \in (X \times X)$$

2. The complementary relation of $\triangle$, denoted by $\triangle_c$ or $\triangledown$ (depending on the context) is defined as:

$$\mu_\triangledown(x, y) = \begin{cases} 0 \text{ if } x = y \\ 1 \text{ if } x \neq y \end{cases}, \nu_\triangledown(x, y) = \begin{cases} 1 \text{ if } x = y \\ 0 \text{ if } x \neq y \end{cases}, \forall(x, y) \in (X \times X)$$

We have in particular the following result.

Theorem 6.1 *Let $\alpha, \beta, \lambda, \rho$ be any t-norms or t-conorms and $E \in (X \times X)$ then:*

1. *$E \overset{\alpha,\beta}{\underset{\lambda,\rho}{\circ}} \triangle = \triangle \overset{\alpha,\beta}{\underset{\lambda,\rho}{\circ}} E = E$ if and only if α, ρ are t-conorms, and β, λ are t-norms.*
2. *$E \overset{\lambda,\rho}{\underset{\alpha,\beta}{\circ}} \triangledown = \triangledown \overset{\lambda,\rho}{\underset{\alpha,\beta}{\circ}} E = E$ if and only if α, ρ are t-conorms, and β, λ are t-norms.*

A deep study of identity relations is made in [14].

From now on, and unless otherwise stated, we take $\alpha = \vee$, β a t-norm, $\lambda = \wedge$ and ρ a t-conorm. β^* will denote the dual t-conorm of the t-norm β.

Definition 6.4 We will say that $E \in$ A-IFR$(X \times X)$ is:

1. *Reflexive*, if for every $x \in X$, $\mu_E(x, x) = 1$. Notice that for every $x \in X$, $\nu_E(x, x) = 0$.
2. *Antireflexive*, if for every $x \in X$, $\nu_E(x, x) = 1$, and then $\mu_E(x, x) = 0$. That is to say, E_c is reflexive.
3. *Symmetric*, if for every $(x, y) \in X \times X$, $\mu_E(x, y) = \mu_E(y, x)$ and $\nu_E(x, y) = \nu_E(y, x)$. That is to say, $E = E^{-1}$. Otherwise we will say that it is asymmetric.
4. *Antisymmetric*, if for every $(x, y) \in X \times X$,

$$x \neq y, then \begin{cases} \mu_E(x, y) \neq \mu_E(y, x) \\ \nu_E(x, y) \neq \nu_E(y, x) \\ \pi_E(x, y) = \pi_E(y, x) \end{cases}$$

5. *Perfect antisymmetric*, if for every $(x, y) \in X \times X$,

$$x \neq y \text{ and } \begin{cases} \mu_E(x, y) > 0 \\ or \\ \mu_E(x, y) = 0 \text{ and } \nu_E(x, y) < 1, \end{cases}$$

then $\mu_E(y, x) = 0$ and $\nu_E(y, x) = 1$.

6. *Transitive*, if $E \geq E \overset{\vee,\beta}{\underset{\wedge,\rho}{\circ}} E$ for any β and ρ.

7. *C-transitive*, if $E \leq E \overset{\wedge,\rho}{\underset{\vee,\beta}{\circ}} E$ for any β and ρ.

Remark 6.1 In items (6) and (7), apart from the direction of the inequality, also the order of $\vee$, β, $\wedge$ and ρ changes.

Remark 6.2 The definition of Atanassov's intuitionistic antisymmetry does not recover the fuzzy antisymmetry for the case in which the relation E considered is fuzzy. However, the definition considered here is justified because, for every $x, y \in X$, the relation

$$x \preceq_E y \text{ if and only if } \begin{cases} \mu_E(y, x) \leq \mu_E(x, y) \\ \nu_E(y, x) \geq \nu_E(x, y), \end{cases}$$

is an order in the referential X if the Atanassov's intuitionistic fuzzy relation $E \in$ A-IFR($X \times X$) is reflexive, transitive and antisymmetric fuzzy [16]. This fact does not hold if, instead of the antisymmetric previous property, the definition of antisymmetric fuzzy property given by Kaufmann [26] is considered.

However, the definition of Atanassov's intuitionistic perfect antisymmetry does recover the definition of perfect fuzzy antisymmetry given by Zadeh [31] when the considered relation is fuzzy.

Definition 6.5 Let $E \in$ A-IFR($X \times X$). Let β be a t-conorm and ρ a t-norm.

1. The *Transitive closure* of E is the minimal A-IFR $\hat{E}$ on $X \times X$ which contains E and it is transitive. Besides:

$$\hat{E} = E \vee E \overset{\vee,\wedge}{\underset{\wedge,\vee}{\circ}} E \vee E \overset{\vee,\wedge}{\underset{\wedge,\vee}{\circ}} E \overset{\vee,\wedge}{\underset{\wedge,\vee}{\circ}} E \vee \cdots \vee E \overset{\vee,\wedge}{\underset{\wedge,\vee}{\circ}} \ldots \overset{n-times}{} \overset{\vee,\wedge}{\underset{\wedge,\vee}{\circ}} E$$

2. The *C-transitive closure* of E is the maximal A-IFR $\check{E}$ on $X \times X$ contained in E and it is c-transitive. Besides:

$$\check{E}_c = E_c \wedge E_c \overset{\wedge,\vee}{\underset{\vee,\wedge}{\circ}} E_c \wedge E_c \overset{\wedge,\vee}{\underset{\vee,\wedge}{\circ}} E_c \overset{\wedge,\vee}{\underset{\vee,\wedge}{\circ}} E_c \wedge \cdots \wedge E_c \overset{\wedge,\vee}{\underset{\vee,\wedge}{\circ}} \ldots \overset{n-times}{} \overset{\wedge,\vee}{\underset{\vee,\wedge}{\circ}} E_c .$$

Next the relation that exists between the properties of a relation and the properties of its complementary relation is studied.

Theorem 6.2 *Let* $E, L \in A - IFR(X \times X)$, β *be a t-norm and* ρ *be a t-conorm. Then,* $(E \overset{\vee,\beta}{\underset{\wedge,\rho}{\circ}} L)_c = E_c \overset{\wedge,\rho}{\underset{\vee,\beta}{\circ}} L_c$

Theorem 6.3 *Let $E \in A - IFR(X \times X)$ and E_c be its complementary relation. Take $\beta = \wedge$ and $\rho = \vee$. Then $\check{E}_c = (\hat{E})_c$.*

Next theorem relates the properties of an Atanassov's intuitionistic fuzzy relation and those of its complementary.

Theorem 6.4 *Let $E \in A - IFR(X \times X)$*

1. *E is reflexive if and only if E_c is antireflexive.*
2. *E is symmetric if and only if E_c is symmetric.*
3. *E is antisymmetric if and only if E_c is antisymmetric.*
4. *E is transitive if and only if E_c is c-transitive.*

Definition 6.6 An A-IFR $E \in A - IFR(X \times X)$, is called:

1. A tolerance relation on $X \times X$ if E is reflexive and symmetric;
2. An atolerance relation on $X \times X$ if E is symmetric and antireflexive;
3. A preorder if it is reflexive and transitive;
4. An order if it is reflexive, transitive and antisymmetric;
5. A perfect ordering if it is reflexive, transitive and perfect antisymmetric;
6. A strict order if E is antireflexive, transitive and antisymmetric;
7. A similarity relation on $X \times X$ if E is reflexive, symmetric and transitive;
8. A dissimilarity relation on $X \times X$ if E is symmetric, antireflexive and transitive.

It is possible to provide a definition similar to the previous one using c-transitivity, where the structures of similarity, preorder, order, strict order, dissimilarity and perfect order will have to fulfill the same properties as the ones given in the definition, but substituting the transitive property by the c-transitive one. Whenever we do so, it will be specified.

The first result to note about the reflexive property is the following:

Theorem 6.5 *For every $E \in$ A-IFR $(X \times X)$, it holds that*

1. *If E is reflexive, then $\triangle \leq E$.*
2. *If E is antireflexive, then $\triangledown \geq E$.*

In [16] it is shown that, if $E \in$ A-IFR$(X \times X)$ is reflexive, then

$$E^{(n)} = \overbrace{E \overset{\vee,\wedge}{\underset{\wedge,\vee}{\circ}} E \overset{\vee,\wedge}{\underset{\wedge,\vee}{\circ}} E \cdots \overset{\vee,\wedge}{\underset{\wedge,\vee}{\circ}} E}^{n-times}$$

$\mu_A(x) \in [0, 1], \qquad \nu_A(x) \in [0, 1],$

with $n = 1, 2, \ldots$ it is reflexive. Besides we also have the following result.

Theorem 6.6 *Let E be a reflexive A-IFR in $X \times X$. Then*

1. *$(E)^{-1}$ is reflexive,*
2. *$E \vee L$ is reflexive for every $L \in$ A-IFR$(X \times X)$,*
3. *$E \wedge L$ is reflexive if and only if $L \in$ A-IFR$(X \times X)$ is reflexive.*

Theorem 6.7 *If α, β, λ and ρ are t-norms or t-conorms and $E, L \in A\text{-}IFR(X \times X)$ are symmetric then $E \overset{\alpha,\beta}{\underset{\lambda,\rho}{\circ}} L = (L \overset{\alpha,\beta}{\underset{\lambda,\rho}{\circ}} E)^{-1}$*

Corollary 6.1 *If $E, L \in A\text{-}IFR(X \times X)$ are tolerance relations, then*

1. *$E \vee L$ is tolerance relation;*
2. *$E \wedge L$ is tolerance relation;*
3. *$\hat{E}$ is tolerance relation;*

Theorem 6.8 *Let $E, L \in A\text{-}IFR(X \times X)$, with E an tolerance relation, L an similarity relation and $E \leq L$ then $\hat{E} \leq L$.*

In [16], it is proven that if $E, L \in A\text{-}IFR(X \times X)$ and $E \leq L$, then $\hat{E} \leq \hat{L}$ and $\check{E} \leq \check{L}$. Besides, it is known [16] that if E is reflexive and transitive, then $E = E \overset{\vee,\beta}{\underset{\wedge,\rho}{\circ}} E$. Next theorem can be proved from the previous results.

Theorem 6.9 *In every Atanassov's intuitionistic preorder:*

1. *$E = E \overset{\vee,\wedge}{\underset{\wedge,\vee}{\circ}} E$*
2. *$E = E^k$ for every $k = 1, 2, \ldots$ are verified.*

The next theorem establishes the existing relation between the structures of $E \in A\text{-}IFR(X \times X)$ and the ones of its complementary relation.

Theorem 6.10 *Let E be an A-IFR on $X \times X$*

1. *E is an order if and only if E_c is an strict order with respect to the c-transitive property;*
2. *E is an tolerance relation if and only if E_c is an atolerance relation;*
3. *E is an tolerance relation if and only if $\check{E}_c$ is an dissimilarity relation with respect to the c-transitive property;*

Theorem 6.11 *$E \in A\text{-}IFR(X \times X)$ is an similarity relation if and only if E_c is an dissimilarity relation with respect to the c-transitive property.*

The properties of the A-IFRs presented in this section show that the definition of these properties does not always coincide with the definition of the properties of fuzzy relations. It happens that there are properties like perfect Atanassov's intuitionistic antisymmetry that recover Zadeh's perfect antisymmetry for the case fuzzy, while Atanassov's intuitionistic antisymmetry does not recover Kaufmann's antisymmetry for the fuzzy case.

To finish this section, it is important to introduce a new type of relations [14], which will be relevant in our subsequent developments.

Definition 6.7 A relation $E \in A\text{-}IFR(X \times X)$ is partially included, if for every $x, y, z \in X$ with $\mu_E(x, y) \neq \mu_E(y, z)$, $Sign(\mu_E(x, y) - \mu_E(y, z)) = Sign(v_E(y, z) - v_E(x, y))$

Theorem 6.12 *If $R \in FR(X \times X)$, then R is partially included.*

Theorem 6.13 *Let $E \in A\text{-}IFR(X \times X)$, if $\pi_E(x, y) = K = constant$ for every (x, y) $\in X \times X$, then E is partially included.*

To close this section is important to note that new A-IFR have recently been defined and studied, for example preference relations in [20, 32]. This field of knowledge is in expansion and has a huge potential because applications are now less expensive in performance reduction.

6.3 Construction of A-IFRs from Fuzzy Relations with Predetermined Properties

In this section the way to build A-IFRs with predetermined properties from fuzzy relations is studied, i.e., the opposite of the problem studied in the previous section. The Atanassov's intuitionistic fuzzy set construction theorems studied in [12] will be used in order to build Atanassov's intuitionistic fuzzy relations from fuzzy relations.

The section explores different ways to construct Atanassov's intuitionistic fuzzy relations that are reflexive, symmetric, antisymmetric, perfect antisymmetric, partially included and transitive, from fuzzy relations, like it is showed in [17].

The first construction theorem of Atanassov's Intuitionistic Fuzzy Sets is:

Theorem 6.14 [12] *Let $\Phi_{\alpha_x} \colon FS(X) \to A\text{-}IFS(X)$ with $\Phi_{\alpha_x}(A) = \{\langle x, \mu_{\Phi_{\alpha_x}}(x),$ $\nu_{\Phi_{\alpha_x}}(x)\rangle | x \in X\}$ such that*

1. *$\mu_{\Phi_{\alpha_x}(A)}(x) = a + b\mu_A(x) - \alpha_x c$ with fixed $a, b, c \in \mathbb{R}$ for all A in FS.*
2. *$\pi_{\Phi_{\alpha_x}(A)}(x) = c$ for all $x \in X$.*
3. *If $\pi_A(x) = 0, \forall x \in X$, then $\Phi_\alpha = A$.*

Then

1. *$\mu_{\Phi_{\alpha_x}(A)}(x) = \mu_A(x) - \alpha_x c$.*
2. *$\nu_{\Phi_{\alpha_x}(A)}(x) = 1 - \mu_A(x) + \alpha_x c - c$ for all $x \in X$*

Conversely if the last two conditions hold, so do the first three ones.

In the construction developed in the theorem above for each x it is taken an α_x, so that the A-IFS constructed is such that with Atanassov's D_α operator it is not possible to recover the initial fuzzy set, since this operator requires a constant value of $\alpha \in [0, 1]$.

From this theorem it is known that:

Theorem 6.15 *Let $E \in A\text{-}IFR(X \times X)$ be built with Theorem 6.14, from a fuzzy relation $R \in FR(X \times X)$, i.e. $E = \Phi_\alpha(R)$. Then*

1. E is reflexive if and only if R is reflexive and $\pi_E(x, x) = 0$ for all $x \in X$.
2. E is symmetric if and only if R is symmetric and $\pi_E(x, y) = \pi_E(y, x)$ for all $(x, y) \in X \times X$.
3. E is antisymmetric if and only if R is antisymmetric and $\pi_E(x, y) = \pi_E(y, x)$ for all $(x, y) \in X \times X$.
4. E is perfect antisymmetric if and only if R is perfect fuzzy antisymmetric.

Theorem 6.16 Let $R \in FR(X \times X)$ such that $Inf\{\mu_R(x, y) \mid (x, y) \in X \times X\} > 0$ and let $E \in A\text{-}IFR(X \times X)$ be constructed from R with Theorem 6.14 such that for all $(x, y) \in X \times X$, $\pi_E(x, y) = K = constant$. Then

1. E is partially included.
2. If $\beta = \wedge$, E is transitive if and only if R is transitive.

Regarding condition $Inf(\mu_R(x, y)) > 0$ it is noteworthy that for all $(x, y) \in X \times X$,

$$\pi_E(x, y) \leq \frac{\mu_R(x, y)}{\alpha} .$$

If the infimum is equal to zero, since in this theorem it is considered that $\pi_E(x, y) = K$, then $\pi_E(x, y) = 0$ and therefore

$$\mu_E(x, y) = \mu_R(x, y), \qquad \nu_E(x, y) = 1 - \mu_R(x, y) .$$

Thus $E = R \in FR(X \times X)$, i.e. no A-IFR is constructed. Thus, the above theorem is not very useful for relations $R \in FR(X \times X)$ where there is some $(x, y) \in X \times X$ such that $\mu_R(x, y) = 0$. So as to avoid this limitation the following theorem is presented. It allows an analogous result to the one above but with the index $\pi_E(x, y)$ chosen variable.

Theorem 6.17 Let $E \in A\text{-}IFR(X \times X)$ be constructed from $R \in FR(X \times X)$ by Theorem 6.14 and let $\gamma \in (0, 1)$. If

$$\pi_E(x, y) = \gamma \frac{\mu_R(x, y)}{\alpha}$$

for all $(x, y) \in X \times X$, then

1. E is partially included.
2. If $\beta = \wedge$, E is transitive if and only if R is transitive.

Theorem 6.18 Let $L \in A\text{-}IFR(X \times X)$, $L \neq 0$ and $p \in [0, 1]$ fixed. The Atanassov's intuitionistic fuzzy set E built from $D_p(L) \in FR(X \times X)$, such that $D_1(E) = D_p(L)$, by means of Theorem 6.14. If E is transitive, then $D_p(L)$ is transitive.

Theorem 6.19 For all relations $E \in A\text{-}IFR(X \times X)$ built from the fuzzy relation $R \in FR(X \times X)$ by Theorem 6.14, $E \overset{\vee,\beta}{\underset{\wedge,\beta^*}{\circ}} E \preceq R \overset{\vee,\beta}{\underset{\wedge,\beta^*}{\circ}} R$ holds

As it has been said with Theorem 6.14, as $\alpha > 0$, it is not possible to recover the fuzzy set A as the first element of the family of fuzzy sets $\{D_\alpha(\Phi_\alpha(A))\}_{\alpha \in [0,1]}$.

This justifies the following construction theorem, wich allows to generate Atanassov's intuitionistic fuzzy sets so that with Atanassov's D_p operator it is easy to recover the fuzzy set used in the construction, taking $p = \alpha = 0$.

Theorem 6.20 [12] *Let $A \in FS(X)$, let the mapping $\pi_A : X \to [0, 1]$, with $\pi_A(x) \leq 1 - \mu_A(x)$, $\forall x \in X$ and let $\varphi : FS \to A\text{-}IFS$, with $\varphi(A) = \{\langle x, \mu_{\varphi(A)}(x), \nu_{\varphi(A)}(x)\rangle | x \in X\}$ such that*

1. $\mu_{\varphi(A)}(x) = a + b\mu_{(A)}(x)$ *with fixed $a, b \in \mathbb{R}$, $\forall A \in FS$.*
2. $\pi_{\varphi(A)}(x) = \pi_{(A)}(x)$, $\forall x \in X$.
3. *If $\pi_{(A)}(x) = 0$, $\forall x \in X$, then $\varphi(A) = A$.*

Then

1. $\mu_{\varphi(A)}(x) = \mu_{(A)}(x)$
2. $\nu_{\varphi(A)}(x) = 1 - \mu_{(A)}(x) - \pi_{(A)}(x)$, *for all $x \in X$.*

Conversely, the last two statements imply the first three.

From this theorem it is deduced the following one.

Theorem 6.21 *Let $E \in A\text{-}IFR(X \times X)$ be built with Theorem 6.20, from a fuzzy relation $R \in FR(X \times X)$, i.e. $E = \varphi(R)$. Then*

1. *E is reflexive if and only if R is reflexive.*
2. *E is symmetric if and only if R is symmetric and $\pi_E(x, y) = \pi_E(y, x)$ for all $(x, y) \in X \times X$.*
3. *E is antisymmetric if and only if R is antisymmetric and $\pi_E(x, y) = \pi_E(y, x)$ for all $(x, y) \in X \times X$*
4. *If E is perfect antisymmetric, then R is perfect fuzzy antisymmetric.*

Theorem 6.22 *Let $E \in A\text{-}IFR(X \times X)$ be built with Theorem 6.20, from a fuzzy relation $R \in FR(X \times X)$. E is partially included if for all $x, y, z \in X$, $Sign(\mu_R(x, y) - \mu_R(y, z)) = Sign(\pi_E(x, y) - \pi_E(y, z))$ holds.*

Theorem 6.23 *Let $R \in FR(X \times X)$ such that $Sup\{\mu_R(x, y)) < 1 \mid (x, y) \in X \times X\}$ and let $E \in A\text{-}IFR(X \times X)$ be constructed from R with Theorem 6.20 such that $\pi_E(x, y) = constant = K \leq 1 - \mu_R(x, y)$ for all $(x, y) \in X \times X$. Then*

1. *E is partially included.*
2. *If $\beta = \wedge$, E is transitive if and only if R is transitive.*

Requiring that the supremum is less than 1 is due to the fact that if it is equal to 1, then $\pi_E(x, y) = K = 0$ for all $(x, y) \in X \times X$, so, $E = R \in FR(X \times X)$ and it would not be Atanassov's intuitionistic fuzzy.

Therefore the above theorem is not valid for relations R with some $\mu_R(x, y) = 1$. So as to avoid this limitation the following theorem is proposed.

Theorem 6.24 *Let $E \in A\text{-}IFR(X \times X)$ be constructed from $R \in FR(X \times X)$ by Theorem 6.20 and let $\gamma_{x,y} \in (0, 1]$. If $\pi_E(x, y) = \gamma_{x,y}(1 - \mu_R(x, y)) \leq 1 - \mu_R(x, y)$, then*

1. *E is partially included.*
2. *If $\beta = \wedge$, E is transitive if and only if R is transitive.*

Theorem 6.25 *For all relations $E \in A\text{-}IFR(X \times X)$ built from a fuzzy relation $R \in FR(X \times X)$ by Theorem 6.20, $E \overset{\vee,\beta}{\underset{\wedge,\beta^*}{\circ}} E \preceq R \overset{\vee,\beta}{\underset{\wedge,\beta^*}{\circ}} R$ holds.*

Theorem 6.26 *Let $L \in A\text{-}IFR(X \times X)$ and let $p \in [0, 1]$ be fixed. From $D_p(L) \in FR(X \times X)$ the Atanassov's intuitionistic fuzzy set E such that $D_0(E) = D_p(L)$ is built by means of Theorem 6.20. If E is transitive, then $D_p(L)$ is transitive.*

The degree of membership $\mu_A(x)$ coincides, for all Atanassov's intuitionistic fuzzy sets A, with the degree of membership of fuzzy set $D_0(A)$ and the degree of non-membership $\nu_A(x)$ with the degree of non-membership of $D_1(A)$. This fact allows to infer that there is a strong connection between an A-IFS and two fuzzy sets. This observation lead the study naturally to analyze the construction of an A-IFS from two *FS*.

Theorem 6.27 *[12] Let $A, B \in FS, A \leq B$. For each $x \in X$ a s_x is taken such that $s_x \in (0, 1]$ and $\mu_B(x) - \mu_A(x) \leq s_x \mu_B(x)$. Let $\pi_{(A,B)} : X \to [0, 1]$, be given by $\pi_{(A,B)}(x) = \frac{\mu_B(x) - \mu_A(x)}{s_x}$, and let $\Psi_{s_x} : FS \times FS \to A\text{-}IFS$, with*

$$\Psi_{s_x}(A, B) = \{(x, \mu_{\Psi_{s_x}(A,B)}(x), \nu_{\Psi_{s_x}(A,B)}(x)) | x \in X\}$$

such that

1. *$\mu_{\Psi_{s_x}(A,B)}(x) = a + b\mu_{(A)}(x) + c\mu_{(B)}(x)$ with fixed $a, b, c \in \mathbb{R}, \forall A, B \in FS$.*
2. *$\pi_{\Psi_{s_x}(A,B)}(x) = \pi_{(A,B)}(x), \forall x \in X$.*
3. *If $A = B$, then $\Psi_{s_x}(A, A) = A$.*
4. *If $B = 1$, then $\nu_{\Psi_{s_x}(A,B)}(x) = 0, \forall x \in X$.*

Then

1. *$\mu_{\Psi_{s_x}(A,B)}(x) = (\frac{1}{s_x})\mu_{(A)}(x) - (\frac{(1-s_x)}{s_x})\mu_{(B)}(x)$*
2. *$\nu_{\Psi_{s_x}(A,B)}(x) = 1 - \mu_{(B)}(x)$, for all $x \in X$.*

Conversely, the last two statements imply the first four ones.

It is useful to explain that in the following conditions it is possible to recover the *FRs* A and B. If $s_x = s(A, B) \in (0, 1], \forall x \in X$ such that $Sup_{x \in X}\mu_B(x) - Inf_{x \in X}\mu_A(x) \leq s(A, B)Sup_{x \in X}\mu_B(x)$ then $D_1(\Psi_{s(A,B)}(A, B)) = B$ and $D_{1-s(A,B)}(\Psi_{s(A,B)}(A, B)) = A$

Theorem 6.28 *Let $E \in A\text{-}IFR(X \times X)$ be built with Theorem 6.27, from fuzzy relations $R, P \in FR(X \times X)$ such that $R < P$, i.e. $E = \Psi(R, P)$. Then*

1. *E is reflexive if and only if R and P are reflexive.*
2. *E is symmetric if and only if R and P are symmetric.*
3. *E is antisymmetric if and only if R and P are antisymmetric and $\pi_E(x, y) = \pi_E(y, x)$ for all $(x, y) \in X \times X$*
4. *E is perfect antisymmetric if and only if P is perfect fuzzy antisymmetric.*

Note that last statement only requires the perfect antisymmetry of P. Hence, it is easy to prove that if $R \leq P$ and P is perfect antisymmetric, then R also is.

Theorem 6.29 *Let $E \in A\text{-}IFR(X \times X)$ be constructed from $R, P \in FRs(X \times X)$ with Theorem 6.27 taking $s = 1$. Then E is transitive if and only if R and P are transitive.*

Theorem 6.30 *Let $E \in A\text{-}IFR(X \times X)$ be built with Theorem 6.27, from $R, P \in FR(X \times X)$. E is partially included if for all $x, y, z \in X$,*
$$Sign(\pi_E(x, y) - \pi_E(y, z)) = Sign(\mu_P(y, z) - \mu_P(x, y)) .$$

Theorem 6.31 *Let $L \in A\text{-}IFR(X \times X)$ be transitive such that for a $p \in [0, 1]$ fixed, $D_p(L) \in FR(X \times X)$ is not transitive. Thus Atanassov's intuitionistic fuzzy relation $E \in A\text{-}IFR(X \times X)$ built by Theorem 6.27 from $R \leq P = D_p(L) \in FR(X \times X)$ is not transitive.*

Theorem 6.32 *Let $R, P \in FR(X \times X)$ such that $R < P$, satisfying for all $(x, y) \in X \times X$ the condition $\mu_P(x, y) - \mu_R(x, y) = constant = K$, and let $E \in A\text{-}IFR(X \times X)$ be built from R and P by Theorem 6.27. Then*

1. *E is partially included.*
2. *if $\beta = \wedge$ then the following statements are equivalent*

   a. *E is partially included,*
   b. *P is transitive,*
   c. *R is transitive.*

Note that K in this theorem is always different from zero, i.e. $\pi_E(x, y) \neq 0$, because $R \neq P$.

Theorem 6.33 *Let $R, P \in FR(X \times X)$ such that $R < P$ and $R = \gamma P$ with $\gamma \in [0, 1)$. If $E \in A\text{-}IFR(X \times X)$ is built from R and P by Theorem 6.27, then*

1. *E is partially included.*
2. *if $\beta = \wedge$ then the following statements are equivalent*

   a. *E is transitive,*
   b. *P is transitive,*
   c. *R is transitive.*

Theorem 6.34 *Let $E \in A\text{-}IFR(X \times X)$ built from $R, P \in FR(X \times X)$ such that $R \leq P$ by Theorem 6.27, then*

1. $E \overset{\vee,\beta}{\underset{\wedge,\beta^*}{\circ}} E \leq P \overset{\vee,\beta}{\underset{\wedge,\beta^*}{\circ}} P$

2. $E \overset{\vee,\beta}{\underset{\wedge,\beta^*}{\circ}} E \preceq P \overset{\vee,\beta}{\underset{\wedge,\beta^*}{\circ}} P$

3. $E \overset{\vee,\beta}{\underset{\wedge,\beta^*}{\circ}} E \preceq (\frac{1}{s})R \overset{\vee,\beta}{\underset{\wedge,\beta^*}{\circ}} R$

4. $E \overset{\vee,\beta}{\underset{\wedge,\beta^*}{\circ}} E \preceq R \overset{\vee,\beta}{\underset{\wedge,\beta^*}{\circ}} R$

Corollary 6.2 *If $R, P \in FR(X \times X)$, with $R \leq P$, are reflexive and transitive, then*

1. $E \leq P$,
2. $E \preceq (\frac{1}{s})R$

6.4 Conclusions

In this chapter we have made a review of the main concepts related to Atanassov's intuitionistic fuzzy relations. In particular, we have focused in the construction of A-IFRs which preserve some specific properties, specially by using Atanassov's operators, which allow to reduced Atanassov's intuitionistic fuzzy sets to fuzzy sets.

Acknowledgments This work has been supported by projects TIN2013-40765-P and TIN2012-32482 of the Spanish Ministry of Science.

References

1. K. Atanassov, Intuitionistic fuzzy sets. Fuzzy Sets Syst. **20**, 87–96 (1986)
2. K. Atanassov, Review and new results on intuitionistic fuzzy sets. IM-MFAIS **1** (1988)
3. K. Atanassov, Two operators on intuitionistic fuzzy sets. Comptes rendus de l'Academie bulgare des Sciences, Tome 41 (1988)
4. K. Atanassov, *Intuitionistic Fuzzy Sets. Theory and Applications* (Physica-Verlag, Heidelberg, 1999)
5. K. Atanassov, Intuitionistic fuzzy sets, VII ITKR's Session, Deposed in Central Science and Technology. Library of the Bulgarian academy of sciences, Sofia, 1983, pp. 1684–1697
6. K.T. Atanassov, More on intuitionistic fuzzy sets. Fuzzy Sets Syst. **33**, 37–45 (1989)
7. G. Beliakov, H. Bustince, S. James, T. Calvo, J. Fernandez, Aggregation for Atanassovs intuitionistic and interval valued fuzzy sets: the median operator. IEEE Trans. Fuzzy Syst. **20**(3), 487–498 (2012)
8. U. Bentkowska, H. Bustince, A. Jurio, M. Pagola, B. Pekala, Decision making with an interval-valued fuzzy preference relation and admissible orders. Appl. Soft Comput. **35**, 792–801 (2015)
9. T. Buhaescu, Some observations on intuitionistic fuzzy relations. Itimerat Seminar on Functional Equations, 1989, 111–118
10. P. Burillo, H. Bustince, Estructuras algebraicas en conjuntos IFS, II Congreso Nacional ESTYLF, Boadilla del Monte, Madrid, Spain, 1992, pp. 135–147
11. P. Burillo, H. Bustince, Structures on intuitionistic fuzzy relations. Fuzzy Sets Syst. **78**, 293–303 (1996)

12. P. Burillo, H. Bustince, Construction theorems for intuitionistic fuzzy sets. Fuzzy Sets Syst. **84**, 271–281 (1996)
13. P. Burillo, H. Bustince, Entropy on intuitionistic fuzzy sets and on interval-valued fuzzy sets. Fuzzy Sets Syst. **78**, 305–316 (1996)
14. P. Burillo, H. Bustince, Intuitionistic fuzzy relations (Part I). Mathware Soft Comput. **2**, 5–38 (1995)
15. P. Burillo, H. Bustince, Orderings in the referential set induced by an intuitionistic fuzzy relation. Notes IFS **I**, 93–103 (1995)
16. H. Bustince, Conjuntos Intuicionistas e Intervalo valorados Difusos: Propiedades y Construcción, Relaciones Intuicionistas Fuzzy, Thesis, Universidad Pública de Navarra (1994)
17. H. Bustince, Construction of intuitionistic fuzzy relations with predetermined properties. Fuzzy Sets Syst. **109**, 379–403 (2000)
18. H. Bustince, P. Burillo, Intuitionistic fuzzy relations (Part II). Mathware Soft Comput. **2**, 117–148 (1995)
19. H. Bustince, P. Burillo, Perturbation of intuitionistic fuzzy relations. Int. J. Uncertain. Fuzziness Knowl. Based Syst. **9**(1), 81–103 (2001)
20. L. De Miguel, H. Bustince, J. Fernandez, E. Induráin, A. Kolesárová, R. Mesiar, Construction of admissible linear orders for interval-valued Atanassov intuitionistic fuzzy sets with an application to decision making. Inf. Fusion **27**, 189–197 (2016)
21. G. Deschrijver, E.E. Kerre, On the composition of intuitionistic fuzzy relations. Fuzzy Sets Syst. **136**, 333–361 (2003)
22. J. Drewniak, U. Dudziak, Aggregations preserving classes of fuzzy relations. Kybernetika **41**(3), 265–284 (2005)
23. J. Drewniak, U. Dudziak, Preservation of properties of fuzzy relations during aggregation processes. Kybernetika **43**(2), 115–132 (2007)
24. U. Dudziak, B. Pekala, Equivalent bipolar fuzzy relations. Fuzzy Sets Syst. **161**, 234–253 (2010)
25. U. Dudziak, Weak and graded properties of fuzzy relations in the context of aggregation process. Fuzzy Sets Syst. **161**, 216–233 (2010)
26. A. Kaufmann, Introduction a la Théorie des Sous-Ensembles Flous, vols. I–IV (Masson. Paris, 1977)
27. P. Melo-Pinto, P. Couto, H. Bustince, E. Barrenechea, M. Pagola, J. Fernandez, mage segmentation using Atanassovs intuitionistic fuzzy sets. Expert Syst. Appl. **40**(1), 15–26 (2013)
28. N.R. Pal, H. Bustince, M. Pagola, U.K. Mukherjee, D.P. Goswami, G. Beliakov, Uncertainties with Atanassovs intuitionistic fuzzy sets: fuzziness and lack of knowledge. Inf. Sci. **228**, 61–74 (2013)
29. B. Pekala, Properties of Atanassov's intuitionistic fuzzy relations and Atanassov's operators. Inf. Sci. **213**, 84–93 (2012)
30. E. Szmidt, I. Kacprzyk, Intuitionistic fuzzy sets in group decision making. Notes IFS **2**, 15–32 (1996)
31. L.A. Zadeh, Similarity relations and fuzzy orderings. Inf. Sci. **3**, 177–200 (1971)
32. X. Zeshui, Intuitionistic preference relations and their application in group decision making. Inf. Sci. **177**, 2363–2379 (2007)

Chapter 7
On Weighting Triangles Using Fuzzy Relations and Its Application to Aggregation Functions

Tomasa Calvo Sánchez, Ramón Fuentes-González and Pilar Fuster-Parra

Abstract In this work, a new lattice L determined by the class of weighting triangles as a base of L-fuzzy subsets is proposed. Furthermore, extended orders and operators which are obtained by means of fuzzy binary relations $F_\triangle$ associated to a weighting triangle are included. Moreover, some new expressions have been defined for Extended Ordered Weighted Averaging operators, and Extended Aggregation functions.

7.1 Introduction

Let us denote $I[0, 1]^X$ as the interval-valued fuzzy sets determined by the functions from X to the set of intervals $I[0, 1] = \{[\alpha, \beta] \mid 0 \le \alpha \le \beta \le 1\}$. The class $I[0, 1]^X$ constitutes an important example of the class of L-fuzzy subsets $A \in L^X$ of a set X. In this case, the order that represents the inclusion in $I[0, 1]^X$ is based on the order between intervals $I[0, 1]$ that gives the structure of a complete lattice. This order is built using the usual chain $[0, 1]$. In $I[0, 1]^X$ laws such as union, intersection, negation, etc., may be defined as specific extensions of t-norms, uninorms, etc., which are again extensions of similar operators in $[0, 1]$. This mechanism of extending is also used to introduce other operators in $I[0, 1]^X$, as the associated to implications, quantifiers, etc., in $I[0, 1]$. This type of extensions, together with the identification of an element $\alpha \in [0, 1]$ with the interval $[\alpha, \alpha] \in I[0, 1]$, constitutes the consideration of the class $[0, 1]^X$ of fuzzy sets of X as a part of the interval-valued fuzzy subsets $I[0, 1]^X$.

T. Calvo Sánchez
Universidad de Alcalá, Alcalá de Henares, Madrid, Spain
e-mail: tomasa.calvo@uah.es

R. Fuentes-González
Universidad Pública de Navarra, Pamplona, Spain
e-mail: rfuentes@unavarra.es

P. Fuster-Parra (✉)
Universitat Illes Balears, Palma de Mallorca, Balears, Spain
e-mail: pilar.fuster@uib.es

© Springer International Publishing Switzerland 2016
T. Calvo Sánchez and J. Torrens Sastre (eds.), *Fuzzy Logic and Information Fusion*,
Studies in Fuzziness and Soft Computing 339, DOI 10.1007/978-3-319-30421-2_7

In this work we propose a new complete lattice $(L, \preccurlyeq) = (\mathcal{W}, \preccurlyeq)$ as a structure to assess the L-fuzzy subsets: the one determined by the class of weighting triangles $\mathcal{W}$, which are elements $\triangle \in \mathcal{W}$ which appear in the study of some aggregation operators [1–5]. The order $\preccurlyeq$ in $\mathcal{W}$ is also related with the one of the chain [0, 1]. Furthermore, a mechanism to generate operators like negation, t-norms, uninorms, implications, etc., through extensions of others of the same type in [0, 1] is also included. Every element $\alpha \in [0, 1]$ can be consider as a triangle $\triangle_{(\alpha)} \in \mathcal{W}$, i.e., the algebra determined by the lattice $\mathcal{W}$ and its operations is an extension of the chain [0, 1] together with the operations t-norms, uninorms, implications, quantifiers, etc.

These extensions (order, and operations) appear when the weighting triangles $\triangle$ are reinterpreted as specific fuzzy binary relations $F_\triangle \in [0, 1]^{\mathbb{N}^* \times \mathbb{N}^*}$ in the set of natural numbers $\mathbb{N}^* = \{1, 2, \ldots\}$. The study of these fuzzy relations is the core of this work. Finally, they are used to find new expressions to some aggregation operators linked to the weighting triangles $\triangle$: the ones of type EOWA (Extended Ordered Weighted Averaging), and the extended aggregation functions.

The paper is organized as follows. Section 7.2 presents some necessary concepts that relies on subsequent sections. Section 7.3 introduces the relation of distribution $F_\triangle$ associated to a weighting triangle $\triangle$. Section 7.4 presents induced structures in the classes of triangles $\mathcal{W}, \mathcal{RW}, \ldots$ by operations in some classes $\mathcal{DF}, \mathcal{RDF}, \ldots$ of fuzzy relations. Section 7.5 EOWA operators and extended aggregated functions through relations of distribution are presented. Section 7.6 shows closure and interior operators in lattices. Finally, Sect. 7.7 concludes the paper.

7.2 Preliminaires

In [1], the authors characterized the extended aggregation operators, (EOWA, EAF, EQWLM, ...), using weighting lists $\triangle$ of type:

$$\triangle = (1, (w_1^2, w_2^2), (w_1^3, w_2^3, w_3^3), \ldots, (w_1^n, w_2^n, \ldots w_i^n, \ldots, w_n^n), \ldots) \in \prod_{n \geq 1} [0, 1]^n$$

where the numerical values $w_i^n \in [0, 1]$ are such that the sum of the components of each weighting list $(w_1^n, w_2^n, \ldots, w_n^n)$ verifies: $\sum_{i=1}^{n} w_i^n = 1$, $\forall n \in \mathbb{N}^* - \{1\}$, where $\mathbb{N}^* = \{1, 2, 3, \ldots\}$.

As it is usual, we will represent these weighting lists in a triangular shape:

$$\triangle = \begin{matrix} & & & 1 & & & \\ & & w_1^2 & & w_2^2 & & \\ & w_1^3 & & w_2^3 & & w_3^3 & \\ w_1^n & w_2^n & & w_3^n & & w_4^n & \ldots & w_n^n \\ & & & \ldots & & & \end{matrix} \quad .$$

The authors [1] called this shape a *weighting triangle*. We will represent the class of triangles by $\mathscr{W}$. Some examples of weighting triangles are:

(a) For $\alpha \in [0, 1]$, let us consider the succession $\triangle_{(a)}$:

$$(1, (\alpha, 1 - \alpha), (\alpha, 0, 1 - \alpha), \ldots, (\alpha, \underbrace{0, \ldots, 0}_{n-2}, 1 - \alpha), \ldots),$$

that is

$$\triangle_{(a)} = \begin{matrix} & & & 1 & & & \\ & & \alpha & & (1-\alpha) & & \\ & \alpha & & 0 & & (1-\alpha) & \\ \alpha & & 0 & & 0 & & (1-\alpha) \\ & & & \cdots & & & \end{matrix}$$

In particular, for $\alpha \in \{0, 1\}$, the weighting triangles $\triangle_{(0)}$, $\triangle_{(1)}$ have the shape:

$$\triangle_{(0)} = \begin{matrix} & & 1 & & \\ & 0 & & 1 & \\ 0 & & 0 & & 1 \\ 0 & 0 & & 0 & 1 \\ & & \cdots & & \end{matrix} \quad , \quad \triangle_{(1)} = \begin{matrix} & & 1 & & \\ & 1 & & 0 & \\ 1 & & 0 & & 0 \\ 1 & 0 & & 0 & 0 \\ & & \cdots & & & . \end{matrix}$$

(b) For all $\alpha \in]0, 1[$, we consider the weighting triangle $\triangle_{p(\alpha)}$ such that: $w_m^n = \binom{n-1}{m-1} \alpha^{n-m} (1 - \alpha)^{m-1}$, $n \in \mathbb{N}^*$, $m \leq n$. In particular, $\triangle_{p(\frac{1}{2})}$ is the normalized Pascal triangle [1]:

$$\triangle_{p(\frac{1}{2})} = \begin{matrix} & & & 1 & & & \\ & & 1/2 & & 1/2 & & \\ & 1/4 & & 2/4 & & 1/4 & \\ 1/8 & & 3/8 & & 3/8 & & 1/8 \\ & & & \cdots & & & . \end{matrix}$$

(c) The weighting triangles $\triangle_{median}$ and $\triangle_{mean}$ associated with median and mean operators, with values $median_m^n \in [0, 1]$ and $mean_m^n \in [0, 1]$ such that:

$$median_i^{2k} = \frac{1}{2}, \text{ if } i \in \{k, k + 1\},$$
$$median_i^{2k-1} = 1, \text{ if } i = k,$$
$$median_i^n = 0, \text{ otherwise,}$$

and,

$$mean_i^n = \frac{1}{n}, \quad n \in \mathbb{N}^*, \quad i \leq n :$$

are

$$\Delta_{median} = \begin{matrix} & & 1 & & \\ & \frac{1}{2} & & \frac{1}{2} & \\ 0 & & 1 & & 0 \\ 0 & \frac{1}{2} & & \frac{1}{2} & 0 \\ & & \cdots & & \end{matrix}, \quad \Delta_{mean} = \begin{matrix} & & 1 & & \\ & \frac{1}{2} & & \frac{1}{2} & \\ \frac{1}{3} & & \frac{1}{3} & & \frac{1}{3} \\ \frac{1}{4} & \frac{1}{4} & & \frac{1}{4} & \frac{1}{4} \\ & & \cdots & & \end{matrix}.$$

Now we introduce some definitions associated to weighting triangles:

Definition 7.1 [1, 2] The *reverse weighting triangle* Δ^r of a triangle $\Delta = (w_i^n)$, $i = 1, \ldots, n, n \geq 1$ is the new triangle: $\Delta^r = (u_i^r)$, $i = 1, \ldots, n, n \geq 1$ such that: $u_j^n = w_{n-j+1}^n$ for $n > 1$ and $j \in \{1, 2, \ldots, n\}$.

For instance:

$$\Delta_{(\alpha)}^r = \Delta_{(1-\alpha)}, \quad \Delta_{p(\alpha)}^r = \Delta_{p(1-\alpha)}, \quad \Delta_{median}^r = \Delta_{median}, \quad \cdots$$

The operator $^r : \mathscr{W} \to \mathscr{W}$ that associates Δ with its reverse Δ^r, is involutive: $(\Delta^r)^r = \Delta$ for all $\Delta \in \mathscr{W}$.

Definition 7.2 [1, 2] A weighting triangle $\Delta \in \mathscr{W}$ is *left (right) regular* if for all $n > 1$ and $p = 1, 2, \ldots, n$ it is verified:

$$\sum_{i=1}^{p} w_i^{n+1} \leq \sum_{i=1}^{p} w_i^n \quad \left(\sum_{i=1}^{p} w_i^n \leq \sum_{i=1}^{p+1} w_i^{n+1}\right). \tag{7.1}$$

The weighting triangle Δ is *regular*, whenever it is on the left and right:

$$\sum_{i=1}^{p} w_i^{n+1} \leq \sum_{i=1}^{p} w_i^n \leq \sum_{i=1}^{p+1} w_i^{n+1}. \tag{7.2}$$

Let $\overleftarrow{\mathscr{RW}} \subset \mathscr{W}$ ($\overrightarrow{\mathscr{RW}} \subset \mathscr{W}$) be the class of regular weighting triangles to the left (right) and let $\mathscr{RW} = (\overleftarrow{\mathscr{RW}}) \cap (\overrightarrow{\mathscr{RW}})$ the class of regular weighting triangles. In this work, we will analyze the set $\mathscr{RW}$ and its properties, some of them could be proven for elements of $\overleftarrow{\mathscr{RW}}$ and $\overrightarrow{\mathscr{RW}}$.

7.3 Relation of Distribution F_Δ Associated to a Weighting Triangle Δ

Each weighting triangle $\Delta \in \mathscr{W}$ is characterized by a fuzzy binary relation $F_\Delta \in [0, 1]^{\mathbb{N}^* \times \mathbb{N}^*}$ which be called *relation of distribution* of Δ.

Definition 7.3 Let $\mathbb{N}^* = \{1, 2, \ldots\}$. Given the weighting triangle $\triangle = (w_i^n)$, $i = 1, \ldots, n$, $n \geq 1$, a *relation of distribution* of $\triangle$ is a fuzzy relation $F_\triangle : \mathbb{N}^* \times \mathbb{N}^* \longrightarrow [0, 1]$ such that

$$F_\triangle(n, m) = \begin{cases} w_1^n, & if \ m = 1 \\ \sum_{i=1}^m w_i^{n+m-1}, & if \ m > 1. \end{cases}$$

The relation $F_\triangle$ is well defined, because for all $(n, m) \in \mathbb{N}^* \times \mathbb{N}^*$:

$$0 \leq \sum_{i=1}^m w_i^{n+m-1} \leq \sum_{i=1}^{n+m-1} w_i^{n+m-1} = 1 \implies F_\triangle(n, m) \in [0, 1].$$

Particularly, for all $m \in \mathbb{N}^*$: $F_\triangle(1, m) = \sum_{i=1}^m w_i^m = 1$.

Let us observe that the elements of $F_\triangle$ relation can be represented as a non limited table:

$$F_\triangle = \begin{matrix} 1 & 1 & 1 & \cdots & 1 & \cdots \\ F_\triangle(2, 1) & F_\triangle(2, 2) & F_\triangle(2, 3) & \cdots & F_\triangle(2, m) & \cdots \\ F_\triangle(3, 1) & F_\triangle(3, 2) & F_\triangle(3, 3) & \cdots & F_\triangle(3, m) & \cdots \\ & & \cdots & & & \\ F_\triangle(n, 1) & F_\triangle(n, 2) & F_\triangle(n, 3) & \cdots & F_\triangle(n, m) & \cdots \\ \vdots & \vdots & \vdots & & \vdots & \end{matrix}.$$

Example 7.1 The case of $\triangle_{median}$ and $\triangle_{mean}$, are:

$$F_{median}(n, m) = \begin{cases} 0, & if \ m \leq n - 2 \\ \frac{1}{2}, & if \ m = n - 1 \\ 1, & if \ \ m \geq n \end{cases},$$

$F_{mean}(n, m) = \frac{m}{n+m-1}$, $\forall(n, m) \in \mathbb{N}^* \times \mathbb{N}^*$:

$$F_{median} = \begin{matrix} 1 & 1 & 1 & 1 \\ \frac{1}{2} & 1 & 1 & 1 \\ 0 & \frac{1}{2} & 1 & 1 \\ 0 & 0 & \frac{1}{2} & 1 \\ & & \cdots & \end{matrix} .. \quad F_{mean} = \begin{matrix} 1 & 1 & 1 & 1 \\ \frac{1}{2} & \frac{2}{3} & \frac{3}{4} & \frac{4}{5} \\ \frac{1}{3} & \frac{2}{4} & \frac{3}{5} & \frac{4}{6} \\ \frac{1}{4} & \frac{2}{5} & \frac{3}{6} & \frac{4}{7} \\ & & \cdots & \end{matrix} ..$$

Proposition 7.1 *Every $\triangle \in \mathscr{W}$, verifies:*
(i) $F_\triangle(1, m) = 1$, $\forall m \in \mathbb{N}^*$;
(ii) $F_\triangle(n + 1, m) \leq F_\triangle(n, m + 1)$, $\forall(n, m) \in \mathbb{N}^* \times \mathbb{N}^*$;
If a weighting triangle $\triangle$ is left (right) regular then
(iii) $F_\triangle(n + 1, m) \leq F_\triangle(n, m) \big(F_\triangle(n, m) \leq F_\triangle(n, m + 1)\big) \forall(n, m) \in \mathbb{N}^* \times \mathbb{N}^*$;

If the weighting triangle $\triangle$ is regular then
(iv) $F_\triangle(n+1, m) \leq F_\triangle(n, m) \leq F_\triangle(n, m+1), \ \forall(n, m) \in \mathbb{N}^* \times \mathbb{N}^*$.

Proof (i) By definition of $F_\triangle(1, m)$.

(ii) $F_\triangle(n+1, m) = \sum_{i=1}^{m} w_i^{n+1+m-1} \leq \sum_{i=1}^{m+1} w_i^{n+m} = \sum_{i=1}^{m+1} w_i^{n+(m+1)-1} = F_\triangle(n, m+1)$.

(iii) Suppose $\triangle$ is left regular. Then,

$$F_\triangle(n+1, m) = \sum_{i=1}^{m} w_i^{n+1+m-1} = \sum_{i=1}^{m} w_i^{n+(m-1)+1} \leq \sum_{i=1}^{m} w_i^{n+m-1} = F_\triangle(n, m).$$

Similarly, if $\triangle$ is right regular:

$$F_\triangle(n, m) = \sum_{i=1}^{m} w_i^{n+m-1} \leq \sum_{i=1}^{m+1} w_i^{n+m} = \sum_{i=1}^{m+1} w_i^{n+(m+1)-1}$$
$$= F_\triangle(n, m+1).$$

(iv) can be deduced by *(iii)* taking into account that $\triangle$ is regular, so it is left and right regular. $\square$

The weights w_k^n of $\triangle$ are determined by $F_\triangle$ as follows:

Proposition 7.2 *For all weighting triangle* $\triangle = (w_i^n)$, $i = 1, \ldots, n$, $n \geq 1$, *then* $w_1^n = F_\triangle(n, 1)$, $\forall n \in \mathbb{N}^*$ *and,*
$$w_m^n = F_\triangle(n-m+1, m) - F_\triangle(n-m+2, m-1), \text{ if } n \geq m > 1.$$

Proof Suppose $n \geq m > 1$, then:

$$F_\triangle(n-m+1, m) = \sum_{i=1}^{m} w_i^{n-m+1+m-1} = \sum_{i=1}^{m} w_i^n,$$

$$F_\triangle(n-m+2, m-1) = \sum_{i=1}^{m-1} w_i^{n-m+2+m-1-1} = \sum_{i=1}^{m-1} w_i^n$$

so, $w_m^n = F_\triangle(n-m+1, m) - F_\triangle(n-m+2, m-1)$. $\square$

Remark 7.1 Let us consider $\mathscr{DF} \subset [0, 1]^{\mathbb{N}^* \times \mathbb{N}^*}$ the class of fuzzy relations G that verifies *(i)* and *(ii)* ; $\overleftarrow{\mathscr{RDF}} \subset \mathscr{DF}$ those that verify the property that characterized weighting triangles left regular; $\overrightarrow{\mathscr{RDF}} \subset \mathscr{DF}$ those that verify the property that characterized weighting triangles right regular and, finally $\mathscr{RDF} \subset \mathscr{DF}$ the class that verifies all properties related to regular triangles:

$$\mathscr{DF} = \{G \in [0, 1]^{\mathbb{N}^* \times \mathbb{N}^*} / (G(1, m) = 1) \& (G(n+1, m) \leq G(n, m+1))\},$$

$$\overleftarrow{\mathscr{RDF}} = \{G \in [0, 1]^{\mathbb{N}^* \times \mathbb{N}^*} / (G(1, m) = 1) \&$$
$$(G(n+1, m) \leq G(n, m+1)) \& (G(n+1, m) \leq G(n, m))\},$$

$$\overrightarrow{\mathscr{RDF}} = \{G \in [0, 1]^{\mathbb{N}^* \times \mathbb{N}^*} / (G(1, m) = 1) \&$$
$$(G(n+1, m) \leq G(n, m+1)) \& (G(n, m) \leq G(n, m+1)),$$
$$\forall(n, m) \in \mathbb{N}^* \times \mathbb{N}^*\},$$

$$\mathscr{RDF} = \{G \in [0, 1]^{\mathbb{N}^* \times \mathbb{N}^*} / (G(1, m) = 1) \&$$
$$(G(n+1, m) \leq G(n, m) \leq G(n, m+1)), \ \forall(n, m) \in \mathbb{N}^* \times \mathbb{N}^*\}.$$

Proposition 7.3 *(i) Let $G \in \mathscr{DF}$ and $\triangle_G = (w_i^n)$, $i = 1, \ldots, n, n \geq 1$ be the weighting list obtained by G:*

$$w_k^n = \begin{cases} G(n, k), & k = 1 \\ G(n - k + 1, k) - G(n - k + 2, k - 1), & k > 1 \end{cases}.$$

Then $\triangle_G$ is a weighting triangle with distribution relation $F_{\triangle_G} = G$.
(ii) If $G \in (\mathscr{RDF})$ then the triangle $\triangle_G$ is regular.

Proof (i) The element $w_k^n \in [0, 1]$ for all $n \geq k$, so $w_1^n = G(n, 1) \in [0, 1]$ and if $n \geq k > 1$, of $0 \leq F(n - m + 2, m - 1) \leq F(n - m + 1, m) \leq 1$, it follows $w_k^n \in [0, 1]$. It verifies $w_1^1 = G(1, 1) = 1$. Let us prove $\sum_{i=1}^{n} w_i^n = 1$, for $n \geq 2$:

$$\sum_{i=1}^{n} w_i^n = w_1^n + \sum_{i=2}^{n} w_i^n = w_1^n + \sum_{i=2}^{n}(G(n - i + 1, i) - G(n - i + 2, i - 1))$$
$$= G(n, 1) + (G(n - 1, 2) - G(n, 1)) + \cdots + ((G(1, n) - G(2, n - 1))$$
$$= G(1, n) = 1.$$

Let us show $F_{\triangle_G} = G$. Let $(n, m) \in \mathbb{N}^{*2}$:

$$F_{\triangle_G}(n, m) = \sum_{i=1}^{m} w_i^{n+m-1} = w_1^{n+m-1} + \sum_{i=2}^{m} w_i^{n+m-1}$$
$$= G(n + m - 1, 1) + \sum_{i=2}^{m}(G(n + m - 1 - i + 1, i)$$
$$- G(n + m - 1 - i + 2, i - 1))$$
$$= G(n + m - 1, 1) + G(n + m - 2, 2) - G(n + m - 1, 1)$$
$$+ \cdots + G(n, m) - G(n + 1, m - 1) = G(n, m).$$

(ii) Suppose $G \in (\mathscr{RDF})$. It verifies:

$$\sum_{i=1}^{m} w_i^{n+1} = \sum_{i=1}^{m} w_i^{(n-m+2)+m-1} = F_{\triangle_G}(n - m + 2, m) = G(n - m + 2, m),$$
$$\sum_{i=1}^{m} w_i^{n} = \sum_{i=1}^{m} w_i^{(n-m+1)+m-1} = F_{\triangle_G}(n - m + 1, m) = G(n - m + 1, m),$$
$$\sum_{i=1}^{m+1} w_i^{n} = \sum_{i=1}^{m+1} w_i^{(n-m+1)+(m+1)-1}$$
$$= F_{\triangle_G}(n - m + 1, m + 1) = G(n - m + 1, m + 1),$$

and hypothesis:

$$G(n - m + 2, m) \leq G(n - m + 1, m) \leq G(n - m + 1, m + 1),$$

we obtain the conditions to regularity $\triangle_G$:

$$\sum_{i=1}^{p} w_i^{n+1} \leq \sum_{i=1}^{p} w_i^{n} \leq \sum_{i=1}^{p+1} w_i^{n+1}.$$

$\square$

7.4   Induced Structures in the Class of Triangles $\mathscr{W}$, $\mathscr{RW}$ by Operations $\mathscr{DF}$, $\mathscr{RDF}$

The following Proposition will be used to give a partial order to weighting triangles.

Proposition 7.4 *(i) Ordered sets* $(\mathscr{DF}, \leq)$ *and* $(\mathscr{RDF}, \leq)$ *determined by fuzzy relations in Remark 7.1, with restriction of usual order* $\leq$ *in* $[0, 1]^{\mathbb{N}^* \times \mathbb{N}^*}$ *are sublattices of lattice* $([0, 1]^{\mathbb{N}^* \times \mathbb{N}^*}, \leq, \wedge, \vee)$, *with operations* $\wedge$ *(infimum) and* $\vee$ *(supreme) defined.*

(ii) The minimal and maximal elements of $(\mathscr{DF}, \leq, \wedge, \vee)$ *and* $(\mathscr{RDF}, \leq, \wedge, \vee)$ *are the fuzzy relations* $F_{(0)}$ *and* $F_{(1)}$.

(iii) The lattices $(\mathscr{DF}, \leq)$ *and* $(\mathscr{RDF}, \leq)$ *are complete, and supreme operator* sup *and infimum* inf *of a family* $\mathscr{F}$ *of elements of* $\mathscr{DF}$ *are given by:*

If $\mathscr{H} \neq \emptyset$, *are the usual supreme operation* $\bigvee \mathscr{H}$ *and infimum* $\bigwedge \mathscr{H}$ *of the family* $\mathscr{H}$ *in the complete lattice* $([0, 1]^{\mathbb{N}^* \times \mathbb{N}^*} \leq)$: sup $\mathscr{H} = \bigvee \mathscr{H}$, inf $\mathscr{H} = \bigwedge \mathscr{H}$, *and in the case* $\mathscr{H} = \emptyset$: sup $\emptyset = F_{(0)}$, inf $\emptyset = F_{(1)}$.

(iv) If $\mathfrak{I}$ *represents the subset:* $\mathfrak{I} = \{F_{(\alpha)}/\alpha \in [0, 1]\} \subset \mathscr{DF}$ *of distribution considered in (a'), then the chain* $(\mathfrak{I}, \leq)$ *included in* $(\mathscr{RDF}, \leq)$ *is isomorphic to the chain* $([0, 1], \leq)$.

Proof (i) Let F, G *be elements of* $\mathscr{DF} \subset [0, 1]^{\mathbb{N}^* \times \mathbb{N}^*}$. *It is verified:* $(F \vee G)$ $(n + 1, m) = \max\{F(n + 1, m), G(n + 1, m)\} \leq \max\{F(n, m + 1), G(n, m + 1)\}$ $= (F \vee G)(n, m + 1)$, *it shows that* $F \vee G \in \mathscr{DF}$. *A similar reasoning shows that* $F \wedge G \in \mathscr{DF}$.

If F *and* G *belong to* $\mathscr{RDF}$, *again max and min, together the definition* $F \vee G$ *and of* $F \wedge G$, *show that the last ones verify (7.1), i.e., belong to* $\mathscr{RDF}$.

(ii) Because $F_{(0)}(n, m) = 0$ *if* $n > 1$ *and* $F_{(1)}(n, m) = 1$, *for all* $(n, m) \in \mathbb{N}^{*2}$ *and belong to* $\mathscr{RDF}$, *evidently, it verifies* $F_{(0)} \leq F \leq F_{(1)}$ *for all* $F \in \mathscr{DF}$, *(and* $\mathscr{RDF}$).

(iii) Let $\mathscr{H} \neq \emptyset$ *be a family of* $(\mathscr{DF}, \leq)$ *and let* $\bigvee \mathscr{H}$ *its supreme in* $([0, 1]^{\mathbb{N}^* \times \mathbb{N}^*} \leq)$. *It verifies:* $(\bigvee \mathscr{H})(n + 1, m) = \sup\{F(n + 1, m)/F \in \mathscr{H}\} \leq \sup\{F(n, m + 1)/ F \in \mathscr{H}\} = (\bigvee \mathscr{H})(n, m + 1)$, *it shows* sup $\mathscr{H} = \bigvee \mathscr{H} \in \mathscr{DF}$. *A similar reasoning shows that* inf $\mathscr{H} = \bigwedge \mathscr{H} \in \mathscr{DF}$. *As it is bounded,* sup $\emptyset = F_{(0)}$ *and* inf $\emptyset = F_{(1)}$, *so,* $(\mathscr{DF}, \leq)$ *is complete.*

If $\mathscr{H} \neq \emptyset$ is a family of $(\mathscr{R}\mathscr{D}\mathscr{F}, \leq)$, a similar reasoning shows $\bigvee \mathscr{H}$ as $\bigwedge \mathscr{H}$ belongs to $\mathscr{R}\mathscr{D}\mathscr{F}$ and consequently sup $\mathscr{H}$ e inf $\mathscr{H}$. As $\{F_{(0)}, F_{(1)}\} \subset \mathscr{R}\mathscr{D}\mathscr{F}$, it verifies that $(\mathscr{R}\mathscr{D}\mathscr{F}, \leq)$ is complete.

(iv) As $F_{(\alpha)}(n, m) = \alpha$ if $n > 1$, so $\alpha \leq \beta \Longleftrightarrow F_{(\alpha)} \leq F_{(\beta)}$. □

Note that $(\mathscr{R}\mathscr{D}\mathscr{F}, \leq)$ is a complete sublattice of complete lattice $(\mathscr{D}\mathscr{F}, \leq)$. However, although this is a sublattice of $([0, 1]^{\mathbb{N}^* \times \mathbb{N}^*}, \leq)$, is not a complete sublattice in it, because $F_{(0)} = \sup \emptyset \neq \bigvee \emptyset = \emptyset$.

A consequence of last Proposition 7.4, it is to obtain structures of lattices of weighting triangles $(\mathscr{W}, \preccurlyeq, \curlywedge, \curlyvee)$ and $(\mathscr{R}\mathscr{W}, \preccurlyeq, \curlywedge, \curlyvee)$ where the order relation $\preccurlyeq$ and the operations $\curlywedge$ (infimum) and $\curlyvee$ (supreme) are:

Proposition 7.5 *(i) The relation $\preccurlyeq$ defined in $\mathscr{W}$, (and its restriction to $\mathscr{R}\mathscr{W}$), by*

$$\triangle_1 \preccurlyeq \triangle_2 \Longleftrightarrow F_{\triangle_1} \leq F_{\triangle_2}, \tag{7.3}$$

is of order and $(\mathscr{W}, \preccurlyeq, \curlywedge, \curlyvee)$, $(\mathscr{R}\mathscr{W}, \preccurlyeq, \curlywedge, \curlyvee)$ are lattices with minimal element $\triangle_{(0)}$ and maximal element $\triangle_{(1)}$ in which the operations inf ($\curlywedge$) and sup ($\curlyvee$) verify:

$$\triangle_1 \curlywedge \triangle_2 = \triangle_{F_{\triangle_1} \wedge F_{\triangle_2}}, \tag{7.4}$$

$$\triangle_1 \curlyvee \triangle_2 = \triangle_{F_{\triangle_1} \vee F_{\triangle_2}}. \tag{7.5}$$

(ii) The lattices $(\mathscr{W}, \preccurlyeq, \curlywedge, \curlyvee)$, $(\mathscr{R}\mathscr{W}, \preccurlyeq, \curlywedge, \curlyvee)$ are complete such as infimum $\curlywedge \mathscr{H}$ and supreme $\curlyvee \mathscr{H}$ of a family $\mathscr{H}$ of weighting triangles are the weighting triangles obtained respectively by

$$\curlywedge \mathscr{H} = \triangle_{\wedge \{F_\triangle / \triangle \in \mathscr{H}\}}, \quad \curlyvee \mathscr{H} = \triangle_{\vee \{F_\triangle / \triangle \in \mathscr{H}\}}.$$

(iii) Let us consider the chain $([0, 1], \leq)$ immerse in $(\mathscr{R}\mathscr{W}, \preccurlyeq)$ identifying $\alpha \in [0, 1]$ with the triangle $\triangle_{(\alpha)}$.

Proof (i) Because $F_\triangle$ determined uniquely a $\triangle$, the relation $\preccurlyeq$ is an order relation. Let us $\triangle_1$ and $\triangle_2$ elements of $\mathscr{W}$. As $F_{\triangle_i} \leq F_{\triangle_1} \vee F_{\triangle_2}$ for $i \in \{1, 2\}$, of (7.3) it follows: $\triangle_i = \triangle_{F_{\triangle_i}} \preccurlyeq \triangle_{F_{\triangle_1} \vee F_{\triangle_2}}$, which shows that this last one is an upper bound of $\{\triangle_1, \triangle_2\}$. Let us $\triangle$ be other upper bound. Of $\triangle_i \preccurlyeq \triangle$ we deduce that $F_{\triangle_i} \leq F_\triangle$, so $F_{\triangle_1} \vee F_{\triangle_2} \leq F_\triangle$, where we obtain $\triangle_{F_{\triangle_1} \vee F_{\triangle_2}} \preccurlyeq \triangle_{F_\triangle} = \triangle$, which shows that the supreme exists $\triangle_1 \curlyvee \triangle_2$ and it is equal to $\triangle_{F_{\triangle_1} \vee F_{\triangle_2}}$. A similar demonstration justifies that $\triangle_1 \curlywedge \triangle_2 = \triangle_{F_{\triangle_1} \wedge F_{\triangle_2}}$. In addition, if $\triangle_1$ and $\triangle_2$ belong to $\mathscr{R}\mathscr{W}$ then $F_{\triangle_1} \vee F_{\triangle_2}$ and $F_{\triangle_1} \wedge F_{\triangle_2}$ belong to $\mathscr{R}\mathscr{D}\mathscr{F}$ and in consequence $\triangle_{F_{\triangle_1} \vee F_{\triangle_2}}$ and $\triangle_{F_{\triangle_1} \wedge F_{\triangle_2}}$ are also regular. We have shown $(\mathscr{W}, \preccurlyeq)$ is a lattice and that $(\mathscr{R}\mathscr{W}, \preccurlyeq)$ is a sublattice of the one before. It is clear that $\triangle_{(0)}$ and $\triangle_{(1)}$, (belong to $\mathscr{R}\mathscr{W}$), are elements minimum and maximum in both lattices.

(ii) Let $\mathcal{H}$ be a family of weighting triangles included in $\mathcal{W}$ or in $\mathcal{RW}$. If $\mathcal{H} = \emptyset$, let us write $\curlyvee\emptyset = \triangle_{(0)}$ and $\curlywedge\emptyset = \triangle_{(1)}$. Let $\mathcal{H} \neq \emptyset$. It is verified for each $\triangle_s$ of $\mathcal{H}$: $F_{\triangle_s} \leq \bigvee\{F_\triangle \mid \triangle \in \mathcal{H}\}$, so $\triangle_s \preccurlyeq \triangle_{\vee F_\triangle}$ which shows that $\triangle_{\vee F_\triangle}$ is an upper bound of $\mathcal{H}$. Let $\widehat{\triangle}$ be other upper bound of $\mathcal{H}$, of $\triangle_s \preccurlyeq \widehat{\triangle}$, $\forall \triangle_s \in \mathcal{H}$ it is obtained $\bigvee\{F_\triangle \mid \triangle \in \mathcal{H}\} \leq F_{\widehat{\triangle}}$, so $\triangle_{\vee F_\triangle} \preccurlyeq \triangle_{F_{\widehat{\triangle}}} = \widehat{\triangle}$, which shows that there exists the supreme $\mathcal{H}$ in $\mathcal{W}$ and it is the weighting triangle whose distribution relation is $\bigvee\{F_\triangle \mid \triangle \in \mathcal{H}\}$. If $\mathcal{H}$ is included in $\mathcal{RW}$, so $\bigvee\{F_\triangle \mid \triangle \in \mathcal{H}\}$ belongs to $\mathcal{RDF}$ it is shown that the supremum of $\mathcal{H}$ is also in $\mathcal{RW}$.

A similar reasoning shows that there exists the infimum of $\mathcal{H}$, in $\mathcal{W}$ and in $\mathcal{RW}$, and it is weighting triangle with relation of distribution: $\bigwedge\{F_\triangle \mid \triangle \in \mathcal{H}\}$.

(iii) From the last paragraph of Proposition 7.1 it is deduced:
$$\alpha \leq \beta \iff \triangle_{(\alpha)} \preccurlyeq \triangle_{(\beta)}. \qquad \qquad \square$$

Remark 7.2 The expression of the order relation $\preccurlyeq$ (7.3) depending on the weights of triangles is the following: Given $\triangle_1 = (w_i^n)$, and $\triangle_2 = (u_i^n)$, $i = 1, \ldots, n, n \geq 1$, taking into account Definition 7.3:

$$\triangle_1 \preccurlyeq \triangle_2 \iff \left(\sum_{i=1}^{m} w_i^{n+m-1} \leq \sum_{i=1}^{m} u_i^{n+m-1}, \ \forall(n, m) \in \mathbb{N}^{*2} \right), \qquad (7.6)$$

i.e., (taking into account $\sum_{i=1}^{n} w_i^n = \sum_{i=1}^{n} u_i^n = 1$ for all $n \in \mathbb{N}^*$), $\triangle_1 \preccurlyeq \triangle_2$ if and only if:

$$w_1^2 \leq u_1^2$$

$$w_1^3 \leq u_1^3, \ (w_1^3 + w_2^3) \leq (u_1^3 + u_2^3)$$

$$\ldots$$

$$w_1^n \leq u_1^n, \ (w_1^n + w_2^n) \leq (u_1^n + u_2^n), \ldots, (w_1^n + w_2^n + \cdots + w_{n-1}^n)$$
$$\leq (u_1^n + u_2^n + \cdots + u_{n-1}^n)$$

Example 7.2 Using Definition 7.3 and Proposition 7.5, together with the triangles $\triangle_{median}$ and $\triangle_{mean}$ included in Sect. 1, we $\triangle_1 = \triangle_{median} \curlywedge \triangle_{mean}$ and $\triangle_2 = \triangle_{median} \curlyvee \triangle_{mean}$, such as the initial weighting appear in the following non limited tables:

$$\triangle_1 = \begin{array}{ccccc} & & 1 & & \\ & \frac{1}{2} & & \frac{1}{2} & \\ 0 & & \frac{2}{3} & & \frac{1}{3} \\ 0 & & \frac{1}{2} & \frac{1}{4} & \frac{1}{4} \\ & & \ldots & & \end{array}, \qquad \triangle_2 = \begin{array}{ccccc} & & 1 & & \\ & \frac{1}{2} & & \frac{1}{2} & \\ \frac{1}{3} & & \frac{2}{3} & & 0 \\ \frac{1}{4} & \frac{1}{4} & & \frac{1}{2} & 0 \\ & & \ldots & & \end{array} .$$

The before triangles are examples of elements related to the order $\preccurlyeq$, because:

$$\triangle_1 \preccurlyeq \triangle_{median} \preccurlyeq \triangle_2,$$
$$\triangle_1 \preccurlyeq \triangle_{mean} \preccurlyeq \triangle_2.$$

The relation of order $\preccurlyeq$ is partial, for instance:

$$\triangle_{median} \npreccurlyeq \triangle_{mean} \text{ and } \triangle_{mean} \npreccurlyeq \triangle_{median}.$$

The completeness of lattice $(\mathcal{RDF}, \leq)$, together with elements minimum $F_{(0)}$ and maximum $F_{(1)}$ of lattice $(\mathcal{DF}, \leq)$, allows the introduction of a closure operator $^{-} : \mathcal{DF} \longrightarrow (\mathcal{RDF})$ and an interior operator $^{\circ} : \mathcal{DF} \longrightarrow (\mathcal{RDF})$:

Definition 7.4 *Let $F \in (\mathcal{DF}, \leq)$. (i) We call closure of F in $\mathcal{RDF}$, to the infimum $\overline{F}$ of the elements of $\mathcal{RDF}$ bigger or equal to F:*

$$\overline{F} = \bigwedge \{G \in (\mathcal{RDF})/F \leq G\} \in (\mathcal{RDF}).$$

(ii) We will call interior of F in $\mathcal{RDF}$, to the supremum $\overset{\circ}{F}$ of the elements of $\mathcal{RDF}$ smaller or equal to F:

$$\overset{\circ}{F} = \bigvee \{H \in (\mathcal{RDF})/H \leq F\} \in (\mathcal{RDF}).$$

These operators verify the properties of closure and interior in partially ordered sets:

(iii) $\overset{\circ}{F} \leq F \leq \overline{F}$ $\forall F \in \mathcal{DF}$,

(iv) $F \leq G \Longrightarrow (\overset{\circ}{F} \leq \overset{\circ}{G}) \& (\overline{F} \leq \overline{G})$,

(v) $(\overset{\circ}{\overset{\circ}{F}}) = \overset{\circ}{F} \& \overline{\overline{(F)}} = \overline{F}$.

The closure and interior operator induce other triangles $^{-} : \mathcal{W}(\mathcal{TRI}) \longrightarrow (\mathcal{RW}(\mathcal{TRI}))$, and $^{\circ} : \mathcal{W}(\mathcal{TRI}) \longrightarrow (\mathcal{RW}(\mathcal{TRI}))$ as follows:

Definition 7.5 *Let $\triangle \in \mathcal{W}(\mathcal{TRI})$ and $F_\triangle$ its associated relation of distribution. (i) We will call closure of $\triangle$ in $\mathcal{RW}(\mathcal{TRI})$, to the weighting triangle $\overline{\triangle} \in (\mathcal{RW}(\mathcal{TRI}))$ such that :*

$$\overline{\triangle} = \triangle_{\overline{F}_\triangle} \in (\mathcal{RW}(\mathcal{TRI})).$$

(ii) We will call interior of $\triangle$ in $\mathcal{RW}(\mathcal{TRI})$, to the weighting triangle $\overset{\circ}{\triangle} \in (\mathcal{RW}(\mathcal{TRI}))$ such that:

$$\overset{\circ}{\triangle} = \triangle_{\overset{\circ}{F}_\triangle} \in (\mathcal{RW}(\mathcal{TRI})).$$

And they are such that

(iii) $\overset{\circ}{\Delta} \preccurlyeq \Delta \preccurlyeq \overline{\Delta}$,

(iv) $\Delta \preccurlyeq \Delta \Longrightarrow (\overset{\circ}{\Delta}_1 \preccurlyeq \overset{\circ}{\Delta}_2)\&(\overline{\Delta}_1 \preccurlyeq \overline{\Delta}_2)$,

(v) $(\overset{\circ}{\Delta}) = \overset{\circ}{\Delta}, \quad \overline{(\overline{\Delta})} = \overline{\Delta}$.

Example 7.3 Let $\{\alpha, \beta\} \subset [0, 1]$ and $F_{<\alpha,\beta>} \in \mathcal{DF}$ such that, for $n > 1$:

$$F_{<\alpha,\beta>}(n, m) = \begin{cases} \alpha & (n = 2k, m = 2r - 1) \text{ ó } (n = 2k - 1, m = 2r) \\ \beta & (n = 2k - 1, m = 2r - 1) \text{ ó } (n = 2k, m = 2r) \end{cases} :$$

$$F_{<\alpha,\beta>} = \begin{matrix} 1 & 1 & 1 & 1 \\ \alpha & \beta & \alpha & \beta \\ \beta & \alpha & \beta & \alpha \\ \alpha & \beta & \alpha & \beta \end{matrix} \dots.$$
$$\dots$$

It is verified $F_{<\alpha,\beta>} \in (\mathcal{RDF}) \Longleftrightarrow \alpha = \beta$. It is shown that $\overset{\circ}{F}_{<\alpha,\beta>} = F_{(\min(\alpha,\beta))}$, and $\overline{F}_{<\alpha,\beta>} = F_{(\max(\alpha,\beta))}$.

As a consequence, the interior $\overset{\circ}{\Delta}_{<\alpha,\beta>}$ and the closure $\overline{\Delta}_{<\alpha,\beta>}$ in $\mathcal{RW}(\mathcal{TRI})$ of weighting triangle $\Delta_{<\alpha,\beta>} \in \mathcal{W}(\mathcal{TRI})$:

$$\Delta_{<\alpha,\beta>} = \begin{matrix} 1 & & & \\ \alpha & 1 - \alpha & & \\ \beta & 0 & 1 - \beta & \\ \alpha & 0 & 0 & 1 - \alpha \end{matrix}$$
$$\dots \qquad .$$

are, respectively, the weighting triangles

$\overset{\circ}{\Delta}_{<\alpha,\beta>} = \Delta_{(\min(\alpha,\beta))}$, and $\overline{\Delta}_{<\alpha,\beta>} = \Delta_{(\max(\alpha,\beta))}$ of $\mathcal{RW}(\mathcal{TRI})$:

$$\overset{\circ}{\Delta}_{<\alpha,\beta>} = \begin{matrix} 1 & & & \\ \min(\alpha, \beta) & 1 - \min(\alpha, \beta) & & \\ \min(\alpha, \beta) & 0 & 1 - \min(\alpha, \beta) & \\ \min(\alpha, \beta) & 0 & 0 & 1 - \min(\alpha, \beta) \end{matrix} ,$$
$$\dots$$

$$\overline{\Delta}_{<\alpha,\beta>} = \begin{matrix} 1 & & & \\ \max(\alpha, \beta) & 1 - \max(\alpha, \beta) & & \\ \max(\alpha, \beta) & 0 & 1 - \max(\alpha, \beta) & \\ \max(\alpha, \beta) & 0 & 0 & 1 - \max(\alpha, \beta) \end{matrix} .$$
$$\dots$$

7.5 Extended Operators Through Distribution Relations

In [4], the concept of EOWA operator extends the n-dimensional OWA operator [5] of Yager:

Definition 7.6 [4] *A mapping* $A : \bigcup_{n \geq 1} [0, 1]^n \to [0, 1]$ *is an EOWA operator (Extended Ordered Weighted Averaging operator) if the restriction of A to each $[0, 1]^n$ is an n-dimensional OWA operator. In order words, if for each n there exist $(w_1^n, w_2^n, \ldots, w_n^n) \in [0, 1]^n$ with $\sum_{i=1}^{n} w_i = 1$, such that $A(x_1, \ldots, x_n) = \sum_{i=1}^{n} w_i y_i$ for all $(x_1, \ldots, x_n) \in [0, 1]^n$ and $(y_1, \ldots, y_n)$ being a list obtained from the previous rearranging in descending order.*

Let $\mathscr{EOWA}$ be the set of EOWA operators.

Proposition 7.6 *Let $F_\triangle$ be the relation of distribution of the weighting triangle associated to operator A. Then for $n \geq 2$:*

$$A(x_1, x_2, \ldots, x_n) = y_n + \sum_{i=1}^{n-1} F_\triangle(n - i + 1, i)(y_i - y_{i+1}). \tag{7.7}$$

Proof If $n \geq 2$, and w_m^n represents a generic weight of $\triangle$, considering that $F_\triangle(1, n) = 1$:

$$\begin{aligned}
A(x_1, x_2, \ldots, x_n) &= \sum_{i=1}^{n} w_i^n y_i = w_1^n y_1 + \sum_{i=2}^{n} w_i^n y_i \\
&= F_\triangle(n, 1)y_1 + \sum_{i=2}^{n} (F_\triangle(n - i + 1, i) \\
&\quad - F_\triangle(n - i + 2, i - 1))y_i \\
&= F_\triangle(n, 1)y_1 + (F_\triangle(n - 1, 2) - F_\triangle(n, 1))y_2 \\
&\quad + \cdots + (F_\triangle(1, n) - F_\triangle(2, n - 1))y_n \\
&= y_n + \sum_{i=1}^{n-1} F_\triangle(n - i + 1, i)(y_i - y_{i+1}).
\end{aligned}$$
$\square$

Note that if n is fixed, an OWA operator [5] can be expressed by:

$$A(x_1, x_2, \ldots, x_n) = y_n + \sum_{i=1}^{n-1} s_i(y_i - y_{i+1}),$$

with $0 \leq s_1 \leq \cdots \leq s_{n-1} \leq 1$.

For instance, the expression for *mean operator* $\mathfrak{M}_1 = A_{mean}$ is:

$$A_{mean}(x_1, x_2, \ldots, x_n) = y_n + \sum_{i=1}^{n-1} \frac{i}{n}(y_i - y_{i+1}). \tag{7.8}$$

In the set $\mathcal{EOWA}$ the order relation can be introduced using the natural order in the initial set $[0, 1]$ of mappings $A \in \mathcal{EOWA}$. This relation, which is usually used to compare fuzzy subsets, is given by:

Definition 7.7 Let $\leq$ be *a partial order in* $\mathcal{EOWA}$ *induced by the extension of order in the set* $[0, 1]$:

$$A_1 \leq A_2 \iff A_1(\overline{x}) \leq A_2(\overline{x}), \quad \forall \overline{x} \in \bigcup_{n \geq 1} [0, 1]^n. \tag{7.9}$$

We obtain a partially ordered set $(\mathcal{EOWA}, \leq)$ included in the lattice $([0, 1]^{\cup [0,1]^n}, \leq)$. We characterize this order between $\mathcal{EOWA}$ operators using the order $\leq$ between relations of distribution F belonging to lattice $(\mathcal{DF}, \leq, \wedge, \vee, F_{(0)}, F_{(1)})$.

Proposition 7.7 *Let A_1, A_2 be elements of $\mathcal{EOWA}$. Let $\triangle_1$, $\triangle_2$ be their weighting triangles and let $F_{\triangle_1}$, $F_{\triangle_2}$ be the relations of distribution of them. Then:*

$$A_1 \leq A_2 \iff F_{\triangle_1} \leq F_{\triangle_2}.$$

Proof First, let us suppose $F_{\triangle_1} \leq F_{\triangle_2}$. Considering $\overline{x} = (x_1, x_2, \ldots, x_n) \in [0, 1]^n$, then for all $i = 1, \ldots, n - 1$ the inequalities are verified $F_{\triangle_1}(n - i + 1, i) \leq F_{\triangle_2}(n - i + 1, i)$ and $(y_i - y_{i+1}) \geq 0$, then

$$A_1(\overline{x}) = y_n + \sum_{i=1}^{n-1} F_{\triangle_1}(n - i + 1, i)(y_i - y_{i+1})$$

$$\leq y_n + \sum_{i=1}^{n-1} F_{\triangle_2}(n - i + 1, i)(y_i - y_{i+1}) = A_2(\overline{x})$$

and consequently: $A_1 \leq A_2$.

Let us now consider $A_1 \leq A_2$. Let $(p, q) \in \mathbb{N}^{*2}$. If $p = 1$, then $F_{\triangle_1}(1, q) \leq 1 = F_{\triangle_1}(1, q)$. If $p > 1$, let us consider a $(p + q - 1)$-uple $\overline{x} = (x_1, x_2, \ldots, x_{q-1}, x_q, x_{q+1}, \ldots, x_{p+q-1})$ such that $x_1 = x_2 = \cdots = x_{q-1} = x_q > x_{q+1} = \cdots = x_{p+q-1}$. Then: $x_{p+q-1} + F_{\triangle_1}(p, q)(x_q - x_{q+1}) = A_1(\overline{x}) \leq A_2 (\overline{x}) = x_{p+q-1} + F_{\triangle_2}(p, q)(x_q - x_{q+1})$, it proves the inequality $F_{\triangle_1}(p, q) \leq F_{\triangle_2}(p, q)$. $\square$

This proposition shows that the partially ordered set $(\mathcal{EOWA}, \leq)$ is isomorphic to $(\mathcal{DF}, \leq)$ and, as a consequence, to $(\mathcal{W}, \preccurlyeq)$.

Remark 7.3 (1) The partially ordered set $(\mathcal{EOWA}, \leq)$ is a lattice with minimal element the operator min and maximal element the operator max such that, if $\overline{x} = (x_1, x_2, \ldots, x_n)$ is a generic element of $\bigcup_{n \geq 1} [0, 1]^n$,

$$\min(\overline{x}) = y_n + \sum_{i=1}^{n-1} 0(y_i - y_{i+1}) = y_n,$$

$$\max(\overline{x}) = y_n + \sum_{i=1}^{n-1} (y_i - y_{i+1}) = y_1.$$

(2) $(\mathscr{EOWA}, \leq)$ is a complete lattice where the minimal and maximal elements are given by:

If $\mathscr{A} = \varnothing : \bigwedge \varnothing = MAX, \bigvee \varnothing = \min,$ and

if $\mathscr{A} \neq \varnothing$ is a family of EOWA operators such that for $A \in \mathscr{A}$, F_A represents the relation of distribution of its weighting triangle $\triangle_A$ then:

$$(\bigwedge \mathscr{A})(\overline{x}) = y_n + \sum_{i=1}^{n-1} \inf\{F_A(n-i+1, i)/A \in \mathscr{A}\}(y_i - y_{i+1}),$$

$$(\bigvee \mathscr{A})(\overline{x}) = y_n + \sum_{i=1}^{n-1} \sup\{F_A(n-i+1, i)/A \in \mathscr{A}\}(y_i - y_{i+1}).$$

In particular, if A_1 and, A_2 are EOWA operators:

$$(A_1 \wedge A_2)(\overline{x}) = y_n + \sum_{i=1}^{n-1} \min\{F_{A_1}(n-i+1, i), F_{A_2}(n-i+1, i)\}$$
$(y_i - y_{i+1}),$

$$(A_1 \vee A_2)(\overline{x}) = y_n + \sum_{i=1}^{n-1} \max\{F_{A_1}(n-i+1, i), F_{A_2}(n-i+1, i)\}$$
$(y_i - y_{i+1}).$

(3) If $'$ is a negation in $[0, 1]$, then the mapping $' : \mathscr{EOWA} \longrightarrow \mathscr{EOWA}$ such that for all $A \in \mathscr{EOWA}$ associates the element $A' \in \mathscr{EOWA}$ defined by:

$$A'(\overline{x}) = y_n + \sum_{i=1}^{n-1} (F_{\triangle}(i+1, n-i))'(y_i - y_{i+1})$$

is a negation in $(\mathscr{EOWA}, \leq)$ which is strong if $'$ is strong in $[0, 1]$.

(4) For Zadeh negation, given $A \in \mathscr{EOWA}$, the inverse $A^r \in \mathscr{EOWA}$ is such that $A^r = A'$, i.e.,

$$A^r(\overline{x}) = y_n + \sum_{i=1}^{n-1} (1 - F_{\triangle}(i+1, n-i))(y_i - y_{i+1}).$$

(5) We consider the chain $([0, 1], \leq)$ inside $(\mathscr{EOWA}, \leq)$ identifying $\alpha \in [0, 1]$ with the EOWA operator represented by $A_{(\alpha)}$ and defined by:

$$A_{(\alpha)}(\overline{x}) = y_n + \sum_{i=1}^{n-1} \alpha(y_i - y_{i+1}) = \alpha y_1 + (1 - \alpha)y_n,$$

the equivalences are verified:

$$A_{(\alpha_1)} \le A_{(\alpha_2)} \iff F_{(\alpha_1)} \le F_{(\alpha_2)} \iff \alpha_1 \le \alpha_2.$$

In particular: $A_{(0)} = \min$ and $A_{(1)} = \max$.

(6) They are extended to non-decreasing mappings (fuzzy quantifiers) $\gamma : [0, 1] \to [0, 1]$ to other $\widehat{\gamma} : \mathscr{EOWA} \to \mathscr{EOWA}$, such that

$$\widehat{\gamma}(A) = \gamma \circ A,$$

i.e.,

$$(\widehat{\gamma}(A))(\overline{x}) = y_n + \sum_{i=1}^{n-1} \gamma(F_A(n-i+1, i))(y_i - y_{i+1}).$$

For instance, $(\widehat{\gamma}(A_{mean}))(\overline{x}) = y_n + \sum_{i=1}^{n-1} \gamma(\frac{i}{n}))(y_i - y_{i+1})$.

(7) The extensions: $'$ and, $\widehat{\gamma}$ of negations $'$ and, the non-decreasing mappings γ are consistent with the immersion of $([0, 1], \le)$ in $(\mathscr{EOWA}, \le)$:

$$(A_{(\alpha)})' = A_{(\alpha')}, \quad \widehat{\gamma}(A_{(\alpha)}) = A_{(\gamma(\alpha))}.$$

(8) It is extended to binary laws $\eta : [0, 1]^2 \to [0, 1]$ such that $\eta(1, 1) = 1$ to other $\widehat{\eta} : \mathscr{EOWA}^2 \to \mathscr{EOWA}$:

- If η is non-decreasing in its two arguments (t-norms, t-conorms, ...) and A, B belong to $\mathscr{EOWA}$:

$$\widehat{\eta}(A, B)(\overline{x}) = y_n + \sum_{i=1}^{n-1} \eta(F_A(n-i+1, i), F_B(n-i+1, i))(y_i - y_{i+1}).$$

- If η is non-increasing in the first and non-decreasing in the second, (implication, etc.):

$$\widehat{\eta}(A, B)(\overline{x}) = y_n + \sum_{i=1}^{n-1} \eta(F_A(i+1, n-i), F_B(n-i+1, i))(y_i - y_{i+1}).$$

(9) In both cases we obtain consistent extensions with the immersion of $([0, 1], \le)$ in $(\mathscr{EOWA}, \le)$:

$$\widehat{\eta}(A_{(\alpha_1)}, A_{(\alpha_2)}) = A_{(\eta(\alpha_1, \alpha_2))}.$$

(10) We obtain an EOWA operator as combination of other:

If $(\alpha_1, \alpha_2, \ldots, \alpha_p) \in [0, 1]^p$ is a p-uple such that $\sum_{j=1}^{p} \alpha_j = 1$, and $A_1, A_2, \ldots,$ A_p $(p \geq 2)$ belong to $\mathscr{EOWA}$ and $F_{\triangle_1}, F_{\triangle_2}, \ldots, F_{\triangle_p}$ are the relations of distribution, the mapping

$$\sum_{j=1}^{p} \alpha_j A_j : \bigcup_{n \geq 1} [0, 1]^n \to [0, 1] \text{ defined for all } \overline{x} \text{ by:}$$

$$\left(\sum_{j=1}^{p} \alpha_j A_j \right)(\overline{x}) = y_n + \sum_{i=1}^{n-1} \left(\sum_{j=1}^{p} \alpha_j F_{\triangle j}(n - i + 1, i) \right)(y_i - y_{i+1}),$$

is an EOWA operator.

In the literature [4], given a lattice L, the extended aggregation functions are defined using two orders $\leqslant_1$ and $\leqslant_2$ in $\bigcup_{n \geq 1} L^n$:

$(\alpha_1, \alpha_2, \ldots, \alpha_n) \leqslant_1 (\beta_1, \beta_2, \ldots, \beta_m)$ if and only if

$$(n \leq m) \& (\alpha_1 \leq \beta_1, \ldots, \alpha_n \leq \beta_n) \Longrightarrow$$
$$(\sup\{\alpha_1, \ldots, \alpha_n\} \leq \inf\{\beta_{n+1}, \ldots, \beta_m\}).$$

$(\alpha_1, \alpha_2, \ldots, \alpha_n) \leqslant_2 (\beta_1, \beta_2, \ldots, \beta_m)$ if and only if

$$(n \geq m) \& (\alpha_1 \leq \beta_1, \ldots, \alpha_m \leq \beta_m) \Longrightarrow$$
$$(\sup\{\alpha_{m+1}, \ldots, \alpha_n\} \leq \inf\{\beta_1, \ldots, \beta_m\}).$$

Definition 7.8 [4]   *An extended aggregation function on L is a mapping $A : \bigcup_{n \geq 1} L^n \to L$ that verifies the following two conditions:*

(a_1) The mapping A is monotone with respect to the orders $\leqslant_1$ and $\leqslant_2$, is if $\overline{x}$ and $\overline{y}$ are elements of $\bigcup_{n \geq 1} L^n$:

$$\overline{x} \leqslant_1 \overline{y} \Longrightarrow A(\overline{x}) \leq A(\overline{y}), \text{ and } \overline{x} \leqslant_2 \overline{y} \Longrightarrow A(\overline{x}) \leq A(\overline{y}).$$

(a_2) The mapping A is idempotent:

$$A(\underbrace{\alpha, \alpha, \ldots, \alpha}_{n}) = \alpha, \quad \forall \alpha \in L, \text{ and } \forall n \geq 1.$$

Let $\mathscr{EAF}(L)$ be the class of extended aggregation functions in L. In [4] the EOWA operators are characterized in $L = [0, 1]$ which are extended aggregation functions:

Proposition 7.8 [4]   *Let $A \in \mathscr{EOWA}$ with the weighting triangle associated $\triangle$. Then $A \in \mathscr{EAF}([0, 1])$ if and only if $\triangle$ is regular.*

Corollary 7.1 *Let $A \in \mathcal{EOWA}$ and $F_\triangle$ the relation of distribution of its weighting triangle $\triangle$. Then $A \in \mathcal{EAF}([0, 1])$ if and only if $F_\triangle \in (\mathcal{RDF})$, i.e., if and only if for all (n, m):*

$$F_\triangle(n + 1, m) \leq F_\triangle(n, m) \leq F_\triangle(n, m + 1). \tag{7.10}$$

As a consequence, the partially ordered set $(\mathcal{EOWA} \cap \mathcal{EAF}([0, 1]), \leq)$ is isomorphic to the partially ordered set $(\mathcal{RDF}, \leq)$ of the relations of distribution that verify (7.10). The first one is a complete lattice, sublattice of $(\mathcal{EOWA}, \leq)$.

7.6 Closure and Interior Operators in Lattices $(\mathcal{EOWA}, \leq)$ and $(\mathcal{EOWA} \cap \mathcal{EAF}([0, 1]), \leq)$

From Definition 7.4 and Definition 7.5, we obtain the following result:
Given $A \in \mathcal{EOWA}$ associated to the relation of distribution F, there exist dimensions of the same in $\mathcal{EOWA} \cap \mathcal{EAF}([0, 1])$, (its interior $\mathring{A}$ associated to $\mathring{F}$ and its closure $\overline{A}$ associated to $\overline{F}$), such that

$$\mathring{A} \leq A \leq \overline{A}.$$

Remark 7.4 We could consider closure and interior operators in a more general case. Let M be a mapping $M : \bigcup_{n \geq 1} [0, 1]^n \to [0, 1]$ such that $M \leq \max$ with the usual order extended from the order in the arrival set $[0, 1]$.

We define the *closure* of M: $Cl(M) \in \mathcal{EOWA}$, and $\overline{M} \in \mathcal{EOWA} \cap \mathcal{EAF}([0, 1])$ by:

$$Cl(M) = \bigwedge \{A \in \mathcal{EOWA} / M \leq A\},$$

$$\overline{M} = \bigwedge \{A \in \mathcal{EOWA} \cap \mathcal{EAF}([0, 1]) / M \leq A\}.$$

These definitions make sense, because
$M \leq A_{(1)} = \max \in \mathcal{EOWA} \cap \mathcal{EAF}([0, 1])$, and $\mathcal{EOWA}$, and $\mathcal{EOWA} \cap \mathcal{EAF}([0, 1])$ are closing by the operation $\wedge$.

It is verified: $Cl(M) \leq \overline{M}$.

Similarly, if M is such that $\min \leq M$, we define the *interior* operators $Ap(M)$ and $\mathring{M}$:

$$Ap(M) = \bigvee \{A \in \mathcal{EOWA} / A \leq M\},$$

$$\mathring{M} = \bigvee \{A \in \mathcal{EOWA} \cap \mathcal{EAF}([0, 1]) / A \leq M\}.$$

Then $\mathring{M} \leq Ap(M)$.

For instance, let $\mathfrak{G} : \bigcup_{n \geq 1} [0, 1]^n \to [0, 1]$ the *geometric mean* operator:

$$\mathfrak{G}(x) = x, \quad \mathfrak{G}(x_1, x_2, \ldots, x_n) = \sqrt[n]{\prod_{i=1}^{n} x_i}, \quad if \ n > 1$$

and let $\mathfrak{M}_{-1} : \bigcup_{n \geq 1} [0, 1]^n \to [0, 1]$ the *harmonic mean* operator such that, for $\overline{x} = (x_1, x_2, \ldots, x_n)$:

$$\mathfrak{M}_{-1}(\overline{x}) = \begin{cases} x_1 & if & n = 1 \\ 0 & if \ (n > 1)\&(\exists x_i : x_i = 0) \\ \frac{n}{\sum_{i=1}^{n} \frac{1}{x_i}} & if \ (n > 1)\&(\forall x_i : x_i \neq 0) \end{cases} .$$

Then [6]: $\min \leq \mathfrak{M}_{-1} \leq \mathfrak{G} \leq \max$, so, there exist $Cl(\mathfrak{M}_{-1})$, $\overline{\mathfrak{M}_{-1}}$, $Ap(\mathfrak{M}_{-1})$, and $\overset{\circ}{\mathfrak{M}}_{-1}$ associated to $\mathfrak{M}_{-1}$ as $Cla(\mathfrak{G})$, $\overline{\mathfrak{G}}$, $Ap(\mathfrak{G})$, and $\overset{\circ}{\mathfrak{G}}$ associated to $\mathfrak{G}$.

Let us see, in the case of closing operators, all matching with the mean operator A_{mean}, i.e., A_{mean} is the smallest EOWA operator, and the smallest extended aggregation function which is bigger that geometric mean, and harmonic mean.

Proposition 7.9 *Let (n, r) be natural numbers such that $1 \leq r < n$ and let $S_{n,r}$ and $T_{n,r}$ subsets of $\mathbb{R}$ such that $S_{n,r} = \{\frac{\varepsilon^{n-r} - \varepsilon^n}{1 - \varepsilon^n} / 0 < \varepsilon < 1\}$ and $T_{n,r} = \{\frac{r\varepsilon}{r\varepsilon + n - r} / 0 < \varepsilon < 1\}$. Then:*

$$\inf S_{n,r} = \inf T_{n,r} = 0 \ y \ \sup S_{n,r} = \sup T_{n,r} = \frac{r}{n}.$$

Proof Let us consider the function $f_{n,r}$ with domain $]0, 1[$ and defined by $f_{n,r}(x) = \frac{x^{n-r} - x^n}{1 - x^n}$. We will prove that it is an increasing monotone function. Its derivative in $]0, 1[$ is given by:

$$\frac{d(f_{n,r})}{dx}(x) = \frac{x^{n-r-1}(n - r + rx^n - nx^r)}{(1 - x^n)^2}$$

$$= \frac{x^{n-r-1}(1 - x)^2 (r \sum_{i=1}^{n-r} i x^{n-i-1} + (n - r) \sum_{j=1}^{r-1} (r - j)x^{r-j-1})}{(1 - x^n)^2},$$

which is a positive value for all $x \in]0, 1[$, so, $f_{n,r}$ is increasing in the considered domain.

As a consequence, we obtain

$\inf S_{n,r} = \lim_{\varepsilon \to 0+} \frac{\varepsilon^{n-r} - \varepsilon^n}{1 - \varepsilon^n} = 0,$

$\sup S_{n,r} = \lim_{\varepsilon \to 1-} \frac{\varepsilon^{n-r} - \varepsilon^n}{1 - \varepsilon^n} = \lim_{\varepsilon \to 1-} \frac{\varepsilon^{n-r}(1 - \varepsilon)(\varepsilon^{r-1} + \varepsilon^{r-2} + \cdots + \varepsilon + 1)}{(1 - \varepsilon)(\varepsilon^{n-1} + \varepsilon^{n-2} + \cdots + \varepsilon + 1)} = \frac{r}{n}.$

Let us consider now the function $g_{n,r}$ whose domain is the open interval $]0, 1[$ and defined by $g_{n,r}(x) = \frac{rx}{rx+n-r}$. Its derivative is $\frac{n-r}{(rx+n-r)^2}$, so $g_{n,r}$ is increasing. As a consequence: $\inf T_{n,r} = \lim\limits_{\varepsilon \to 0+} \frac{r\varepsilon}{r\varepsilon+n-r} = 0$, and $\sup T_{n,r} = \lim\limits_{\varepsilon \to 1-} \frac{r\varepsilon}{r\varepsilon+n-r} = \frac{r}{n}$. $\square$

Proposition 7.10 *It is verified:*

(i) $Cl(\mathfrak{G}) = \overline{\mathfrak{G}} = A_{mean}, \ Cl(\mathfrak{M}_{-1}) = \overline{\mathfrak{M}_{-1}} = A_{mean};$

(ii) $Ap(\mathfrak{G}) = \overset{\circ}{\mathfrak{G}} = \min, \ Ap(\mathfrak{M}_{-1}) = \overset{\circ}{\mathfrak{M}}_{-1} = \min.$

Proof (i) A classical result [6] shows that for all $\overline{x} \in \bigcup\limits_{n \geq 1} [0, 1]^n$ then: $\min(\overline{x}) \leq \mathfrak{M}_{-1}(\overline{x}) \leq \mathfrak{G}(\overline{x}) \leq A_{mean}(\overline{x}) \leq \max(\overline{x})$. So, A_{mean} is an upper bound of $\mathfrak{M}_{-1}$, and of $\mathfrak{G}$ in $\mathscr{EOWA}$, and also in $\mathscr{EOWA} \cap \mathscr{EAF}([0, 1])$:

$$A_{mean} \in \{A \in \mathscr{EOWA} / \mathfrak{G} \leq A\},$$
$$A_{mean} \in \{A \in \mathscr{EOWA} \cap \mathscr{EAF}([0, 1]) / \mathfrak{G} \leq A\},$$
$$A_{mean} \in \{A \in \mathscr{EOWA} / \mathfrak{M}_{-1} \leq A\},$$
$$A_{mean} \in \{A \in \mathscr{EOWA} \cap \mathscr{EAF}([0, 1]) / \mathfrak{M}_{-1} \leq A\}.$$

We will prove that in all cases A_{mean} is the smallest of these upper bounds, and as a consequence, it is the closing operator of $\mathfrak{M}_{-1}$, and of $\mathfrak{G}$ in $\mathscr{EOWA}$, and in $\mathscr{EOWA} \cap \mathscr{EAF}([0, 1])$.

Let A be other upper bound and let $\overline{x} = (x_1, x_2, \ldots, x_n)$ with $n > 1$ be a generic element. According to (7.7), and according to hypothesis, if $F_\triangle$ represents the relation of distribution associated to A, then, in the case of geometric mean $\mathfrak{G}$:

$$\sqrt[n]{\prod_{i=1}^{n} x_i} \leq y_n + \sum_{i=1}^{n-1} F_\triangle(n-i+1, i)(y_i - y_{i+1}). \tag{7.11}$$

Let us suppose $\varepsilon \in]0, 1[$, $n > 1$ and $k < n$, and n-uple

$$\overline{x}_{n,k,\varepsilon} = (\underbrace{1, 1, \ldots, 1}_{k}, \underbrace{\varepsilon^n, \varepsilon^n, \ldots, \varepsilon^n}_{n-k}) \in [0, 1]^n.$$

According to (7.11), for this $\overline{x}_{n,k,\varepsilon}$ we obtain the inequality:

$$\varepsilon^{n-k} \leq \varepsilon^n + F_\triangle(n-k+1, k)(1 - \varepsilon^n),$$

so, for all $\varepsilon \in]0, 1[$ it is verified:
$F_\triangle(n-k+1, k) \geq \frac{\varepsilon^{n-k} - \varepsilon^n}{(1-\varepsilon^n)}$. That is, $F_\triangle(n-k+1, k)$ is an upper bound of bounded subset of real numbers $S_{n,k} = \{\frac{\varepsilon^{n-k} - \varepsilon^n}{(1-\varepsilon^n)} / \varepsilon \in]0, 1[\}$, so it will be greater or equal than $\sup S_{n,k} = \frac{k}{n}$, so we obtain

$F_\triangle(n-k+1,k) \geq \frac{k}{n} = F_{mean}(n-k+1,k)$, as a consequence $F_\triangle(n,m) \geq \frac{m}{n+m-1} = F_{mean}(n,m)$. It proves that A_{mean} operator associated to F_{mean} is the closing of $\mathfrak{G}$ in $\mathscr{EOWA}$ and in $\mathscr{EOWA} \cap \mathscr{EAF}([0,1])$: $Cl(\mathfrak{G}) = \overline{\mathfrak{G}} = A_{mean}$.

In the case of harmonic mean $\mathfrak{M}_{-1}$, let us consider A, (with relation of distribution $F_\triangle$), an upper bound of $\mathfrak{M}_{-1}$. It is verified for all $\overline{x} = (x_1, x_2, \ldots, x_n)$ such that $n > 1$:

$$\frac{n}{\sum\limits_{i=1}^{n} \frac{1}{x_i}} \leq y_n + \sum\limits_{i=1}^{n-1} F_\triangle(n-i+1,i)(y_i - y_{i+1}).$$

Let $\varepsilon \in]0,1[$ and $\overline{z}_{n,k,\varepsilon} = (\underbrace{1,1,\ldots,1}_{k}, \underbrace{\varepsilon,\varepsilon,\ldots,\varepsilon}_{n-k}) \in [0,1]^n$. The before inequality is reduced to

$$\frac{n}{k + \frac{(n-k)}{\varepsilon}} \leq \varepsilon + F_\triangle(n-k+1,k)(1-\varepsilon).$$

So, for $\varepsilon \in]0,1[$ it is verified that $F_\triangle(n-k+1,k) \geq \frac{k\varepsilon}{k\varepsilon+(n-k)}$, and in consequence $F_\triangle(n-k+1,k)$ is greater than $\sup T_{n,k} = \frac{k}{n}$. This result shows that $A \geq \mathfrak{M}_{-1} \Longrightarrow A \geq A_{mean} \geq \mathfrak{M}_{-1}$, i.e., $Cl(\mathfrak{M}_{-1}) = \overline{\mathfrak{M}_{-1}} = A_{mean}$.

(ii) Clearly, min (belongs to $\mathscr{EOWA}$, and to $\mathscr{EOWA} \cap \mathscr{EAF}([0,1])$), is a lower bound of $\mathfrak{G}$. Let now A be another lower bound of $\mathfrak{G}$, and let $F_\triangle$ be the relation of distribution of A. Using the n-uple $\overline{x}_{n,k,\varepsilon}$, we obtain now

$$\varepsilon^{n-k} \geq \varepsilon^n + F_\triangle(n-k+1,k)(1-\varepsilon^n),$$

so, for all $\varepsilon \in]0,1[$ it is verified: $F_\triangle(n-k+1,k) \leq \frac{\varepsilon^{n-k}-\varepsilon^n}{(1-\varepsilon^n)}$. That is, $F_\triangle(n-k+1,k)$ is an lower bound of the bounded subset of real numbers $S_{n,k} = \{\frac{\varepsilon^{n-k}-\varepsilon^n}{(1-\varepsilon^n)} / \varepsilon \in]0,1[\}$, so it will be smaller or equal than $\inf S_{n,k} = 0$, so we obtain $F_\triangle(n-k+1,k) = 0 = F_{(0)}(n-k+1,k)$, being $F_{(0)}$ the relation of distribution associated to min. As a consequence, if $n > 1$, then $F_\triangle(n,m) = 0 = F_{(0)}(n,m)$, it proves that the only element, both in $\mathscr{EOWA}$, and in $\mathscr{EOWA} \cap \mathscr{EAF}([0,1])$, that makes smaller $\mathfrak{G}$ is min, so $Ap(\mathfrak{G}) = \overset{\circ}{\mathfrak{G}} = $ min. A similar reasoning shows that $Ap(\mathfrak{M}_{-1}) = \overset{\circ}{\mathfrak{M}}_{-1} = $ min. $\qquad\qquad\square$

7.7 Conclusions

These results show how the expression of weighting triangles and their EOWA operators associated through relation of distribution let us define new concepts, relations, operations, etc. For all negation in $[0,1]$, then $(\mathscr{DF}, \leq, \wedge, \vee, ', F_{(0)}, F_{(1)})$, $(\mathscr{W}, \preccurlyeq, \curlywedge, \curlyvee, ', \triangle_{(0)}, \triangle_{(1)})$ and $(\mathscr{EOWA}, \leq, \bigwedge, \bigvee, ', \text{min}, \text{max})$, can be considered as extensions of the chain: $([0,1], \leq, \text{min}, \text{max}, ', 0, 1)$. When the negation is the one of Zadeh, these sublattices are also extensions of the lattice of [0, 1] with negation: $(\mathscr{I}([0, 1]), \leq, \wedge, \vee,', F_{(0)}, F_{(1)})$. On other hand, on the structure of $\mathscr{D}\mathscr{F}$, $\mathscr{W}$ and $\mathscr{E}\mathscr{O}\mathscr{W}\mathscr{A}$ it has been introduced a class of operations as extension of others in [0, 1], which are interesting from the point of view of fuzzy logic, as norms, implications, etc. In addition, the relations of distribution turn out to be a practical tool to show some interesting properties, as evidenced by the results of Sect. 7.6. All these structures constitute a set L of assessment of subset $L-$fuzzy $A : X \rightarrow L$ of a referential X, (for instance, *valued-triangle sets* $A_\triangle \in \mathscr{W}^X$). When the Zadeh negation is considered, this class of $L-$fuzzy includes the structure of usual fuzzy subsets $A : X \rightarrow [0, 1]$.

Acknowledgments Tomasa Calvo Sánchez received the support of the Spanish Government projects TIN2012-32482, TIN2013-42795-P, and TIN2014-56381-REDT (LODISCO). She has contributed to this work during her sabbatical year at UIB.

References

1. T. Calvo, G. Mayor, J. Suñer, M. Mas, C. Carbonell, Generation of weighting triangles associated with aggregation functions. Int. J. Uncertain., Fuzziness Knowl.-Based Syst. **8**(4), 417–451 (2000)
2. G. Beliakov, A. Pradera, T. Calvo, *Aggregation Functions: A Guide for Practitioners*, Studies in Fuzziness and Soft Computing, vol. 221 (Springer, Heidelberg, 2007)
3. G. Beliakov, H. Bustince Sola, T. Calvo Sánchez, *A Practical Guide to Averaging Functions*, Studies of Fuzziness and Soft Computing, vol. 329 (Springer, Heilderberg, 2015)
4. G. Mayor, T. Calvo, On Extended Aggregation Functions, *Proc. IFSA'97*, Praga, pp. 282–285 (1997)
5. R.R. Yager, On ordered weighted averaging aggregation operators in multicriteria decision making. IEEE Trans. Syst., Man Cybern. **18**(1), 183–190 (1988)
6. G.H. Hardy, J.E. Littelwood, G. Pólya, *Inequalities* (Cambridge University Press 1934), Second edition 1952. Reprinted 1978

Chapter 8
New Advances in the Aggregation of Asymmetric Distances. The Bounded Case

Isabel Aguiló, Tomasa Calvo Sánchez, Pilar Fuster-Parra, Javier Martín, Jaume Suñer and Oscar Valero

Abstract In 1981, J. Borsík and J. Doboš studied the problem of how to merge, by means of a function, a family of distances into a single one. To this end, they introduced the notion of distance aggregation function and gave a characterization of such functions. Later on, in 2010, the notion of distance aggregation function was extended to the framework of asymmetric distances by G. Mayor and O. Valero. Thus, asymmetric distance aggregation functions were introduced and a characterization of this new type of functions was also given. Concretely, the aforesaid characterization states that the functions which allow to merge a family of asymmetric distances into a single one are exactly those that are amenable, monotone and subadditive. In the present chapter we consider the problem of aggregating a family of bounded asymmetric distances. To this end, the notion of bounded asymmetric distance aggregation function is introduced and a full description of such functions is provided. The obtained results are illustrated by means of examples. Furthermore, the relationship between asymmetric aggregation functions and the bounded ones is discussed.

I. Aguiló · P. Fuster-Parra · J. Martín · J. Suñer · O. Valero (✉)
Universitat de les Illes Balears, Palma de Mallorca, Illes Balears, Spain
e-mail: o.valero@uib.es

I. Aguiló
e-mail: isabel.aguilo@uib.es

P. Fuster-Parra
e-mail: pilar.fuster@uib.es

J. Martín
e-mail: javier.martin@uib.es

J. Suñer
e-mail: jaume.sunyer@uib.es

T. Calvo Sánchez
Universidad de Alcalá, Alcalá de Henares, Madrid, Spain
e-mail: tomasa.calvo@uah.es

T. Calvo Sánchez and J. Torrens Sastre (eds.), *Fuzzy Logic and Information Fusion*,
Studies in Fuzziness and Soft Computing 339, DOI 10.1007/978-3-319-30421-2_8

8.1 Introduction

The interest in the mathematical theory of information aggregation has grown in
the last years because of its wide range of applications to practical problems that
arise in a natural way in areas as, for instance, image processing, decision making,
control theory, medical diagnosis, robotics or biology. In the aforesaid fields, it is
necessary to process incoming data that comes from sources of a different nature in
order to obtain a working conclusion. In such situations the pieces of information
are symbolized via some numerical values. As a consequence the fusion methods
that are based on numerical aggregation operators play a central role in the theory
of information aggregation. A wide class of techniques of aggregation impose a
constraint in order to select the most suitable aggregation operator for the problem
to be solved. In general this constraint consists of considering only those operators
that provide an output data which preserves relevant properties of the input data. A
typical example of this type of situation is given when one wants to merge some kind
of distances in order to obtain a new one which will provide the numerical values
that will be taken into account in order to make a decision.

The problem of merging a collection of distances into a single one was studied
by Borsík and Doboš in [4]. With this aim, they introduced the so-called distance
aggregation functions (metric preserving functions in [4]) and characterized such
functions via the notion of triangle triplets. In order to introduce the Borsík and
Doboš description of the distance aggregation functions, let us recall a few pertinent
concepts.

From now on, we shall use the symbol $\mathbb{R}_+$ to denote the set of nonnegative
real numbers. Following [11] (see also [23]), we will consider the set $\mathbb{R}_+^n = \{\mathbf{a} = (a_1, \ldots, a_n) : a_i \in \mathbb{R}_+\}$ ordered by the pointwise order relation $\preceq$, i.e., $\mathbf{a} \preceq \mathbf{b} \Leftrightarrow a_i \leq b_i$ for all $i = 1, \ldots, n$. Moreover, given $A \subseteq \mathbb{R}_+^n$, a function $f : A \to \mathbb{R}_+$
will be said to be *monotone* provided that $f(\mathbf{a}) \leq f(\mathbf{b})$ for all $\mathbf{a}, \mathbf{b} \in A$ with $\mathbf{a} \preceq \mathbf{b}$.
In addition, on account of [29], a function $f : A \to \mathbb{R}_+$ will be said to be *subadditive*
if $f(\mathbf{a} + \mathbf{b}) \leq f(\mathbf{a}) + f(\mathbf{b})$ for all $\mathbf{a}, \mathbf{b} \in A$ with $\mathbf{a} + \mathbf{b} \in A$, where $+$ stands for the
usual addition on $\mathbb{R}_+^n$.

According to [9], we will say that a function $f : \mathbb{R}_+^n \to \mathbb{R}_+$ is *amenable* whenever
it holds that: $f(\mathbf{a}) = 0 \Leftrightarrow a_1 = a_2 = \cdots = a_n = 0$.

A *distance* (or metric) on a (nonempty) set X is a function $d : X \times X \to \mathbb{R}_+$ such
that for all $x, y, z \in X$:

(i) $d(x, y) = 0 \Leftrightarrow x = y$.
(ii) $d(x, y) = d(y, x)$
(iii) $d(x, z) \leq d(x, y) + d(y, z)$.

A *distance space* (or metric space) is a pair (X, d) such that X is a (nonempty)
set and d is a distance on X.

Following [4] (see also [9]), a function $\Phi : \mathbb{R}_+^n \to \mathbb{R}_+$ is a *distance aggregation
function* provided that the function $D_\Phi : X \times X \to \mathbb{R}^+$ is a distance for every
arbitrary collection of distance spaces $\{(X_i, d_i)\}_{i=1}^n$, where $X = \prod_{i=1}^n X_i$ and

$$D_\Phi(\mathbf{x}, \mathbf{y}) = \Phi\left(d_1(x_1, y_1), \ldots, d_n(x_n, y_n)\right) \qquad (8.1)$$

for all $\mathbf{x} = (x_1, \ldots, x_n), \mathbf{y} = (y_1, \ldots, y_n) \in X$.

A first approach to the description of distance aggregation functions was made by Borsík and Doboš as follows:

Proposition 8.1 *Let $\Phi : \mathbb{R}_+^n \to \mathbb{R}_+$. If Φ is a distance aggregation function, then Φ is amenable.*

Proposition 8.2 *Let $\Phi : \mathbb{R}_+^n \to \mathbb{R}_+$. If Φ is monotone, subadditive and amenable, then Φ is a distance aggregation function.*

Clearly Proposition 8.2 provides a technique to induce functions which are able to merge a collection of distances into a new one. However, it does not yield a characterization of distance aggregation functions. This last fact is due to the existence of distance aggregation functions which are not monotone (see Example 8.1).

Motivated by aforesaid disadvantage, Borsík and Doboš delved further into the study of distance aggregation functions and they gave those conditions that makes a function becomes a distance aggregation function. The aforementioned conditions were given in terms of the so-called triangle triplets. Let us recall that, given $\mathbf{a}, \mathbf{b}, \mathbf{c} \in \mathbb{R}_+^n$, the triplet $(\mathbf{a}, \mathbf{b}, \mathbf{c})$ forms a *triangle triplet* whenever $a_i \le b_i + c_i$, $b_i \le a_i + c_i$ and $c_i \le a_i + b_i$ for all $i = 1, \ldots, n$.

The following result states the announced conditions.

Theorem 8.1 *Let $\Phi : \mathbb{R}_+^n \to \mathbb{R}_+$. Then the below assertions are equivalent:*

(1) Φ is a distance aggregation function.
(2) Φ holds the following properties:

   (2.1) Φ is amenable.
   (2.2) For all $\mathbf{a}, \mathbf{b}, \mathbf{c} \in \mathbb{R}_+^n$, if $(\mathbf{a}, \mathbf{b}, \mathbf{c})$ is a triangle triplet, then so is $(\Phi(\mathbf{a}), \Phi(\mathbf{b}), \Phi(\mathbf{c}))$.

From Theorem 8.1 we immediately obtain the below consequence:

Corollary 8.1 *Every distance aggregation function is subadditive.*

Corollary 8.1 motivates the question of whether the converse of Proposition 8.2 is true in general and, thus, distance aggregation functions are always monotone and subadditive. However, the next example shows that there are distance aggregation functions which are not monotone.

Example 8.1 Consider the function $\Phi_+ : \mathbb{R}_+^2 \to \mathbb{R}_+$ given by $\Phi(0, 0) = 0$ and

$$\Phi(a, b) = \begin{cases} 2 & \text{if first}(a, b) \in]0, 1[\\ 1 & \text{if first}(a, b) \ge 1 \end{cases},$$

where $(a, b) \neq (0, 0)$ and first (a, b) denotes the first value of (a, b) different from 0. Clearly Φ is amenable. Furthermore, it is easily seen that Φ turns triangle triplets into triangle triplets. Hence, by Theorem 8.1, Φ is a distance aggregation function. However, Φ is not monotone. Indeed, $(\frac{1}{2}, \frac{1}{2}) \preceq (1, 1)$ but $\Phi(\frac{1}{2}, \frac{1}{2}) = 2$ and $\Phi(1, 1) = 1$.

Since Borsík and Doboš solved the problem of merging distances, an intense research activity in this direction has been developed. We refer the reader to [9] (and references therein) for a fuller treatment of the topic. A few advances in the study of aggregation of distances have been also published in [5–7, 23].

Inspired, on the one hand, by the work of Borsík and Doboš and motivated, on the other hand, by the usefulness of asymmetric distances in sundry fields of Artificial Intelligence and Computer Science (see, for instance, [10, 12, 20, 21, 24–27, 30]), G. Mayor and O. Valero studied the aggregation problem in the framework of asymmetric distances. Concretely, they introduced the notion of asymmetric distance aggregation function in [16] and they provided a characterization of such functions in the spirit of Theorem 8.1 in [16, 17]. In order to introduce such characterization, let us recall a few concepts about asymmetric distances.

Following [14], an *asymmetric distance* (quasi-metric in [14]) on a (nonempty) set X is a function $d : X \times X \to \mathbb{R}_+$ such that for all $x, y, z \in X$:

(i) $d(x, y) = d(y, x) = 0 \Leftrightarrow x = y$.
(ii) $d(x, z) \leq d(x, y) + d(y, z)$.

An *asymmetric distance space* (quasi-metric space in [14]) is a pair (X, d) such that X is a (nonempty) set and d is an asymmetric distance on X.

The following is an illustrative example of asymmetric distance space.

Example 8.2 Consider the pair $(\mathbb{R}_+, d_{max})$, where $d_{max} : \mathbb{R}_+ \times \mathbb{R}_+ \to \mathbb{R}_+$ is the function given by

$$d_{max}(x, y) = \max\{y - x, 0\}$$

for all $x, y \in \mathbb{R}_+$. On account of [14], $(\mathbb{R}_+, d_{max})$ is an asymmetric distance space.

As usual an asymmetric distance space is called T_1 provided that $d(x, y) = 0 \Rightarrow x = y$.

Of course, a distance on a set X is an asymmetric T_1 distance d on X satisfying, in addition, the following condition for all $x, y \in X$:

(iii) $d(x, y) = d(y, x)$.

According to [16, 17], a function $\Phi : \mathbb{R}_+^n \to \mathbb{R}_+$ is an *asymmetric distance aggregation function* if the function $D_\Phi : X \times X \to \mathbb{R}_+$ is an asymmetric distance for every arbitrary collection of asymmetric distance spaces $\{(X_i, d_i)\}_{i=1}^n$, where

$$X = \prod_{i=1}^n X_i \text{ and}$$

$$D_\Phi(\mathbf{x}, \mathbf{y}) = \Phi\left(d_1(x_1, y_1), \ldots, d_n(x_n, y_n)\right) \qquad (8.2)$$

for all $\mathbf{x} = (x_1, \ldots, x_n), \mathbf{y} = (y_1, \ldots, y_n) \in X$.

Similar to the metric case the following partial description of asymmetric distance spaces can be provided.

Proposition 8.3 *Let $\Phi : \mathbb{R}_+^n \to \mathbb{R}_+$. If Φ is an asymmetric distance aggregation function, then Φ is amenable.*

Proposition 8.4 *Let $\Phi : \mathbb{R}_+^n \to \mathbb{R}_+$. If Φ is monotone, subadditive and amenable, then Φ is an asymmetric distance aggregation function.*

Taking into account Proposition 8.4, it seems natural to pose the following question: Is every asymmetric distance aggregation function always monotone? In the light of the metric case, i.e., there exists metric aggregation functions that are not monotone, we can hope the existence of asymmetric distance aggregation functions that are not monotone. Nevertheless, and surprisingly, this is not the case. Indeed, the following result provides a positive answer to the posed question [16].

Theorem 8.2 *Let $\Phi : \mathbb{R}_+^n \to \mathbb{R}_+$. Then the below assertions are equivalent:*

(1) Φ is an asymmetric distance aggregation function.
(2) Φ is amenable, monotone and subadditive.

In spite of Theorem 8.2 yields a complete description of those functions that are able to merge a collection of asymmetric distances spaces into a new one, a new characterization was proved in terms of triples in order to obtain an extension of Theorem 8.1 to the asymmetric context in [17]. The aforesaid result states the following:

Theorem 8.3 *Let $\Phi : \mathbb{R}_+^n \to \mathbb{R}_+$. Then the below assertions are equivalent:*

(1) Φ is an asymmetric distance aggregation function.
(2) Φ holds the following properties:

  (2.1) Φ is amenable.
  (2.2) For all $\mathbf{a}, \mathbf{b}, \mathbf{c} \in \mathbb{R}_+^n$, if $\mathbf{a} \preceq \mathbf{b} + \mathbf{c}$, then $\Phi(\mathbf{a}) \leq \Phi(\mathbf{b}) + \Phi(\mathbf{c})$.

Clearly Theorem 8.2 is retrieved as a particular case of Theorem 8.3, since those functions satisfying condition (2.2) in statement of Theorem 8.3 are exactly the monotone and subadditive ones.

It must be stressed that from Theorem 8.3 (and Theorem 8.2) one can derive that every asymmetric distance aggregation function is a distance aggregation function. Nonetheless, Example 8.1 provides an instance of distance aggregation function that is not an asymmetric distance aggregation function.

8.2 The Statement of the Problem of Aggregating Bounded Asymmetric Distances

As we have mentioned before, asymmetric distances have shown to be useful in several fields of Artificial Intelligence and Computer Science. In many problems arising in the aforesaid applied areas the used distances enjoy a distinguished property, namely, they are bounded (see, for instance, [12, 15, 18, 27]).

Let us recall that a distance space (X, d) is called *bounded* whenever there exists $c \in \mathbb{R}_{++}$ such that

$$d(x, y) \leq c \tag{8.3}$$

for all $x, y \in X$, where $\mathbb{R}_{++} = \{a \in \mathbb{R}_+ : a > 0\}$ (see [8]). Of course, the preceding notion can be easily extended to the asymmetric context by means of replacing in (8.3) the distance by an asymmetric distance. From now on, we will call *constant of boundedness* of a bounded asymmetric distance space (X, d) to the greatest lower bound of all constants satisfying (8.3). Besides, if $c \in \mathbb{R}_{++}$ is the constant of boundedness of an asymmetric distance space (X, d), then we will say that (X, d) is c-bounded.

The following example yields an instance of bounded asymmetric distance space that will be very useful later on.

Example 8.3 Let $c \in \mathbb{R}_{++}$. Define $d_c : \mathbb{R}_+ \times \mathbb{R}_+ \to \mathbb{R}_+$ by

$$d_c(x, y) = \begin{cases} \min\{y - x, c\} & \text{if } x \leq y \\ c & \text{if } x > y \end{cases}$$

for all $x, y \in \mathbb{R}_+$. According to [28], the pair $(\mathbb{R}_+, d_c)$ is an asymmetric distance space. Clearly the asymmetric distance space $(\mathbb{R}_+, d_c)$ is c-bounded.

Notice that Example 8.2 gives an instance of an asymmetric distance space which is not bounded.

Inspired by the aforementioned applicability of bounded asymmetric distances and taking into account the exposed study about asymmetric distance aggregation functions, it seems interesting to provide a complete description of those functions that allow to generate a bounded asymmetric distance space by means of the fusion of a collection of arbitrary bounded asymmetric distance spaces. Moreover, it seems interesting to discuss whether there exists any relationship between the new type of asymmetric distance aggregation functions and the old one.

In the following sections we will provide the solutions to the posed problems.

8.3 The Solution to the Problem of Aggregating Bounded Asymmetric Distances

In the following we mastermind the aggregation problem for bounded asymmetric distance spaces. To this end, we need to adapt the definition, introduced in Sect. 8.1, of asymmetric distance aggregation function given by Mayor and Valero as follows:

Definition 8.1 Given $c \in \mathbb{R}_{++}$, a mapping $\mathscr{B} : [0, c]^n \to \mathbb{R}_+$ is a bounded asymmetric distance aggregation function provided that the mapping $D_{\mathscr{B}} : X \times X \to \mathbb{R}_+$ is a bounded asymmetric distance for every arbitrary family of c-bounded asymmetric distance spaces $(X_i, d_i)_{i=1}^n$, where $X = \prod_{i=1}^n X_i$ and

$$D_{\mathscr{B}}(\mathbf{x}, \mathbf{y}) = \mathscr{B}\left(d_1(x_1, y_1), \ldots, d_n(x_n, y_n)\right) \tag{8.4}$$

for all $\mathbf{x} = (x_1, \ldots, x_n), \mathbf{y} = (y_1, \ldots, y_n) \in X$.

Notice that Example 8.4 shows that, for each $c \in \mathbb{R}_{++}$, the class of bounded asymmetric aggregation functions is not empty. Moreover, it is interesting to point out that from Definition 8.1 it immediately follows that every bounded asymmetric distance aggregation function transforms a collection of bounded distance spaces into a bounded distance space.

Once introduced the notion of bounded asymmetric aggregation function, the aggregation problem consists of ascertaining a complete description of such functions. In the rest of the section we provide the aforementioned description.

As usual, given $A \subseteq \mathbb{R}_+^n$, a function $f : A \to \mathbb{R}_+$ is *bounded* provided that there exists $d \in \mathbb{R}_+$ such that $f(\mathbf{a}) \leq d$ for all $\mathbf{a} \in A$ [19].

The fact that a bounded asymmetric distance aggregation function can merge a collection of bounded asymmetric distance spaces into a single bounded asymmetric distance space is mainly explained by the next result.

Proposition 8.5 *Let $c \in \mathbb{R}_{++}$. If $\mathscr{B} : [0, c]^n \to \mathbb{R}_+$ is a bounded asymmetric distance aggregation function, then $\mathscr{B}$ is bounded.*

Proof Assume that $\mathscr{B} : [0, c]^n \to \mathbb{R}_+$ is a bounded asymmetric distance aggregation function. Consider the bounded asymmetric distance space $(\mathbb{R}_+, d_c)$ introduced in Example 8.3. Then the pair $([0, c]^n, D_{\mathscr{B}})$ is a bounded asymmetric distance space, where the function $D_{\mathscr{B}} : [0, c]^n \times [0, c]^n \to \mathbb{R}_+$ is the asymmetric distance given by

$$D_{\mathscr{B}}(\mathbf{x}, \mathbf{y}) = \mathscr{B}(d_c(x_1, y_1), \ldots, d_c(x_n, y_n))$$

for all $\mathbf{x} = (x_1, \ldots, x_n), \mathbf{y} = (y_1, \ldots, y_n) \in [0, c]^n$. It follows that there exists $d \in \mathbb{R}_{++}$ such that $D_{\mathscr{B}}(\mathbf{x}, \mathbf{y}) \leq d$ for all $\mathbf{x}, \mathbf{y} \in [0, c]^n$. Moreover, given $\mathbf{a} = (a_1, \ldots, a_n) \in [0, c]^n$, we have guaranteed the existence of $\mathbf{x}^a, \mathbf{y}^a \in [0, c]^n$ such that $d_c(x_i^a, y_i^a) = a_i$ for all $i = 1, \ldots, n$. We conclude that $\mathscr{B}$ is a bounded function, since

$$\mathscr{B}(\mathbf{a}) = \mathscr{B}(a_1, \dots, a_n) = D_{\mathscr{B}}(\mathbf{x}^a, \mathbf{y}^a) \le d$$

for all $\mathbf{a} \in [0, c]^n$.   □

In the light of Proposition 8.5 we have that for each $c \in \mathbb{R}_{++}$ and each bounded asymmetric distance aggregation function $\mathscr{B} : [0, c]^n \rightarrow \mathbb{R}_+$ there always exists $d \in \mathbb{R}_{++}$ such that $\mathscr{B}([0, c]^n) \subseteq [0, d]$. Note that d is the lowest upper bound of the set $\{d \in \mathbb{R}_{++} : f(\mathbf{a}) \le d$ for all $\mathbf{a} \in [0, c]^n\}$.

Proposition 8.6 *Let $c \in \mathbb{R}_{++}$ and let $\mathscr{B} : [0, c]^n \rightarrow \mathbb{R}_+$ be a bounded asymmetric distance aggregation function. Then $\mathscr{B}$ is amenable.*

Proof Let $\mathbf{a} = (a_1, \dots, a_n) \in [0, c]^n$. Consider the bounded metric space $([0, c], d_E)$, where d_E is the Euclidean metric on $[0, c]$. Clearly there exist $x_i^a, y_i^a \in [0, c]$ such that $d_E(x_i^a, y_i^a) = a_i$ for all $i = 1, \dots, n$. In fact, $x_i^a = 0$ and $y_i^a = a_i$ for all $i = 1, \dots, n$. Thus we have that $D_{\mathscr{B}} : [0, c]^n \times [0, c]^n \rightarrow \mathbb{R}_+$ is a bounded distance and that

$$\mathscr{B}(\mathbf{a}) = \mathscr{B}(d_E(x_1^a, y_1^a), \dots, d_E(x_n^a, y_n^a)) = D_{\mathscr{B}}(\mathbf{x}^a, \mathbf{y}^a)$$

for all $\mathbf{a} \in [0, c]^n$. Assume that $\mathscr{B}(\mathbf{a}) = 0$. Then $D_{\mathscr{B}}(\mathbf{x}^a, \mathbf{y}^a) = 0$. It follows that $\mathbf{x}^a = \mathbf{y}^a$. Therefore $x_i^a = y_i^a$ for all $i = 1, \dots, n$. So $0 = x_i^a = y_i^a = a_i$ for all $i = 1, \dots, n$.

The proof of the converse is straightforward.   □

From Proposition 8.6 we obtain the following result for T_1 asymmetric distance spaces.

Corollary 8.2 *Let $c \in \mathbb{R}_{++}$ and let $\mathscr{B} : [0, c]^n \rightarrow \mathbb{R}_+$ be a bounded asymmetric distance aggregation function. If $\{(X_i, d_i)\}_{i=1}^n$ is an arbitrary collection of T_1 c-bounded asymmetric distance spaces, then the bounded asymmetric distance space $(X, D_{\mathscr{B}})$ is T_1, where $X = \prod_{i=1}^n X_i$ and $D_{\mathscr{B}}$ is given by (8.4).*

Proof Suppose that $D_{\mathscr{B}}(\mathbf{x}, \mathbf{y}) = 0$ for any $\mathbf{x}, \mathbf{y} \in X$. Thus

$$\mathscr{B}(d_1(x_1, y_1), \dots, d_n(x_n, y_n)) = D_{\mathscr{B}}(\mathbf{x}, \mathbf{y}) = 0.$$

By Proposition 8.6 we have that $\mathscr{B}$ is amenable and, hence, we deduce that $d_i(x_i, y_i) = 0$ for all $i = 1, \dots, n$. Since each (X_i, d_i) is a T_1 asymmetric distance space we obtain that $x_i = y_i$ for all $i = 1, \dots, n$. Therefore the bounded asymmetric distance space $(X, D_{\mathscr{B}})$ is T_1.   □

The following result will be useful in order to provide the solution of the aggregation problem, i.e., to state a characterization, in the spirit of Theorems 8.2 and 8.3, of bounded asymmetric distance aggregation functions.

Lemma 8.1 *Let $c \in \mathbb{R}_{++}$. Then, for every $a, b, e \in [0, c]$ such that $a \leq b + e$, there exists a bounded asymmetric distance space (X, D) in such a way that there exist $x^{a,b,e}, y^{a,b,e}, z^{a,b,e} \in X$ with $D(x^{a,b,e}, y^{a,b,e}) = a$, $D(x^{a,b,e}, z^{a,b,e}) = b$ and $D(z^{a,b,e}, y^{a,b,e}) = e$.*

Proof Define the mapping $D_c : \mathbb{R}_+^2 \times \mathbb{R}_+^2 \to [0, c]$ by

$$D_c((x_1, x_2), (y_1, y_2)) = \frac{\min\{\max\{y_1 - x_1, 0\}, c\} + \min\{\max\{y_2 - x_2, 0\}, c\}}{2}.$$

It is not hard to check that D_c is a c-bounded asymmetric distance on $\mathbb{R}_+^2$. Besides, it is easily seen that the points of $\mathbb{R}_+^2$, described below, satisfy the required condition.

Case 1. $b \geq a$. Set $\mathbf{x}^{a,b,e} = (-a, 2e)$, $\mathbf{y}^{a,b,e} = (a, 2e)$ and $\mathbf{z}^{a,b,e} = (-a + 2b, 0)$.
Case 2. $b < a$. Set $\mathbf{x}^{a,b,e} = (-a, 0)$, $\mathbf{y}^{a,b,e} = (a, 0)$ and $\mathbf{z}^{a,b,e} = (-a + 2b, 2a - 2b - 2e)$. $\square$

In the next result we provide the desired characterization of bounded asymmetric distance aggregation functions.

Theorem 8.4 *Let $c \in \mathbb{R}_{++}$ and $\mathscr{B} : [0, c]^n \to \mathbb{R}_+$. Then the following assertions are equivalent:*

(1) $\mathscr{B}$ is a bounded asymmetric distance aggregation function.
(2) $\mathscr{B}$ is amenable, bounded and $\mathscr{B}(\mathbf{a}) \leq \mathscr{B}(\mathbf{b}) + \mathscr{B}(\mathbf{e})$ for all $\mathbf{a}, \mathbf{b}, \mathbf{e} \in [0, c]^n$ such that $\mathbf{a} \preceq \mathbf{b} + \mathbf{e}$.
(3) $\mathscr{B}$ is amenable, bounded, monotone and subadditive.

Proof (1) $\Rightarrow$ (2). Assume that $\mathscr{B}$ is a bounded asymmetric distance aggregation function. By Propositions 8.5 and 8.6 we obtain that $\mathscr{B}$ is bounded and amenable, respectively. Next we show that assertions (2) hold. To this end consider the bounded asymmetric distance space $(\mathbb{R}_+^2, D_c)$ given in Lemma 8.1. Hence the function $D_{\mathscr{B}} : (\mathbb{R}_+^2)^n \to \mathbb{R}_+$ given by

$$D_{\mathscr{B}}(\mathbf{x}, \mathbf{y}) = \mathscr{B}(D_c(x_1, y_1), \dots, D_c(x_n, y_n))$$

for all $\mathbf{x}, \mathbf{y} \in (\mathbb{R}_+^2)^n$ is a bounded asymmetric distance. By Lemma 8.1 there exist $\mathbf{x}_i^{a_i,b_i,e_i}, \mathbf{y}_i^{a_i,b_i,e_i}, \mathbf{z}_i^{a_i,b_i,e_i} \in (\mathbb{R}_+^2)^n$ such that $D_c(\mathbf{x}_i^{a_i,b_i,e_i}, \mathbf{y}_i^{a_i,b_i,e_i}) = a_i$, $D_c(\mathbf{x}_i^{a_i,b_i,e_i}, \mathbf{z}_i^{a_i,b_i,e_i}) = b_i$ and $D_c(\mathbf{z}_i^{a_i,b_i,e_i}, \mathbf{y}_i^{a_i,b_i,e_i}) = e_i$ for all $i = 1, \dots, n$. On the one hand, we have that

$$\mathscr{B}(\mathbf{a}) = \mathscr{B}(D_c(\mathbf{x}_1^{a_1,b_1,e_1}, \mathbf{y}_1^{a_1,b_1,e_1}), \dots, D_c(\mathbf{x}_n^{a_n,b_n,e_n}, \mathbf{y}_n^{a_n,b_n,e_n}))$$
$$= D_{\mathscr{B}}(\mathbf{x}^{a,b,e}, \mathbf{y}^{a,b,e}) \leq D_{\mathscr{B}}(\mathbf{x}^{a,b,e}, \mathbf{z}^{a,b,e}) + D_{\mathscr{B}}(\mathbf{z}^{a,b,e}, \mathbf{y}^{a,b,e}).$$

On the other hand, we have that

$$D_{\mathscr{B}}(\mathbf{x}^{a,b,e}, \mathbf{z}^{a,b,e}) = \mathscr{B}(D_c(\mathbf{x}_1^{a_1,b_1,e_1}, \mathbf{z}_1^{a_1,b_1,e_1}), \ldots, D_c(\mathbf{x}_n^{a_n,b_n,e_n}, \mathbf{z}_n^{a_n,b_n,e_n})) = \mathscr{B}(\mathbf{b})$$

and

$$D_{\mathscr{B}}(\mathbf{z}^{a,b,e}, \mathbf{y}^{a,b,e}) = \mathscr{B}(D_c(\mathbf{z}_1^{a_1,b_1,e_1}, \mathbf{y}_1^{a_1,b_1,e_1}), \ldots, D_c(\mathbf{z}_n^{a_n,b_n,e_n}, \mathbf{y}_n^{a_n,b_n,e_n})) = \mathscr{B}(\mathbf{e}).$$

Therefore

$$\mathscr{B}(\mathbf{a}) \le \mathscr{B}(\mathbf{b}) + \mathscr{B}(\mathbf{e}).$$

$(2) \Rightarrow (1)$. Let $\{(X_i, d_i)\}_{i=1}^{n}$ be a family of c-bounded asymmetric distance spaces. Consider $\mathbf{x}, \mathbf{y} \in X = \prod_{i \in \omega} X_i$ such that $D_{\mathscr{B}}(\mathbf{x}, \mathbf{y}) = D_{\mathscr{B}}(\mathbf{y}, \mathbf{x}) = 0$. It follows that

$$\mathscr{B}(d_1(x_1, y_1), \ldots, d_n(x_n, y_n)) = \mathscr{B}(d_1(y_1, x_1), \ldots, d_n(y_n, x_n)) = 0.$$

Since $\mathscr{B}$ is amenable we deduce that $d_i(x_i, y_i) = d_i(y_i, x_i) = 0$ for all $i = 1, \ldots, n$. Whence we obtain that $x_i = y_i$ for all $i = 1, \ldots, n$ and, thus, $\mathbf{x} = \mathbf{y}$.

Since $d_i(x_i, y_i) \le d_i(x_i, z_i) + d_i(z_i, y_i)$ for all $i = 1, \ldots, n$, we have that

$$\mathscr{B}(d_1(x_1, y_1), \ldots, d_n(x_n, y_n))$$
$$\le \mathscr{B}(d_1(x_1, z_1), \ldots, d_n(x_n, z_n)) + \mathscr{B}(d_1(z_1, y_1), \ldots, d_n(z_n, y_n)).$$

Thus,

$$D_{\mathscr{B}}(\mathbf{x}, \mathbf{y}) \le D_{\mathscr{B}}(\mathbf{x}, \mathbf{z}) + D_{\mathscr{B}}(\mathbf{z}, \mathbf{y})$$

for all $\mathbf{x}, \mathbf{y} \in X$.

It remains to prove that the asymmetric distance space $(X, D_{\mathscr{B}})$ is bounded. Since $\mathscr{B}$ is bounded there exists $d \in \mathbb{R}_{++}$ such that $\mathscr{B}(a_1, \ldots, a_n) \le d$ for all $\mathbf{a} \in [0, c]^n$. Whence we deduce that $D_{\mathscr{B}}(\mathbf{x}, \mathbf{y}) = \mathscr{B}(d_1(x_1, y_1), \ldots, d_n(x_n, y_n)) \le d$ for all $\mathbf{x}, \mathbf{y} \in X$.

$(2) \Rightarrow (3)$. We only need to prove that $\mathscr{B}$ is monotone and subadditive. First of all we show that $\mathscr{B}$ is monotone. Indeed, consider $\mathbf{a}, \mathbf{b} \in [0, c]^n$ such that $\mathbf{a} \preceq \mathbf{b}$. Then taking $\mathbf{e} = \mathbf{0}$ in assertion (2) we have that $\mathbf{a} \preceq \mathbf{b} + \mathbf{0}$ and, thus, that $\mathscr{B}(\mathbf{a}) \le \mathscr{B}(\mathbf{b}) + \mathscr{B}(\mathbf{0}) = \mathscr{B}(\mathbf{b})$. Next assume that $\mathbf{a}, \mathbf{b} \in [0, c]^n$ such that $\mathbf{a} + \mathbf{b} \in [0, c]^n$. Since $\mathbf{a} + \mathbf{b} \preceq \mathbf{a} + \mathbf{b}$ assertion (3) yields that $\mathscr{B}(\mathbf{a} + \mathbf{b}) \le \mathscr{B}(\mathbf{a}) + \mathscr{B}(\mathbf{b})$.

$(3) \Rightarrow (2)$. Consider $\mathbf{a}, \mathbf{b}, \mathbf{e} \in [0, c]^n$ such that $\mathbf{a} \preceq \mathbf{b} + \mathbf{e}$. Then we distinguish two cases:

Case 1. $\mathbf{b} + \mathbf{e} \in [0, c]^n$. The monotonicity of $\mathscr{B}$ gives that $\mathscr{B}(\mathbf{a}) \leq \mathscr{B}(\mathbf{b} + \mathbf{e})$. The subadditivity of $\mathscr{B}$ provides that $\mathscr{B}(\mathbf{b} + \mathbf{e}) \leq \mathscr{B}(\mathbf{b}) + \mathscr{B}(\mathbf{e})$. Thus $\mathscr{B}(\mathbf{a}) \leq \mathscr{B}(\mathbf{b}) + \mathscr{B}(\mathbf{e})$ as claimed.

Case 2. $\mathbf{b} + \mathbf{e} \notin [0, c]^n$. Then there exists $i \in \{1, \ldots, n\}$ such that $c < b_i + e_i$. Define $r = b_i - m$, $s = e_i - m$ and $m = \frac{b_i + e_i - c}{2}$. It is clear that $r, s \leq c$ and that $r + s = c$. Thus, by monotonicity of $\mathscr{B}$ we obtain that

$$\mathscr{B}(\mathbf{a}) = \mathscr{B}(a_1, \ldots, a_n) \leq \mathscr{B}(a_1, \ldots, c, \ldots, a_n)$$
$$= \mathscr{B}(a_1, \ldots, r + s, \ldots, a_n) \leq \mathscr{B}(b_1 + e_1, \ldots, r + s, \ldots, a_n + e_n)$$

The subadditivity of $\mathscr{B}$ yields that

$$\mathscr{B}(b_1 + e_1, \ldots, r + s, \ldots, a_n + e_n)$$
$$\leq \mathscr{B}(b_1, \ldots, r, \ldots, a_n) + \mathscr{B}(e_1, \ldots, s, \ldots, e_n)$$
$$\leq \mathscr{B}(b_1, \ldots, b_i, \ldots, a_n) + \mathscr{B}(e_1, \ldots, e_i, \ldots, e_n).$$

Therefore we conclude that

$$\mathscr{B}(\mathbf{a}) \leq \mathscr{B}(\mathbf{b}) + \mathscr{B}(\mathbf{e}). \qquad \square$$

Next we give a few examples of bounded asymmetric distance aggregation functions.

Example 8.4 Let $c \in \mathbb{R}_{++}$. Then the following functions $\mathbb{B} : [0, c]^n \to \mathbb{R}_+$ are bounded asymmetric distance aggregation functions, where:

(1) $\mathscr{B}(\mathbf{a}) = \sum_{i=1}^n w_i a_i$ for all $\mathbf{a} \in [0, c]^n$ and for all $\mathbf{w} \in \mathbb{R}_{++}^n$ (this kind of functions includes the arithmetic mean for which $w_i = \frac{1}{n}$ and $c = 1$ [1]).

(2) $\mathscr{B}(\mathbf{a}) = \max\{w_1 a_1, \ldots, w_n a_n\}$ for all $\mathbf{a} \in [0, c]^n$ and for all $\mathbf{w} \in \mathbb{R}_{++}^n$.

(3) $\mathscr{B}(\mathbf{a}) = \left(\sum_{i=1}^n a_i^p\right)^{\frac{1}{p}}$ for all $\mathbf{a} \in [0, c]^n$ and $p \in [1, \infty[$ (this kind of functions includes the p-means for which $w_i = \frac{1}{n}$ and $c = 1$).

(4) $\mathscr{B}(\mathbf{a}) = \sum_{i=1}^n w_i a_{(i)}$ for all $\mathbf{a} \in [0, c]^n$ and for all $\mathbf{w} \in [0, 1]^n$ with $w_i \geq w_j$ for $i < j$, where $a_{(i)}$ is the i-th largest of the $a_1, \ldots, a_n$ (this kind of functions includes those OWA operators with nondecreasing weights [2]).

(5) $\mathscr{B}(\mathbf{a}) = \min\{c, \sum_{i=1}^n w_i a_i\}$ for $\mathbf{a} \in [0, c]^n$ and for all $\mathbf{w} \in \mathbb{R}_{++}^n$.

(6) $\mathscr{B}(\mathbf{a}) = \begin{cases} 0 & \text{if } a_1 = \cdots = a_n = 0 \\ d & \text{otherwise} \end{cases}$ for all $d \in \mathbb{R}_{++}$.

In addition to the above examples, Aumann functions (see Sect. 8.4) and those n-ary subadditive triangular conorms are another instances of bounded asymmetric distance aggregation functions. Some families of triangular conorms that are subadditive are the following ones (see [13, 23]):

(1) Yager triangular conorms S_λ^Y such that $\lambda \in [1, \infty[$, where $S_\lambda^Y(a, b) = \min\{(a^\lambda + b^\lambda)^{\frac{1}{\lambda}}, 1\}$ with $\lambda \in]0, \infty[$.

(2) Sugeno-Weber triangular conorms S_λ^{SW} such that $\lambda \in [-1, 0]$, where $S_\lambda^{SW}(a, b) = \min\{a + b + \lambda ab, 1\}$ with $\lambda \in [-1, \infty[$.

Besides, the Choquet integral is another important instance of bounded asymmetric distance aggregation functions whenever the corresponding fuzzy measure, that is a normalized capacity according to [11]), is a metric inducing fuzzy measure (MIFM for short) in the sense of [3]. Observe that OWA operators given in statement (4) are retrieved as a particular case of a Choquet integral induced by a MIFM measure.

The next results, which are inspired by those given in [23], provide a little more information on the description of bounded asymmetric distance aggregation functions.

Let us recall that, according to [23], a function $f : [0, c]^n \to \mathbb{R}_+$ has $e \in [0, c]$ as an *absorbent (or annihilator) element in its i-th variable* provided that

$$f(a_1, \ldots, a_{i-1}, e, a_{i+1}, \ldots, a_n) = e$$

for all $a_1, \ldots, a_{i-1}, a_i, \ldots, a_n \in [0, c]$. Following [11], a function $f : [0, c]^n \to \mathbb{R}^+$ has $e \in [0, c]$ as an *absorbent (or annihilator) element* provided that $f(\mathbf{a}) = e$ for all $\mathbf{a} \in [0, c]^n$ such that $e \in \{a_1, \ldots, a_n\}$.

Proposition 8.7 *Let $c \in \mathbb{R}_{++}$ and let $\mathscr{B} : [0, c]^n \to \mathbb{R}^+$ be a bounded asymmetric distance aggregation function. Then $\mathscr{B}$ has not 0 as an absorbent element in any variable.*

Proof Assume that 0 is an absorbent element of $\mathscr{B}$ in some variable. It follows that $\mathscr{B}$ is not amenable. So, by Theorem 8.4, we obtain that $\mathscr{B}$ is not a bounded asymmetric distance aggregation function, which is a contradiction.   □

Considering the unit interval in the preceding result we obtain that n-ary-triangular norms and n-ary weighted geometrical means are not bounded asymmetric distance aggregation functions.

Proposition 8.8 *Let $c, d \in \mathbb{R}_{++}$. If $\mathscr{B} : [0, c]^n \to \mathbb{R}^+$ is a bounded asymmetric distance aggregation function such that d is a bound of $\mathscr{B}$, then $\mathscr{B}$ has not $e \in [0, c]$ as an absorbent element in any variable provided that $d < e$.*

Proof Assume that e is an absorbent element of $\mathscr{B}$ in any variable. We assume without loss of generality that such variable is the first one. Thus we have that $d < \mathscr{B}(e, a_2, \ldots, a_n) = e$ for all $a_2, \ldots, a_n \in [0, c]$. However, the preceding fact is a contradiction because $\mathscr{B}(\mathbf{a}) \leq d$ for all $\mathbf{a} \in [0, c]^n$.   □

On account of [11], a function $f : [0, c]^n \to \mathbb{R}^+$ has $e \in [0, c]$ as an *idempotent element* if $f(e, \ldots, e) = e$.

Proposition 8.9 *Let $c, d \in \mathbb{R}_{++}$ and let $\mathscr{B} : [0, c]^n \to \mathbb{R}_+$ be a bounded asymmetric distance aggregation function. Then $\mathscr{B}$ has not $e \in [0, c]$ as an absorbent element in at least two variables provided that $\mathscr{B}$ has $d \in [0, c]$ as an idempotent element with $e < \frac{d}{2}$.*

Proof Assume that $\mathscr{B}$ has e as absorbent element in two variables. Of course we can assume without loss of generality that such variables are the first ones. Since $\mathscr{B}$ has d as idempotent element we obtain that

$$\mathscr{B}(e + (d - e), e + (d - e), d, \ldots, d) = \mathscr{B}(d, \ldots, d) = d > 2e.$$

By Theorem 8.4 we have that $\mathscr{B}$ is subadditive and, thus, that

$$2e = \mathscr{B}(e, d - e, d, \ldots, d, \ldots, d) + \mathscr{B}(d - e, e, d, \ldots, d) \geq \mathscr{B}(d, \ldots, d) > 2e,$$

which is impossible. □

When the unit interval is under consideration, it immediately follows that n-ary nullnorms are not bounded asymmetric distance aggregation functions when they have as an absorbent element $a \in]0, \frac{1}{2}[$.

A function $f : [0, c]^n \to \mathbb{R}^+$ is called *internal* provided that $\min\{a_1, \ldots, a_n\} \leq f(\mathbf{a}) \leq \max\{a_1, \ldots, a_n\}$ for all $\mathbf{a} \in [0, c]^n$. Moreover, a function $f : [0, c]^n \to \mathbb{R}^+$ is *conjunctive* whenever $f(\mathbf{a}) \leq \min\{a_1, \ldots, a_n\}$ for all $\mathbf{a} \in [0, c]^n$ (see [11]).

Proposition 8.10 *Let $c \in \mathbb{R}_{++}$ and let $\mathscr{B} : [0, c]^n \to \mathbb{R}^+$ be a bounded asymmetric distance aggregation function. Then $\mathscr{B}$ is not conjunctive.*

Proof Assume that $\mathscr{B}$ is conjunctive. Then $\mathscr{B}(0, a, \ldots, a) \leq \min\{0, a\} = 0$ for all $a \in]0, c]$. So $\mathscr{B}$ is not amenable and, by Theorem 8.4, $\mathscr{B}$ is not a bounded asymmetric aggregation function. However, it contradicts our hypothesis. □

When in statement of the preceding result we consider the unit interval, then we immediately deduce that n-quasi-copulas are not bounded asymmetric distance aggregation function, since they are conjunctive (see Theorem 3.80 in [11]).

Of course if a function $f : [0, c]^n \to \mathbb{R}^+$ has every element belonging to $[0, c]$ as an idempotent element, that is $f(a, \ldots, a) = a$ for all $a \in [0, c]$, then the function f is said to be *idempotent* [11].

The next result characterizes the bounded asymmetric distance aggregation functions that are idempotent.

Proposition 8.11 *Let $c \in \mathbb{R}_{++}$ and let $\mathscr{B} : [0, c]^n \to \mathbb{R}^+$ be a bounded asymmetric distance aggregation function. Then the following assertions are equivalent:*

(1) $\mathscr{B}$ is idempotent.
(2) $\mathscr{B}$ is internal.

Proof Proposition 2.63 in [11] guarantees that every monotone function is idempotent if and only if it is internal. By Theorem 8.4 every bounded asymmetric distance aggregation function is monotone. Then the thesis is obtained from the aforesaid Proposition 2.63.   □

Let us recall that, according to [11], a function $f : [0, c]^n \to \mathbb{R}_+$ has $e \in [0, c]$ as a *neutral element* if $f(\mathbf{a}_i \mathbf{e}) = a$ for all $a \in [0, c]$ and $i = 1, \ldots, n$, where $\mathbf{a}_i \mathbf{e}$ denotes the element of $[0, c]^n$ such that the i-th coordinate is a_i and the j-th coordinate with $j \neq i$ is e.

Proposition 8.12 *Let $c \in \mathbb{R}_{++}$ and let $\mathscr{B} : [0, c]^n \to \mathbb{R}_+$ be a bounded asymmetric distance aggregation function. Then does not exist $e \in]0, c]$ such that $\mathscr{B}$ has e as a neutral element.*

Proof Assume for the purpose of contradiction that e is a neutral element of $\mathscr{B}$ with $0 < e$. Then we have that $\mathscr{B}(0, e, \ldots, e) = 0$. It follows that $\mathscr{B}$ is not amenable and, by Theorem 8.4, that $\mathscr{B}$ is not a bounded asymmetric aggregation function, which contradicts our hypothesis.   □

Taking the unit interval in the above result we obtain that n-copulas and n-ary uninorms are not bounded asymmetric aggregation functions, since they have 1 and $e \in]0, 1[$ as neutral element respectively.

In the light of Proposition 8.12 we give a few results that yield more information about the description of bounded asymmetric distance aggregation functions which has 0 as neutral element. Observe that Example 8.4 provides instances of this type of functions.

Proposition 8.13 *Let $c \in \mathbb{R}_{++}$ and let $\mathscr{B} : [0, c]^n \to \mathbb{R}_+$ be a bounded asymmetric distance aggregation function. If $\mathscr{B}$ has 0 as a neutral element, then $\mathscr{B}(\mathbf{a}) \leq \sum_{i=1}^{n} a_i$ for all $\mathbf{a} \in [0, c]^n$.*

Proof The desired conclusion follows from the subadditivity of $\mathscr{B}$, guaranteed by Theorem 8.4, and by the fact that $\mathscr{B}(\mathbf{b}^i{}_i 0) = \mathbf{b}^i{}_i$ for all $i = 1, \ldots, n$, where $\mathbf{b}^i \in [0, c]^n$ with $\mathbf{b}^i = (a_i, \ldots, a_i)$ for all $i = 1, \ldots, n$.   □

Next let us fix a pertinent notion. We will say that $c \in \mathbb{R}_{++}$ is the *constant of boundedness* of a bounded function $f : [0, c]^n \to \mathbb{R}_+$ provided that c is the greatest lower bound of those constants which bound f. In case of c is the constant of boundedness of f will say that f is *c-bounded*.

As a consequence of the above result we obtain the next one.

Corollary 8.3 *Let $c, d \in \mathbb{R}_{++}$ and let $\mathscr{B} : [0, c]^n \to \mathbb{R}^+$ be a bounded asymmetric distance aggregation function which is d-bounded. If $\mathscr{B}$ has 0 as a neutral element, then $d \leq nc$.*

According to [11], a function $f : [0, c]^n \to \mathbb{R}_+$ satisfies the *Lipschitz condition* with respect to a norm $|| \cdot ||$ on $[0, c]^n$ provided that there exists $s \in]0, \infty[$ such that

$$|f(\mathbf{a}) - f(\mathbf{b})| \le s||\mathbf{a} - \mathbf{b}||$$

for all $\mathbf{a}, \mathbf{b} \in [0, c]^n$. Moreover, the greatest lower bound of constants s holding the preceding inequality is called the *Lipschitz constant* of f with respect to $|| \cdot ||$.

The next result states that every bounded asymmetric distance aggregation function with 0 as neutral element satisfies the Lipschitz condition with Lipschitz constant 1 whenever the norm $|| \cdot ||_1$ is considered on $[0, c]^n$, where the norm $|| \cdot ||_1 : [0, c]^n \to \mathbb{R}_+$ is defined by

$$||\mathbf{a}||_1 = \sum_{i=1}^{n} |a_i|$$

for all $\mathbf{a} \in [0, c]^n$.

Proposition 8.14 *Let $c \in \mathbb{R}_{++}$ and let $\mathscr{B} : [0, c]^n \to \mathbb{R}^+$ be a bounded asymmetric distance aggregation function. If $\mathscr{B}$ has 0 as a neutral element, then*

$$|\mathscr{B}(\mathbf{a}) - \mathscr{B}(\mathbf{b})| \le \sum_{i=1}^{n} |b_i - a_i|$$

for all $\mathbf{a}, \mathbf{b} \in [0, c]^n$.

Proof Let $\mathbf{a}, \mathbf{b} \in [0, c]^n$. It is clear that $a_i \le b_i + |b_i - a_i|$ for all $i = 1, \ldots, n$. By assertion (3) in the statement of Theorem 8.4 we have that

$$\mathscr{B}(\mathbf{a}) \le \mathscr{B}(\mathbf{b}) + \mathscr{B}(|b_1 - a_1|, \ldots, |b_n - a_n|).$$

Thus

$$\mathscr{B}(\mathbf{a}) - \mathscr{B}(\mathbf{b}) \le \mathscr{B}(|b_1 - a_1|, \ldots, |b_n - a_n|).$$

Similarly we obtain that

$$\mathscr{B}(\mathbf{b}) - \mathscr{B}(\mathbf{a}) \le \mathscr{B}(|b_1 - a_1|, \ldots, |b_n - a_n|).$$

Hence we have that

$$|\mathscr{B}(\mathbf{a}) - \mathscr{B}(\mathbf{b})| \le \mathscr{B}(|b_1 - a_1|, \ldots, |b_n - a_n|).$$

By Proposition 8.13 we deduce that

$$\mathscr{B}(|b_1 - a_1|, \ldots, |b_n - a_n|) \le \sum_{i=1}^{n} |b_i - a_i|.$$

Finally, we conclude that

$$|\mathscr{B}(\mathbf{a}) - \mathscr{B}(\mathbf{b})| \le \sum_{i=1}^{n} |b_i - a_i|.$$

$\square$

8.3.1 A Refinement of the Problem of Aggregating Bounded Asymmetric Distances

Observe that Definition 8.1 guarantees that bounded asymmetric distance aggregation functions provide bounded asymmetric distance spaces by means of merging a collection of bounded asymmetric distance spaces. However, from Definition 8.1 does not follow that the constant of boundedness of the asymmetric distance space obtained by aggregation is exactly the same than the constant of boundedness of each bounded asymmetric distance space to be merged. In particular, the next example shows that such constants of boundedness do not coincide in general.

Example 8.5 Let $c \in \mathbb{R}_{++}$. Define the function $\mathscr{B} : [0, c]^n \to \mathbb{R}_+$ by

$$\mathscr{B}(\mathbf{a}) = 2 \max\{a_1, \dots, a_n\}.$$

Clearly $\mathscr{B}$ is an amenable, monotone and subadditive function. Moreover, $\mathscr{B}(\mathbf{a}) \le 2c$ for all $\mathbf{a} \in [0, c]^n$. So $\mathscr{B}$ is bounded. Assertion (3) of the statement of Theorem 8.4 guarantees that $\mathscr{B}$ is a bounded asymmetric distance aggregation function. Next consider the collection of c-bounded asymmetric distance spaces $\{(X_i, d_i)\}_{i=1}^{n}$ with $X_i = \mathbb{R}_+$ and $d_i = d_c$ for all $i = 1, \dots, n$. Then the asymmetric distance induced by aggregation of $\{(X_i, d_i)\}_{i=1}^{n}$, $(\mathbb{R}_+^n, D_{\mathscr{B}})$, is a bounded asymmetric distance space. However, $D_{\mathscr{B}}(\mathbf{x}, \mathbf{y}) \le 2c$ for all $\mathbf{x}, \mathbf{y} \in \mathbb{R}_+^n$ and, in addition, $D_{\mathscr{B}}((2, 2, \dots, 2), (1, 1, \dots, 1)) = 2c$. Therefore $(\mathbb{R}_+^n, D_{\mathscr{B}})$ is a $2c$-bounded asymmetric distance space but it is not c-bounded.

In the light of the preceding example we introduce the following refinement of Definition 8.1 in order to be able to discuss when bounded asymmetric distance aggregation functions preserve the constant of boundedness.

Definition 8.2 Given $c \in \mathbb{R}_{++}$, a mapping $\mathscr{B} : [0, c]^n \to \mathbb{R}_+$ is a c-bounded asymmetric distance aggregation function provided that the mapping $D_{\mathscr{B}} : X \times X \to \mathbb{R}_+$ is a c-bounded asymmetric distance for every arbitrary family of c-bounded asymmetric distance spaces $\{(X_i, d_i)\}_{i=1}^{n}$, where $X = \prod_{i=1}^{n} X_i$ and

$$D_{\mathscr{B}}(\mathbf{x}, \mathbf{y}) = \mathscr{B}(d_1(x_1, y_1), \dots, d_n(x_n, y_n))$$

for all $\mathbf{x} = (x_1, \dots, x_n), \mathbf{y} = (y_1, \dots, y_n) \in X$.

In the next example we provide an instance of those bounded asymmetric distance aggregation functions that are c-bounded.

Example 8.6 Consider the probabilistic sum on $[0, 1]$, $\mathscr{S}_P$, that is, the function $\mathscr{S}_P : [0, 1]^2 \to [0, 1]$ defined by

$$\mathscr{S}_P(a, b) = a + b - ab$$

for all $a, b \in [0, 1]$. It is not hard to check that $\mathscr{S}_P$ is subadditive, monotone and amenable. So, by Theorem 8.4, $\mathscr{S}_P$ is a bounded asymmetric distance aggregation function. In fact, the $\mathscr{S}_P$ is a particular case, obtained when $\lambda = -1$, of Sugeno-Weber t-conorm aforementioned in p. 112 [13]. A straightforward computation shows that $\mathscr{S}_P$ is a 1-bounded asymmetric distance aggregation function.

The following result yields a distinguished property which is key to warrant that the asymmetric distance aggregation function preserves the constants of boundedness.

Proposition 8.15 *Let* $c \in \mathbb{R}_{++}$. *If* $\mathscr{B} : [0, c]^n \to \mathbb{R}_+$ *is a* c-*bounded asymmetric distance aggregation function, then* $\mathscr{B}$ *is* c-*bounded.*

Proof By Proposition 8.5 we have that $\mathscr{B}$ is bounded. Next we prove that c is the constant of boundedness of $\mathscr{B}$. With this aim suppose that there exists $d \in \mathbb{R}_{++}$ such that $d < c$ and $\mathscr{B}(\mathbf{a}) \leq d < c$ for all $\mathbf{a} \in [0, c]^n$. It follows that

$$\mathscr{B}(d_c(x_1, y_1), \ldots, d_c(x_n, y_n)) \leq d$$

for all $\mathbf{x}, \mathbf{y} \in [0, c]^n$. Nevertheless the preceding fact is a contradiction, since $\{(X_i, d_i)\}_{i=1}^n$, with $X_i = [0, c]$ and $d_i = d_c|_{[0,c]}$, is a collection of c-bounded asymmetric distance spaces and, thus, the asymmetric distance space induced from it, $([0, c]^n, D_{\mathscr{B}})$, is c-bounded. $\square$

Of course Proposition 8.15 assures that for each $c \in \mathbb{R}_{++}$ and each c-bounded asymmetric distance aggregation function $\mathscr{B} : [0, c]^n \to \mathbb{R}_+$ we have that $\mathscr{B}([0, c]^n) \subseteq [0, c]$.

When one considers c-bounded asymmetric distance functions the following characterization, whose proof is similar to the proof of Theorem 8.4, can be obtained.

Theorem 8.5 *Let* $c \in \mathbb{R}_{++}$ *and* $\mathscr{B} : [0, c]^n \to [0, c]$. *Then the following assertions are equivalent:*

(1) $\mathscr{B}$ is a c-bounded asymmetric distance aggregation function.
(2) $\mathscr{B}$ is amenable, c-bounded and $\mathscr{B}(\mathbf{a}) \leq \mathscr{B}(\mathbf{b}) + \mathscr{B}(\mathbf{e})$ for all $\mathbf{a}, \mathbf{b}, \mathbf{e} \in [0, c]^n$ such that $\mathbf{a} \preceq \mathbf{b} + \mathbf{e}$.
(3) $\mathscr{B}$ is amenable, c-bounded, monotone and subadditive.

Example 8.4 allows us to obtain instances of c-bounded asymmetric distance aggregation functions.

We end the section with the next property of c-bounded asymmetric distance aggregation functions which will play a relevant role in the next section.

Corollary 8.4 *Let* $c \in \mathbb{R}_{++}$. *If* $\mathscr{B} : [0, c]^n \rightarrow [0, c]$ *is a* c-*bounded asymmetric distance aggregation function, then* $\mathscr{B}$ *has* c *as idempotent element.*

Proof By Proposition 8.15 $\mathscr{B}$ is c-bounded. Whence $\mathscr{B}(c, \ldots, c) \leq c$. Assertion (3) in the statement of Theorem 8.5 provides that $\mathscr{B}$ is monotone and, thus, that $\mathscr{B}(\mathbf{a}) \leq \mathscr{B}(c, \ldots, c)$ for all $\mathbf{a} \in [0, c]^n$. Then $\mathscr{B}(c, \ldots, c)$ is a bound of $\mathscr{B}$. It follows that

$$c \leq \mathscr{B}(c, \ldots, c).$$

Therefore $\mathscr{B}(c, \ldots, c) = c$.   □

8.4 Asymmetric Distance Aggregation Functions and Bounded Asymmetric Distance Aggregation Functions: The Relationship

It seems natural to wonder which is the relationship between asymmetric distance aggregation functions and bounded asymmetric distance aggregation functions. We end the chapter giving the answer to the masterminded enquiry in the following two results.

The first one states the relationship between asymmetric distance aggregation functions and the bounded ones.

Theorem 8.6 *Let* $\Phi : \mathbb{R}_+^n \rightarrow \mathbb{R}_+$. *Then the following assertions are equivalent:*

(1) Φ *is an asymmetric distance aggregation function.*
(2) $\Phi|_{[0,c]}$ *is a bounded asymmetric distance aggregation function for all* $c \in \mathbb{R}_{++}$.

Proof (1) $\Rightarrow$ (2). Let $c \in \mathbb{R}_{++}$. By assertion (2) in the statement of Theorem 8.3 we have that Φ is amenable and satisfies

$$\Phi(\mathbf{a}) \leq \Phi(\mathbf{b}) + \Phi(\mathbf{e})$$

for all $\mathbf{a}, \mathbf{b}, \mathbf{e} \in \mathbb{R}_+^n$ such that $\mathbf{a} \preceq \mathbf{b} + \mathbf{e}$. Then $\Phi|_{[0,c]}$ satisfies

$$\Phi|_{[0,c]}(\mathbf{a}) \leq \Phi|_{[0,c]}(\mathbf{b}) + \Phi|_{[0,c]}(\mathbf{e})$$

for all $\mathbf{a}, \mathbf{b}, \mathbf{e} \in [0, c]^n$ such that $\mathbf{a} \preceq \mathbf{b} + \mathbf{e}$. It remains to prove that $\Phi|_{[0,c]}$ is bounded. By assertion (3) in the statement of Theorem 8.2 Φ is monotone. Then $\Phi|_{[0,c]}$ is monotone and, hence, we have that

$$\Phi|_{[0,c]}(\mathbf{a}) \leq \Phi|_{[0,c]}(c, \ldots, c)$$

for all $\mathbf{a} \in [0, c]^n$. It follows that $\Phi|_{[0,c]}$ is bounded. Consequently, by Theorem 8.4, we conclude that $\Phi|_{[0,c]}$ is a bounded asymmetric distance aggregation function.

(2) $\Rightarrow$ (1). It is clear that if $\Phi|_{[0,c]}$ satisfies $\Phi|_{[0,c]}(\mathbf{a}) \leq \Phi|_{[0,c]}(\mathbf{b}) + \Phi|_{[0,c]}(\mathbf{e})$ for all $\mathbf{a}, \mathbf{b}, \mathbf{e} \in [0, c]^n$ such that $\mathbf{a} \preceq \mathbf{b} + \mathbf{e}$ and for all $c \in \mathbb{R}_{++}$, then $\Phi(\mathbf{a}) \leq \Phi(\mathbf{b}) + \Phi(\mathbf{e})$ for all $\mathbf{a}, \mathbf{b}, \mathbf{e} \in \mathbb{R}_+^n$ such that $\mathbf{a} \preceq \mathbf{b} + \mathbf{e}$. Moreover, if $\Phi|_{[0,c]}$ is amenable for each $c \in \mathbb{R}_{++}$, then Φ is amenable. Consequently, by Theorem 8.3, we have that Φ is an asymmetric distance aggregation function. $\square$

The second result clarifies the relationship between asymmetric distance aggregation functions and the c-bounded ones.

Theorem 8.7 *Let $\Phi : \mathbb{R}_+^n \to \mathbb{R}_+$. Then the following assertions are equivalent:*

(1) Φ is an asymmetric distance aggregation function and $c \leq \Phi(c, \ldots, c)$ for all $c \in \mathbb{R}_{++}$.

(2) $\Phi|_{[0,c]}$ is a c-bounded asymmetric distance aggregation function for all $c \in \mathbb{R}_{++}$.

Proof (1) $\Rightarrow$ (2). Let $c \in \mathbb{R}_{++}$. By Theorem 8.6 we have that $\Phi|_{[0,c]}$ is a bounded asymmetric distance aggregation function for all $c \in \mathbb{R}_{++}$. Next we prove that $\Phi|_{[0,c]}$ is c-bounded. To this end, assume for the purpose of contradiction that there exists $d \in]0, c[$ such that $\Phi|_{[0,c]}(\mathbf{a}) \leq d$ for all $\mathbf{a} \in [0, c]^n$. Then

$$c \leq \Phi|_{[0,c]}(c, \ldots, c) \leq d < c$$

which is a contradiction. So $\Phi|_{[0,c]}$ is c-bounded. Assertion (3) in the statement of Theorem 8.5 guarantees that $\Phi|_{[0,c]}$ is a c-bounded asymmetric distance aggregation function.

(2) $\Rightarrow$ (1). Assume that $\Phi|_{[0,c]}$ is a c-bounded asymmetric distance aggregation function for all $c \in \mathbb{R}_{++}$. Then $\Phi|_{[0,c]}$ is a bounded asymmetric distance aggregation function for all $c \in \mathbb{R}_{++}$. Theorem 8.6 gives that Φ is an asymmetric distance aggregation function. Moreover, by Corollary 8.4, we have that $c \leq \Phi|_{[0,c]}(c, \ldots, c)$. Therefore $c \leq \Phi(c, \ldots, c)$ for all $c \in \mathbb{R}_{++}$. $\square$

A natural question is to wonder whether if there exist asymmetric distance aggregation functions that satisfy the condition "$c \leq \Phi(c, \ldots, c)$" or such a condition is too much restrictive. The answer to the preceding question is, of course, affirmative. In particular, every monotone, subadditive function $f : \mathbb{R}_+^n \to \mathbb{R}_+$ such that $\max\{a_1, \ldots, a_n\} \leq f(\mathbf{a})$ for all $\mathbf{a} \in \mathbb{R}_+^n$ is an asymmetric distance aggregation function that satisfies the condition under consideration. A relevant example of this class of asymmetric distance aggregation functions is provided by the so-called Aumann functions.

Let us recall that a function $f : \mathbb{R}_+^n \to \mathbb{R}_+$ is said to be an *Aumann function* provided that f holds the following properties (see [22]):

(i) f is monotone.
(ii) f is subadditive.
(iii) $f(a, 0 \ldots, 0) = f(0, a, 0 \ldots, 0) = \cdots = f(0, \ldots, 0, a)$ for all $a \in \mathbb{R}_+$.

Of course the weighted maximum and the weighted sum (see Example 8.4) are instances of Aumann functions and, thus, asymmetric distance aggregation functions that satisfy the required property "$c \leq \Phi(c, \ldots, c)$". Nevertheless, the family of Aumann functions gives new nontrivial examples of asymmetric distance aggregation functions satisfying the desired property. Concretely, every function $f : \mathbb{R}_+^n \to \mathbb{R}_+$ given by

$$f(\mathbf{a}) = \max \left\{ \max \{a_1, \ldots, a_n\}, h\left(\frac{1}{n}\sum_{i=1}^n a_i\right) \right\} \tag{8.5}$$

for all $\mathbf{a} \in \mathbb{R}_+^n$, where h is a monotone, subadditive function which holds $a \leq h(a) \leq na$ for all $a \in \mathbb{R}_+$. Examples of those functions obtained by (8.5) are, for instance, the functions $g : \mathbb{R}_+^n \to \mathbb{R}_+$ given by

$$g(\mathbf{a}) = \max \left\{ \max \{a_1, \ldots, a_n\}, a_1 + \min\left\{1, \sum_{i=2}^n a_i\right\} \right\}$$

for all $\mathbf{a} \in \mathbb{R}_+^n$.

Acknowledgments The authors would like to express their gratitude for all the affection and the strong support received from Professor Gaspar Mayor over the years and for all his work as a mentor throughout his career. The authors acknowledge the support of the Spanish Ministry of Economy and Competitiveness, under grants TIN2012-32482, TIN2013-42795-P and TIN2014-56381-REDT (LODISCO).

References

1. G. Beliakov, H. Bustince, T. Calvo, *A Practical Guide to Averaging Functions*. Studies of Fuzziness and Soft Computing, vol. 329 (Springer, Heilderberg, 2015)
2. G. Beliakov, A. Pradera, T. Calvo, *Aggregation Functions: A Guide for Practitioners*. Studies in Fuzziness and Soft Computing, vol. 221 (2007)
3. J. Bolton, P. Gader, J.N. Wilson, Discrete Choquet integral as a distance metric. IEEE T. Fuzzy Syst. **16**, 1107–1110 (2008)
4. J. Borsik, J. Doboš, On a product of metric spaces. Math. Slovaca **31**, 193–205 (1981)
5. J. Casasnovas, F. Roselló, Averaging fuzzy biopolymers. Fuzzy Set. Syst. **152**, 139–158 (2005)
6. E. Castiñeira, A. Pradera, E. Trillas, On distances aggregation, in *Proceedings of the Information Processing and Management of Uncertainty in Knowledge-based Systems International Conference* (2000), pp. 693–700
7. E. Castiñeira, A. Pradera, E. Trillas, On the aggregation of some classes of fuzzy relations, in *Technologies for Constructing Intelligent Systems*, ed. by B. Bouchon-Meunier, J. Gutierrez, L. Magdalena, R. Yager (Springer, 2002), pp. 125–147
8. E.T. Copson, *Metric Spaces* (Cambridge University Press, Cambridge, 1968)
9. J. Doboš, *Metric Preserving Functions* (Štroffek, Košice, 1998)

10. L.M. García-Raffi, S. Romaguera, M.P. Schellekens, Applications of the complexity space to the general probabilistic divide and conquer algorithms. J. Math. Anal. Appl. **348**, 346–355 (2008)
11. M. Grabisch, J.L. Marichal, R. Mesiar, E. Pap, *Aggregation Functions* (Cambridge University Press, New York, 2009)
12. P. Hitzler, A.K. Seda, *Mathematical Aspects of Logic Programming Semantics* (CRC Press, Boca Raton, 2010)
13. E.P. Klement, R. Mesiar, E. Pap, *Triangular Norms* (Kluwer Academic Publishers, Dordrecht, 2000)
14. H.P.A. Künzi, Nonsymmetric distances and their associated topologies: about the origins of basic ideas in the area of asymmetric topology, in *Handbook of the History of General Topology*, vol. 3, ed. by C.E. Aull, R. Lowen (Kluwer Academic Publishers, 2001), pp. 853–968
15. J. Martín, G. Mayor, O. Valero, On quasi-metric aggregation functions and fixed point theorems. Fuzzy Set. Syst. **228**, 188–104 (2013)
16. G. Mayor, O. Valero, Aggregating asymmetric distances in computer sciences, in *Computational Intelligence in Decision and Control*, vol. I, ed. by D. Ruan et al. (World Scientific, 2008), pp. 477–482
17. G. Mayor, O. Valero, Aggregation of asymmetric distances in computer science. Inf. Sci. **180**, 803–812 (2010)
18. S.G. Matthews, Partial metric topology. Ann. New York Acad. Sci. **728**, 183–197 (1994)
19. E.A. Ok, *Real Analysis with Economic Applications* (Princeton University Press, New Jersey, 2007)
20. V. Pestov, A. Stojmirović, Indexing schemes for similarity search: an illustrated paradigm. Fund. Inf. **70**, 367–385 (2006)
21. V. Pestov, A. Stojmirović, Indexing schemes for similarity search in datasets of short protein fragments. Inf. Syst. **32**, 1145–1165 (2007)
22. I. Pokorný, Some remarks on metric preserving functions of several variables. Tatra Mt. Math. Publ. **8**, 89–92 (1996)
23. A. Pradera, E. Trillas, A note on pseudometrics aggregation. Int. J. Gen. Syst. **31**, 41–51 (2002)
24. S. Romaguera, E.A. Sánchez-Pérez, O. Valero, Computing complexity distances between algorithms. Kybernetika **39**, 569–582 (2003)
25. S. Romaguera, P. Tirado, O. Valero, New results on mathematical foundations of asymptotic complexity analysis of algorithms via complexity space. Int. J. Comput. Math. **89**, 1728–1741 (2012)
26. S. Romaguera, M. Schellekens, O. Valero, The complexity space of partial functions: a connection between complexity analysis and denotational semantics. Int. J. Comput. Math. **88**, 1819–1829 (2011)
27. S. Romaguera, O. Valero, Asymptotic complexity analysis and denotational semantics for recursive programs based on complexity spaces, in *Semantics-Advances in Theories and Mathematical Models*, vol. 1, ed. by M. Afzal (InTech Open Science, 2012), pp. 99–120
28. S. Romaguera, O. Valero, E.A. Sánchez-Pérez, Quasi-normed monoids and quasi-metrics. Publ. Math.-Debrecen **62**, 53–69 (2003)
29. S. Saminger, R. Mesiar, U. Bodenhofer, Domination of aggregation operators and preservation of transitivity. Int. J. Uncertain. Fuzz. **10**, 11–35 (2002)
30. M. Schellekens, The Smyth completion: a common foundation for denotational semantics and complexity analysis. Electron. Notes Theor. Comput. Sci. **1**, 211–232 (1995)

Chapter 9
Multidistances and Dispersion Measures

**Miguel Martínez-Panero, José Luis García-Lapresta
and Luis Carlos Meneses**

Abstract In this paper, we provide a formal notion of absolute dispersion measure that is satisfied by some classical dispersion measures used in Statistics, such as the range, the variance, the mean deviation and the standard deviation, among others, and also by the absolute Gini index, used in Welfare Economics for measuring inequality. The notion of absolute dispersion measure shares some properties with the notion of multidistance introduced and analyzed by Martín and Mayor in several recent papers. We compare absolute dispersion measures and multidistances and we establish that these two notions are compatible by showing some functions that are simultaneously absolute dispersion measures and multidistances. We also establish that remainders obtained through the dual decomposition of exponential means, introduced by García-Lapresta and Marques Pereira, are absolute dispersion measures up to sign.

9.1 Introduction

In some simple situations, everybody seems to have an intuition about the notion of dispersion, being able to inform if some objects are or not more scattered than others. However, if we aim to measure the magnitude of their spread in order to provide a representation of such perception, even when dealing with mathematical objects or data, many troubles naturally arise. Of course, there exists a well-known approach from Descriptive Statistics, but recently some interesting attempts have been done by extending the usual binary concept of distance to more general settings. The tittle of the seminal paper by Martín and Mayor [15] introducing the so-called multidistances is significative in this sense: "How separated Palma, Inca and Manacor are?".

M. Martínez-Panero · J.L. García-Lapresta (✉)
PRESAD Research Group, BORDA Research Unit, IMUVa,
Dept. de Economía Aplicada, Universidad de Valladolid, Valladolid, Spain
e-mail: lapresta@eco.uva.es

L.C. Meneses
PRESAD Research Group, IMUVa, Dept. de Economía Aplicada,
Universidad de Valladolid, Valladolid, Spain

© Springer International Publishing Switzerland 2016
T. Calvo Sánchez and J. Torrens Sastre (eds.), *Fuzzy Logic and Information Fusion*,
Studies in Fuzziness and Soft Computing 339, DOI 10.1007/978-3-319-30421-2_9

On the other hand, as suggested (without any formal definition) by García-Lapresta and Marques Pereira [9], the remainders of exponential means can be somehow considered as dispersion measures. We now establish this fact regarding the formal definition of an absolute dispersion measure introduced in this paper.

All in all, the mentioned concepts have common links in terms of closeness among different objects. This is the reason why in this paper we have considered their common background in order to establish their formal connections.

The rest of the paper is organized as follows. Section 9.2 introduces the notation and a comprehensive list of properties which will appear along the paper. Some of them are considered in our definition of absolute dispersion measure, in Sect. 9.3, and the fulfillment of some other properties for this measures is checked or tested. Then, in Sect. 9.4 we relate the notion of multidistance with that of absolute dispersion measure, and this relationship is also analyzed, in Sect. 9.5, in connection with the remainders of exponential means. Section 9.6 presents a synoptic diagram of the mentioned relationships, as well as some conjectures for further research, and our concluding remarks.

9.2 Preliminaries

Let I be $[0, 1]$ or $\mathbb{R}$, and $\mathbb{I} = \bigcup_{n \in \mathbb{N}} I^n$. Vectors in I^n are denoted as $\boldsymbol{x} = (x_1, \ldots, x_n)$, $\boldsymbol{0} = (0, \ldots, 0)$, $\boldsymbol{1} = (1, \ldots, 1)$. Accordingly, $x \cdot \boldsymbol{1} = (x, \ldots, x)$ for every $x \in I$.

Given $\boldsymbol{x}, \boldsymbol{y} \in I^n$, by $\boldsymbol{x} \geq \boldsymbol{y}$ we mean $x_i \geq y_i$ for every $i \in \{1, \ldots, n\}$, and by $\boldsymbol{x} > \boldsymbol{y}$ we mean $\boldsymbol{x} \geq \boldsymbol{y}$ and $\boldsymbol{x} \neq \boldsymbol{y}$. Given $\boldsymbol{x} \in I^n$, the increasing reordering of the coordinates of $\boldsymbol{x}$ is indicated as $x_{(1)} \leq \cdots \leq x_{(n)}$. In particular, $x_{(1)} = \min\{x_1, \ldots, x_n\}$ and $x_{(n)} = \max\{x_1, \ldots, x_n\}$. The arithmetic mean of $\boldsymbol{x}$ is symbolized as usual by $\mu(\boldsymbol{x})$. Given a permutation π on $\{1, \ldots, n\}$, we denote $\boldsymbol{x}_\pi = (x_{\pi(1)}, \ldots, x_{\pi(n)})$. Finally, the cardinality of the set $\{x_1, \ldots, x_n\}$ appears as $\#\{x_1, \ldots, x_n\}$.

We begin by defining standard properties of real functions on $\mathbb{R}^n$. For further details the interested reader is referred to Fodor and Roubens [7], Calvo et al. [6], Beliakov et al. [3], García-Lapresta and Marques Pereira [9], Grabisch et al. [11] and Beliakov et al. [2].

Definition 9.1 Let $A : I^n \longrightarrow \mathbb{R}$ be a function.

1. A is *idempotent* if for every $x \in I$ it holds $A(x \cdot \boldsymbol{1}) = x$.
2. A is *symmetric* if for every permutation π on $\{1, \ldots, n\}$ and every $\boldsymbol{x} \in I^n$ it holds $A(\boldsymbol{x}_\pi) = A(\boldsymbol{x})$.
3. A is *monotonic* if for all $\boldsymbol{x}, \boldsymbol{y} \in I^n$ it holds $\boldsymbol{x} \geq \boldsymbol{y} \Rightarrow A(\boldsymbol{x}) \geq A(\boldsymbol{y})$.
4. A is *strictly monotonic* if for all $\boldsymbol{x}, \boldsymbol{y} \in I^n$ it holds $\boldsymbol{x} > \boldsymbol{y} \Rightarrow A(\boldsymbol{x}) > A(\boldsymbol{y})$.
5. A is *compensative* if for every $\boldsymbol{x} \in I^n$ it holds $x_{(1)} \leq A(\boldsymbol{x}) \leq x_{(n)}$.

6. A is *anti-self-dual* if $I = [0, 1]$ and for every $x \in [0, 1]^n$ it holds $A(1 - x) = A(x)$.
7. A is *even* if $I = \mathbb{R}$ and for every $x \in \mathbb{R}^n$ it holds $A(-x) = A(x)$.
8. A is *stable for translations* if for all $x \in I^n$ and $t \in \mathbb{R}$ such that $x + t \cdot 1 \in I^n$ it holds $A(x + t \cdot 1) = A(x) + t$.
9. A is *invariant for translations* if for all $x \in I^n$ and $t \in \mathbb{R}$ such that $x + t \cdot 1 \in I^n$ it holds $A(x + t \cdot 1) = A(x)$.
10. A is *invariant under positive scaling* (or *positively homogeneous of degree* 0) if for all $x \in I^n$ and $\lambda > 0$ such that $\lambda \cdot x \in I^n$ it holds $A(\lambda \cdot x) = A(x)$.

Definition 9.2 Let $A : \mathbb{I} \longrightarrow \mathbb{R}$ be a function.

1. A is *stable* if for all $x \in I^n$ and $i \in \{1, \ldots, n\}$ it holds $A(x, x_i) = A(x)$.
2. A is *contractive* if for all $x \in \mathbb{I}$ there exists $y \in I$ such that $A(x, y) < A(x)$.
3. A is *invariant for replications* if $A(\overbrace{x, \ldots, x}^{m}) = A(x)$ for every $x \in \mathbb{I}$ and any number $m \in \mathbb{N}$ of replications of x.

9.3 Absolute Dispersion Measures

As far as we know, there is no an established notion of absolute dispersion measure in the literature. In this paper we gather some compelling properties that such concept should fulfill. Next, we show some classic statistic estimators which can be understood from this point of view (on this matter, Calot [4] is still useful for further details).

Definition 9.3 A function $D : \mathbb{I} \longrightarrow \mathbb{R}$ is an *absolute dispersion measure* if it satisfies the following conditions

1. *Positiveness*: $D(x) \geq 0$, for every $x \in \mathbb{I}$.
2. *Identity of indiscernibles*: $D(x) = 0 \Leftrightarrow x_1 = \cdots = x_n$, for all $n \in \mathbb{N}$ and $x \in I^n$.
3. *Symmetry*: $D(x_\pi) = D(x)$, for all $n \in \mathbb{N}$, $x \in I^n$ and permutation π on $\{1, \ldots, n\}$.
4. *Invariance for translations*.
5. *Invariance for replications*.
6. *Anti-self-duality* if $I = [0, 1]$ (*evenness* if $I = \mathbb{R}$).

Remark 9.1 If invariance under positive scaling were imposed instead of invariance for translations, then we would move from the scenario of absolute dispersion measures to that of relative ones. However, in this paper we have focused our attention just on the first approach in order to establish connections with multidistances and remainders of exponential means, as will be shown along the paper.

Next we list some classical absolute dispersion measures commonly appearing in the literature. It is easy to check that they verify the above mentioned properties and hence they can also be examined under our approach. Some of them take into account the degree of clustering of the data considering the mean as reference. However, we do no pretend to be exhaustive. For example, some other less known possibilities taking into account the closeness to the median have been avoided.

Definition 9.4 Let $x \in I^n$.

1. The *range* of x is defined as $r(x) = x_{(n)} - x_{(1)}$.
2. The *variance* of x is defined as

$$\sigma^2(x) = \frac{1}{n} \sum_{i=1}^{n} (x_i - \mu(x))^2.$$

3. The *standard deviation* of x is defined as

$$\sigma(x) = \sqrt{\frac{1}{n} \sum_{i=1}^{n} (x_i - \mu(x))^2}.$$

4. The *mean deviation* of x is defined as

$$\mathrm{md}\,(x) = \frac{1}{n} \sum_{i=1}^{n} |x_i - \mu(x)|.$$

5. The Gini index [10], the most popular measure of inequality in Welfare Economics, was introduced by Corrado Gini in 1912. It is based on the average of the absolute differences between all possible pairs of observations (different formulations can be found in Yitzhaki [19] and Aristondo et al. [1, Sect. 3.1], among others).
   The *relative Gini index* is defined as

$$G(x) = \frac{1}{2n^2 \mu(x)} \sum_{i=1}^{n} \sum_{j=1}^{n} |x_i - x_j|, \quad \text{if } \mu(x) \neq 0.$$

The *absolute Gini index* is defined as

$$G_a(x) = \frac{1}{2n^2} \sum_{i=1}^{n} \sum_{j=1}^{n} |x_i - x_j|.$$

Remark 9.2 For $n = 2$, it is interesting to note that both the standard deviation and the mean deviation of $x = (x_1, \ldots, x_n)$ coincide with the *semi-range*, defined as $\frac{x_{(n)} - x_{(1)}}{2}$. In such case, also the mean coincides with the *mid-range* defined as $\frac{x_{(1)} + x_{(n)}}{2}$, and the range becomes four times the absolute Gini index.

Remark 9.3 In the literature it is usual to find the *coefficient of variation* of $x \in \mathbb{I}$ defined as $c_v(x) = \dfrac{\sigma(x)}{|\mu(x)|}$, if $\mu(x) \neq 0$. It is not properly an absolute dispersion measure according to the previous definition, because it vulnerates invariance for translations, and anti-self-duality when $I = [0, 1]$. However, evenness is fulfilled when $I = \mathbb{R}$.

Remark 9.4 It is easy to check that, in the list above, just the range is a stable absolute dispersion measure. On the other hand, according to computer simulations, the variance, standard deviation and mean deviation behave as contractive absolute dispersion measures. Obviously, an open problem consists on providing formal proofs of these facts. In the case of the absolute Gini index, its contractivity can be formally guaranteed in the next section under a multidistance approach.

9.4 Multidistances

The notion of multidistance was introduced by Martín and Mayor [15] from the classical definition of distance between two points (becoming elements in a metric space). In this way, from a more general point of view, these authors consider multidistances among any finite number of points by generalizying the usual triangle inequality (see also Martín and Mayor [16]).

Definition 9.5 A function $M : \mathbb{I} \longrightarrow \mathbb{R}$ is a *multidistance* if it satisfies the following conditions:

1. *Positiveness*: $M(x) \geq 0$, for every $x \in \mathbb{I}$.
2. *Identity of indiscernibles*: $M(x) = 0 \Leftrightarrow x_1 = \cdots = x_n$, for all $n \in \mathbb{N}$ and $x \in I^n$.
3. *Symmetry*: $M(x_\pi) = M(x)$, for all $n \in \mathbb{N}$, $x \in I^n$ and permutation π on $\{1, \ldots, n\}$.
4. *Generalized triangle inequality*: $M(x) \leq M(x_1, y) + \cdots + M(x_n, y)$, for all $n \in \mathbb{N}$, $x \in I^n$ and $y \in I$.

The first examples of multidistances proposed by Martín and Mayor [15] are the *drastic multidistances*. Among them, it is interesting the one defined as follows:

$$D(x) = \#\{x_1, \ldots, x_n\} - 1,$$

which also trivially fulfills all the conditions of absolute dispersion measures.

Another class of multidistances is that of *sum-based multidistances* (see Martín and Mayor [15, 16]), given by any function $D_\lambda : \mathbb{I} \longrightarrow \mathbb{R}$ such that

$$D_\lambda(x) = \begin{cases} 0, & \text{if } n = 1, \\ \lambda(n) \sum_{i<j} |x_i - x_j|, & \text{if } n \geq 2, \end{cases}$$

where $\lambda : \{2, 3, \dots\} \longrightarrow \mathbb{R}$ is any discrete function such that $\lambda(2) = 1$ and $0 < \lambda(n) \leq \frac{1}{n-1}$ for $n > 2$ (this last condition stands for guaranteeing the generalized triangle inequality).

Notice that the absolute Gini index is just twice a multidistance of this family; even more, it is contractive (see Calvo et al. [5, Prop. 8]). However, not all sum-based multidistances are absolute dispersion measures. For example, considering $\lambda(n) = \frac{1}{n-1}$, invariance for replications fails:

$$D_\lambda(x_1, x_2) = \frac{1}{2-1}|x_2 - x_1| = |x_2 - x_1|,$$

whereas

$$D_\lambda(x_1, x_2, x_1, x_2) = \frac{1}{4-1}4|x_2 - x_1| = \frac{4}{3}|x_2 - x_1|.$$

Similarly, usual distances between all possible $\binom{n}{2}$ couples of n points can be used to achieve a multidistance by means of OWA operators. In order to define such *OWA-based multidistances* (see Martín et al. [17]), consider a weighting triangle where the entries are non-negative in each row and they add up to one, i.e.:

$$
\begin{array}{ccccccccc}
 & & & & w_1^1 & & & & \\
 & & & w_1^2 & & w_2^2 & & & \\
 & & w_1^3 & & w_2^3 & & w_3^3 & & \\
 & w_1^4 & & w_2^4 & & w_3^4 & & w_4^4 & \\
w_1^5 & & w_2^5 & & w_3^5 & & w_4^5 & & w_5^5
\end{array}
$$

$\cdots \quad \cdots \quad \cdots \quad \cdots \quad \cdots \quad \cdots \quad \cdots \quad \cdots \quad \cdots$

where $w_j^i \geq 0$ and $\sum_{i=1}^{j} w_i^j = 1$.

Then we can define the function $D_W : \mathbb{I} \longrightarrow \mathbb{R}$ such that

$$
D_W(\boldsymbol{x}) =
\begin{cases}
0, & \text{if } n = 1, \\
\overbrace{W_n \left(|x_2 - x_1|, \dots, |x_n - x_{n-1}|\right)}^{\binom{n}{2}}, & \text{if } n \geq 2,
\end{cases}
$$

where W_n is the OWA operator whose weights are given by the $\binom{n}{2}$th row and $w_1^{\binom{n}{2}} + \dots + w_{n-1}^{\binom{n}{2}} > 0$ for all $n \geq 3$ (see Martín [14]).

Remark 9.5 If all the left weights in the triangle are unitary, i.e., $w_1^{\binom{n}{2}} = 1$ for all $n \in \mathbb{N}$, we obtain the *maximum multidistance* (notice that this case coincides with the range, and hence it is an absolute dispersion measure). However, not all OWA-based multidistance are absolute dispersion measures. For example, if the weights are equal in each row, the corresponding OWA-based multidistance D_W becomes the sum-based multidistance with $\lambda(n) = \frac{1}{\binom{n}{2}} = \frac{2}{n(n-1)}$, which also vulnerates invariance for replications (in this case, it is easy to check that $D_W(x_1, x_2) = |x_2 - x_1| \neq D_W(x_1, x_2, x_1, x_2) = \frac{2}{3}|x_2 - x_1|$).

And also introduced by Martín and Mayor [15, 16], the *Fermat multidistance* $D_F : \mathbb{I} \longrightarrow \mathbb{R}$ is given by:

$$D_F(\boldsymbol{x}) = \min_{x \in I} \left\{ \sum_{i=1}^{n} |x_i - x| \right\} \qquad \text{where } \boldsymbol{x} \in I^n.$$

Once fixed $\boldsymbol{x} \in \mathbb{I}$, such minimum value is effectively reached by any of the points of the *Fermat set* associated with $\{x_1, \ldots, x_n\}$. In our unidimensional context, a classic result (see, for instance, Jackson [12]) provides that such Fermat set is the singleton $\left\{ x_{\left(\frac{n+1}{2}\right)} \right\}$, i.e., the median, if n is odd; or the interval $\left[x_{\left(\frac{n}{2}\right)}, x_{\left(\frac{n+1}{2}\right)} \right]$, if n is even. In this case, it is usual to consider just any of the extremes: $x_{\left(\frac{n}{2}\right)}$, the lower median, or $x_{\left(\frac{n+1}{2}\right)}$, the higher median, or even the average of these two values. In what follows we will choose this last option, calling $\text{med}(\boldsymbol{x})$ the median of $\{x_1, \ldots, x_n\}$. Notice that, as mentioned above, the argument of the minimum appearing in the expression of the Fermat multidistance can be rewritten as $\sum_{i=1}^{n} |x_i - \text{med}(\boldsymbol{x})|$, which is exactly n times the mean deviation with respect to the median.

Proposition 9.1 *The Fermat multidistance is an absolute dispersion measure. Moreover, it is an iterated sum of ranges of the data, where in each term the extreme values are sequentially withdrawn:*

$$D_F(\boldsymbol{x}) = \begin{cases} \left(x_{(n)} - x_{(1)}\right) + \left(x_{(n-1)} - x_{(2)}\right) + \cdots + \left(x_{\left(\frac{n}{2}+1\right)} - x_{\left(\frac{n}{2}\right)}\right), \\ \quad \textit{if } n \textit{ is even}, \\ \\ \left(x_{(n)} - x_{(1)}\right) + \left(x_{(n-1)} - x_{(2)}\right) + \cdots + \left(x_{\left(\frac{n+1}{2}+1\right)} - x_{\left(\frac{n+1}{2}-1\right)}\right), \\ \quad \textit{if } n \textit{ is odd}. \end{cases}$$

Proof The key idea is the very essence of the median, an intermediate value which divides de data in two sets on its left and right sides, each of them which exactly the same number of terms. Then we have

$$D_F(\boldsymbol{x}) = \sum_{i=1}^{n} |x_i - \text{med}(\boldsymbol{x})|$$

$$= \left(x_{(n)} - \text{med}(\boldsymbol{x})\right) + \left(\text{med}(\boldsymbol{x}) - x_{(1)}\right) + \left(x_{(n-1)} - \text{med}(\boldsymbol{x})\right)$$
$$+ \left(\text{med}(\boldsymbol{x}) - x_{(2)}\right) + \cdots = \left(x_{(n)} - x_{(1)}\right) + \left(x_{(n-1)} - x_{(2)}\right) + \cdots ,$$

where the last terms in the corresponding two sums depend on the parity of n. □

Thus, the Fermat multidistance inherits its condition of absolute dispersion measure from the range. And its is also true a sort of reciprocal: the above mentioned fact that the range is also a multidistance.

Remark 9.6 The variance is not a multidistance because the triangle inequality fails:

$$\sigma^2(0, 1) = 0.25 > \sigma^2(0, 0.5) + \sigma^2(1, 0.5) = 0.0625 + 0.0625 = 0.125.$$

On the other side, according to computer simulations, the standard deviation and the mean deviation behave as multidistances, but proofs of these conjectures are yet to be provided.

9.5 Remainders of Exponential Means

In a similar way to the previous scenarios, in what follows we can consider either aggregation functions with a fixed amount $n \in \mathbb{N}$ of input data in the unit interval, or extended aggregation functions defined for any $n \in \mathbb{N}$ (we will not distinguish the notation). Such number n is called the *arity* of the aggregation function.

Definition 9.6 1. A function $A : [0, 1]^n \longrightarrow [0, 1]$ is called an *n-ary aggregation function* if it is monotonic and satisfies $A(\mathbf{1}) = 1$ and $A(\mathbf{0}) = 0$.
2. A function $A : \bigcup_{n \in \mathbb{N}} [0, 1]^n \longrightarrow [0, 1]$ is called an *extended aggregation function* if $A|_{[0,1]^n}$ is an *n*-ary aggregation function for every $n \in \mathbb{N}$.

For the sake of simplicity, the *n*-arity is omitted whenever it is clear from the context.

It is easy to see that for aggregation functions idempotency and compensativeness are equivalent concepts.

9.5.1 The Remainder of an Aggregation Function

We now briefly recall the remainder of an aggregation function appearing in the dual decomposition as introduced by García-Lapresta and Marques Pereira [8, 9].

Definition 9.7 Given an aggregation function $A : [0, 1]^n \longrightarrow [0, 1]$, the function $\widetilde{A} : [0, 1]^n \longrightarrow \mathbb{R}$ defined as

$$\widetilde{A}(\mathbf{x}) = \frac{A(\mathbf{x}) + A(\mathbf{1} - \mathbf{x}) - 1}{2}$$

is called the *remainder* of A.

Clearly, $\widetilde{A}$ is not an aggregation function: $\widetilde{A}(\mathbf{1}) = 0$.

The following result can be found in García-Lapresta and Marques Pereira [9] (excepting that invariance for replications is inherited by the remainder; the proof is immediate).

Proposition 9.2 *The remainder $\widetilde{A}$ inherits from the aggregation function A the properties of continuity, symmetry, invariance for replications, whenever A has these properties.*   □

The following result provide two more properties of the remainder (see García-Lapresta and Marques Pereira [9]).

Proposition 9.3 *Let $A : [0, 1]^n \longrightarrow [0, 1]$ be an aggregation function.*

1. *If A is idempotent, then $\widetilde{A}(x \cdot \mathbf{1}) = 0$ for every $x \in [0, 1]$.*
2. *If A is stable for translations, then $\widetilde{A}$ is invariant for translations.*   □

The first statement establishes that remainders of idempotent aggregation functions are null on the main diagonal. The second statement applies to aggregation functions satisfying stability for translations. In such case, remainders are invariant for translations. These properties of the remainder $\widetilde{A}$ suggest that it may give some information about the dispersion of the coordinates of a vector in $[0, 1]^n$.

9.5.2 Exponential Means

Quasiarithmetic means are the only aggregation functions satisfying continuity, idempotency, symmetry, strict monotonicity and decomposability (see Kolmogoroff [13], Nagumo [18] and Fodor and Roubens [7, pp. 112–114]).

Exponential means are the only quasiarithmetic means satisfying stability for translations.

Given $\alpha \neq 0$, the *exponential mean* $A_\alpha : [0, 1]^n \longrightarrow [0, 1]$ is the aggregation function defined as

$$A_\alpha(\mathbf{x}) = \frac{1}{\alpha} \ln \frac{e^{\alpha x_1} + \cdots + e^{\alpha x_n}}{n}.$$

We now focus on the remainders of exponential means. Given $\alpha \neq 0$, the *remainder* of A_α is the mapping $\widetilde{A}_\alpha : [0, 1]^n \longrightarrow \mathbb{R}$ defined as

$$\widetilde{A}_\alpha(\mathbf{x}) = \frac{1}{2\alpha} \ln \frac{(e^{\alpha x_1} + \cdots + e^{\alpha x_n})(e^{-\alpha x_1} + \cdots + e^{-\alpha x_n})}{n^2}.$$

For every $\alpha \neq 0$, $\widetilde{A}$ satisfies identity of indiscernibles. Moreover, $\widetilde{A}_\alpha$ is continuous, symmetric, anti-self-dual, invariant for translations and invariant for replications (see García-Lapresta and Marques Pereira [9, Sect. 6] for details).

The following result presents the parameter limits of the remainders of exponential means (see García-Lapresta and Marques Pereira [9, Prop. 35]).

Proposition 9.4 *For every* $x \in [0, 1]^n$, *the following statements hold:*

1. $\lim\limits_{\alpha \to \infty} \tilde{A}_\alpha(x) = \dfrac{x_{(n)} - x_{(1)}}{2}$.

2. $\lim\limits_{\alpha \to -\infty} \tilde{A}_\alpha(x) = -\dfrac{x_{(n)} - x_{(1)}}{2}$.

3. $\lim\limits_{\alpha \to 0} \tilde{A}_\alpha(x) = 0$. □

Proposition 9.5 $\tilde{A}_\alpha$ *is an absolute dispersion measure for every* $\alpha > 0$.

Proof It will be shown the only remaining condition, i.e., that A_α is non-negative for every $\alpha > 0$. To this aim, it suffices to check that the argument of the logarithm appearing in the expression of the remainder is not less than. This happens because its numerator is greater than or equal to the denominator:

$$(e^{\alpha x_1} + \cdots + e^{\alpha x_n})(e^{-\alpha x_1} + \cdots + e^{-\alpha x_n})$$

$$= \sum_{i=1}^{n} e^{\alpha(x_i - x_i)} + \sum_{i<j}(e^{\alpha(x_i - x_j)} + e^{\alpha(x_j - x_i)})$$

$$= n + \sum_{i<j} 2\cosh(\alpha(x_i - x_j)) \geq n + 2\frac{(n-1)n}{2} = n + (n-1)n = n^2,$$

where the well known property $\cosh z = \frac{e^z + e^{-z}}{2} \geq 1$ of the hyperbolic cosine has been taken into account in each of the $1 + 2 + \cdots + (n-1) = \frac{(n-1)n}{2}$ terms of the last summation. Hence, the logarithm is always non-negative and the exponential remainder stands as an absolute dispersion measure for $\alpha > 0$. □

Notice that, if $\alpha < 0$, then $\tilde{A}_{-\alpha}(x) = -\tilde{A}_\alpha(x)$, and hence positiveness (required both for absolute dispersion measures and multidistances) is vulnerated. In fact, even for $\alpha > 0$, it is is easy to check that remainders of exponential means are not multidistances. For example, if $\alpha = 0.5$, we obtain:

$$\tilde{A}_{0.5}(0, 1) = 0.06186 > \tilde{A}_{0.5}(0, 0.5) + \tilde{A}_{0.5}(1, 0.5)$$
$$= 0.01358 + 0.01358 = 0.02716$$

and, consequently, the triangle inequality does not hold.

9.6 Concluding Remarks

First, in Fig. 9.1 we show a panoramic diagram connecting the concepts appeared along the paper.

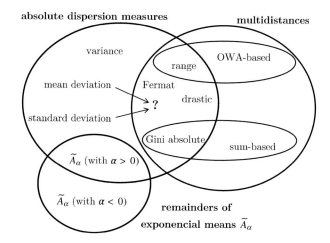

Fig. 9.1 Relationship among concepts

Next, we show a comprehensive table on the fulfillment of several properties considered above. Conjectures appear enclosed in parentheses.

	Symm.	Anti-self-dual.	Evenness	Trans. inv.	Replic. inv.	Contract.	Stabil.
Range	✓	✓	✓	✓	✓	X	✓
Variance	✓	✓	✓	✓	✓	(✓)	X
Standard deviation	✓	✓	✓	✓	✓	(✓)	X
Mean deviation	✓	✓	✓	✓	✓	(✓)	X
Absolute Gini ind.	✓	✓	✓	✓	✓	✓	X
Coef. var.	✓	X	X	X	✓	(✓)	X
Exp. remainder	✓	✓	–	✓	✓	(✓)	X

As commented above, although some open problems remain formally unsolved due to the complexity in their treatment, we consider that our approach opens up wider perspectives and sheds new light into the dispersion analysis.

Acknowledgments This paper is dedicated to Gaspar Mayor, an outstanding researcher and highly excellent person. The authors gratefully acknowledge the comments of Javier Martín, Ana Pérez and Mercedes Prieto, as well as the funding support of the Spanish *Ministerio de Economía y Competitividad* (project ECO2012-32178) and *Consejería de Educación de la Junta de Castilla y León* (project VA066U13).

References

1. O. Aristondo, J.L. García-Lapresta, C. Lasso de la Vega, R.A. Marques Pereira, Classical inequality indices, welfare and illfare functions, and the dual decomposition. Fuzzy Sets Syst. **228**, 114–136 (2013)
2. G. Beliakov, H. Bustince Sola, T. Calvo Sánchez, *A Practical Guide to Averaging Functions* (Springer, Heidelberg, 2015)

3. G. Beliakov, A. Pradera, T. Calvo, *Aggregation Functions: A Guide for Practitioners* (Springer, Heidelberg, 2007)
4. G. Calot, *Cours de Statistique Descriptive* (Dunod, Paris, 1965)
5. T. Calvo, J. Martín, G. Mayor, Measures of disagreement and aggregation of preferences based on multidistances, in *IPMU Part IV, Communications in Computer and Information Science 300*, ed. by S. Greco, et al. (Springer, Berlin, 2012), pp. 549–558
6. T. Calvo, A. Kolesárova, M. Komorníková, R. Mesiar, Aggregation operators: properties, classes and construction methods, in *Aggregation Operators: New Trends and Applications*, ed. by T. Calvo, et al. (Physica-Verlag, Heidelberg, 2002), pp. 3–104
7. J. Fodor, M. Roubens, *Fuzzy Preference Modelling and Multicriteria Decision Support* (Kluwer Academic Publishers, Dordrecht, 1994)
8. J.L. García-Lapresta, R.A. Marques Pereira, Constructing reciprocal and stable aggregation operators, in *Proceedings of International Summer School on Aggregation Operators and their Applications, AGOP 2003* (Alcalá de Henares, Spain, 2003), pp. 73–78
9. J.L. García-Lapresta, R.A. Marques Pereira, The self-dual core and the anti-self-dual remainder of an aggregation operator. Fuzzy Sets Syst. **159**, 47–62 (2008)
10. C. Gini, *Variabilità e Mutabilità* (Tipografia di Paolo Cuppini, Bologna, 1912)
11. M. Grabisch, J.L. Marichal, R. Mesiar, E. Pap, *Aggregation Functions* (Cambridge University Press, Cambridge, 2009)
12. D. Jackson, Note on the median of a set of numbers. Bull. Am. Math. Soc. **27**, 160–164 (1921)
13. A. Kolmogoroff, Sur la notion de la moyenne. Atti della R. Academia Nazionale del Lincei, Rendiconti della Classe di Scienze Fisiche, Mathematiche e Naturali **12**, 388–391 (1930)
14. J. Martín, *Distancias Multiargumento: Tratado General y Aplicaciones*. Ph.D. Thesis, Universitat de les Illes Balears, Palma de Mallorca, Spain
15. J. Martín, G. Mayor, How separated Palma, Inca and Manacor are? in *Proceedings of the International Summer School on Aggregation Operators, AGOP 2009*, (Palma, Spain, 2009), pp. 195–200
16. J. Martín, G. Mayor, Multi-argument distances. Fuzzy Sets Syst. **167**, 92–100 (2011)
17. J. Martín, G. Mayor, O. Valero, Functionally expressible multidistances, in *Proceedings of EUSFLAT-LFA 2011*, pp. 41–46. Atlantis Press (2011)
18. M. Nagumo, Uber eine Klasse der Mittelwerte. Japan. J. Math. **7**, 71–79 (1930)
19. S. Yitzhaki, More than a dozen alternative ways of spelling Gini. Res. Econ. Inequal. **8**, 13–30 (1998)

Chapter 10
Soft Consensus Models in Group Decision Making

Ignacio Javier Perez, Francisco Javier Cabrerizo, Sergio Alonso, Francisco Chiclana and Enrique Herrera-Viedma

Abstract In group decision making problems, when a consensual solution is required, a natural question is how to measure the closeness among experts' opinions in order to obtain the consensus level. To do so, different approaches have been proposed. Following this research line, several authors have introduced hard consensus measures varying between 0 (no consensus or partial consensus) and 1 (full consensus or complete agreement). However, consensus as a full and unanimous agreement is far from being achieved in real situations. So, in practice, a more realistic approach is to use some softer consensus measures, which assess the consensus degree in a more flexible way reflecting better all possible partial agreements obtained through the process. The aim of this chapter is to identify and describe the different existing approaches to compute soft consensus measures in fuzzy group decision making problems. Additionally, we analyze the current models and new challenges on this field.

I.J. Perez
Department of Computer Sciences and Engineering, University of Cadiz, 11519
Puerto Real, Spain
e-mail: ignaciojavier.perez@uca.es

F.J. Cabrerizo
Department of Software Engineering and Computer Systems, Distance Learning
University of Spain, 28040 Madrid, Spain
e-mail: cabrerizo@issi.uned.es

S. Alonso
Department of Software Engineering, University of Granada, 18071 Granada, Spain
e-mail: zerjioi@decsai.ugr.es

F. Chiclana
Faculty of Technology, Centre for Computational Intelligence, De Montfort University,
Leicester LE1 9BH, UK
e-mail: chiclana@dmu.ac.uk

E. Herrera-Viedma (✉)
Department of Computer Science and Artificial Intelligence, University of Granada,
18071 Granada, Spain
e-mail: viedma@decsai.ugr.es

© Springer International Publishing Switzerland 2016

T. Calvo Sánchez and J. Torrens Sastre (eds.), *Fuzzy Logic and Information Fusion*,
Studies in Fuzziness and Soft Computing 339, DOI 10.1007/978-3-319-30421-2_10

10.1   Introduction

In a classical Group Decision Making (GDM) situation there is a problem to solve, a solution set of possible alternatives, and a group of two or more experts, characterized by their own ideas, attitudes, motivations and knowledge, who express their opinions about this set of alternatives in order to achieve a common solution [47, 51, 52].

In an ideal world, each GDM process should be solved in a participatory way, where each individual opinion is taken into account. This idea motivates the concepts of majority and consensus. The term "consensus" has been used for years, even centuries, in a variety of context and areas. It is clear that, ideally, consensus should refer to unanimity of individuals because the option or course of action attained will be best representative for the whole group. Obviously, unanimity may be difficult to obtain.

Ideally, all consensus reaching processes should proceed in a convergent multi-stage way, i.e., the individuals change their opinions step by step until the required consensus level is reached [7, 63]. Of course, this presupposes that the individuals are committed to those changes. Usually, that process is driven by a moderator, who is responsible for driving the consensus reaching session in question by persuading the individuals to change their preferences by rational argument, persuasion, etc., and keeping the process within a period of time considered [7, 30, 31, 41, 42].

We can easily observe that there are several problems related to the existing consensus reaching process. First, a crucial point is the meaning of consensus. Consensus has been traditionally defined as a full and unanimous agreement. For example, several authors have introduced consensus measures assuming values in-between 0 (no consensus or partial consensus) and 1 (full consensus) [5, 64]. However, it has been considered unreachable in the most of real world situations. Due to some inherent differences, individuals rarely arrive at that unanimous agreement, and even if this were the case, the consensus reaching process could be too long for practical purposes. In such a way, the consensus meaning can be softer and it can be viewed not necessarily as a full and unanimous agreement. It can be admitted that the individuals are not willing to fully change their preferences so that consensus will not be a unanimous agreement.

Other relevant problem is the modeling of the consensus reaching processes. Basically, the starting point is here a set of preferences provided by the particular individuals which can be represented and processed using any preferences representation structure and computing tools. In general, they have been assumed to be in matrix form [17, 27, 28]. More recently, preferences have been more and more popular as a more flexible and less formally restrictive representation [41, 42].

The last problem is related with the management of human subjectivity in the consensus process. Since the process of decision making, in particular of GDM, is centered on human beings, coming with their inherent subjectivity, imprecision and vagueness in the articulation of opinions, the theory of fuzzy sets, introduced by Zadeh in [74], has delivered new tools in this field for a long time, as it is a more adequate tool to represent often not clear-cut human preferences encountered in most practical cases. The arrival of fuzzy logic and fuzzy preference relations has changed the field of GDM and consensus. Fuzzy logic can play here a considerable role by providing means for the representation and processing of imprecise information and preferences [25].

Given the importance of obtaining an accepted solution by the whole group, the consensus has had a great attention and it is one of the major goals of GDM problems. The objective of this chapter is to present a comprehensive presentation of the state of the art of all kinds of consensus related problems along the above mentioned situations, with a deep analysis of the respective problems and solutions as well as more relevant new challenges. In particular, we focus on the consensus models based on the concept of fuzzy majority, which is more human-consistent and suitable for reflecting human perceptions of the meaning of consensus.

The chapter is set out as follows. In Sect. 10.2, we describe the usual fuzzy GDM framework and the consensus reaching processes based on moderator. In Sect. 10.3, the main fuzzy consensus models are described. The new challenges in the development of consensus models are shown in Sect. 10.4. Finally, in Sect. 10.5, some concluding remarks are pointed out.

10.2 Background

In this section, we introduce some concepts and approaches about GDM problems and the consensus reaching process.

10.2.1 Group Decision Making Problem

A decision making process, consisting in deriving the best option from a feasible set, is present in just about every conceivable human task. It is obvious that the comparison of different actions according to their desirability in decision problems, in many cases, it cannot be done by using a single criterion or an unique person. Thus, we interpret the decision process in the framework of GDM.

There have been several efforts in the specialized literature to create different models to correctly address and solve GDM situations. Some of them make use of fuzzy theory as it is a good tool to model and deal with vague or imprecise opinions [26, 30, 33, 42, 61, 75].

As we have already mentioned, in a classical GDM situation there is a problem to solve, a solution set of possible alternatives, $X = \{x_1, x_2, \ldots, x_n\}$, $(n \geq 2)$ and a group of two or more experts, $E = \{e_1, e_2, \ldots, e_m\}$, $(m \geq 2)$ characterized by their own ideas, attitudes, motivations and knowledge, who express their opinions about this set of alternatives to achieve a common solution [47, 51, 52]. To do this, each expert has to express his/her preferences on the set of alternatives by means of any preference representation format.

There exist several preference representation structures that can be used by experts to express their opinions about the alternatives in a GDM problem. The most common ones that have been widely used in the literature are:

- *Preference orderings*. The preferences of an expert $e_l \in E$ about a set of feasible alternatives X are described as a preference ordering $O^l = \{o^l(1), \ldots, o^l(n)\}$ where $o^l(\cdot)$ is a permutation function over the indexes set $\{1, \ldots, n\}$ for this expert [65]. Thus, an expert gives an ordered vector of alternatives from best to worst.
- *Utility values*. An expert $e_l \in E$ provides his/her preferences about a set of feasible alternatives X by means of a set of n utility values $U^l = \{u^l_1, \ldots, u^l_n\}, u^l_i \in [0, 1]$, the higher the value for an alternative, the better it satisfies experts' objective [35].
- *Preference relations*. In this case, expert's preferences on X are described by means of a $P^l \subset X \times X$ characterized by a function $\mu_{P^l} : X \times X \to D$ where $\mu_{P^l}(x_i, x_j) = p^l_{ij}$ can be interpreted as the preference degree or intensity of the alternative x_i over x_j expressed in the information representation domain D. Preference relations are the representation format most used in GDM. Different types of preference relations can be used according to the domain used to evaluate the intensity of the preference:

1. *Fuzzy preference relations* [39]: If $D = [0, 1]$ every value p^l_{ij} in the matrix P^l represents the preference degree or intensity of preference of the alternative x_i over x_j: $p^l_{ij} = 1/2$ indicates indifference between x_i and x_j, $p^l_{ij} = 1$ indicates that x_i is absolutely preferred to x_j, and $p_{ij} > 1/2$ indicates that x_i is preferred to x_k. It is usual to assume the additive reciprocity property $p^l_{ij} + p^l_{ji} = 1 \ \forall i, j$.
2. *Multiplicative preference relations* [62]: If $D = [1/9, 9]$ and then every value p^l_{ij} in the matrix P^l represents a ratio of the preference intensity of the alternative x_i to that of x_j, i.e., it is interpreted as x_i is p^l_{ij} times good as x_j: $p^l_{ij} = 1$ indicates indifference between x_i and x_j, $p^l_{ij} = 9$ indicates that x_i is unanimously preferred to x_j, and $p^l_{ij} \in \{2, 3, \ldots, 8\}$ indicates intermediate evaluations. It is usual to assume the multiplicative reciprocity property $p^l_{ij} \cdot p^l_{ji} = 1 \ \forall i, j$ too.
3. *Linguistic preference relations* [30, 31]: If $D = S$, where S is a linguistic term set $S = \{s_0, \ldots, s_g\}$ with odd cardinality $(g + 1)$, $s_{g/2}$ being a neutral label (meaning "equally preferred") and the rest of labels distributed homogeneously around it, then every value p^l_{ij} in the matrix P^l represents the linguistic preference degree or linguistic intensity of preference of the alternative x_i over x_j.

GDM problems are roughly classified into two groups: homogeneous and heterogeneous [14, 32]. A GDM problem is heterogeneous when the opinions of the experts are not equally important. On the contrary, if every opinion is treated equally, we face an homogeneous GDM problem. A way to implement experts' heterogeneity is to assign a weight to every individual. Weights are qualitative or quantitative values that can be assigned in several different ways [14]: a moderator can assign them directly, or the weights can be obtained automatically from the preference expressed by the experts (for example, the most consistent experts could received a higher weight that inconsistent one). The weights can be interpreted as a fuzzy subset, I, with a

Directive	Individual Dominance
↑↓	Minority Influence
	Majority Rules
Participatory	Consensus

Fig. 10.1 GDM methods

membership function, $\mu_I : E \rightarrow [0, 1]$, in such a way that $\mu_I(e_l) \in [0, 1]$ denotes the importance degree of the expert within the group, or how relevant is the person in relation with the problem to be solved [20, 21].

Moreover, there are several methods that can be applied in order to solve GDM problems. These methods can be classified along a spectrum, from directive to participatory decision making (see Fig. 10.1). The methods that are closer to the directive range, mean that the decision is made by a limited, small number of decision makers in the group. For example, individual dominance method, where one person in the group has the authority or power to make the final decision, or minority influence method, that usually takes the form of decisions delegated from larger groups and made by sub-committees [1]. On the contrary, the methods that are lower on the spectrum, towards the participatory range, mean that the decision is made by all the involved parties. For example, majority rules method usually involves the group voting on the alternatives and the alternative receiving the most votes, wins, or consensus method where the consensual agreement is achieved through group discussion of the alternatives, where every group member can agree on an option and commit to the unanimous outcome.

10.2.2   The Consensus Reaching Process

In this chapter we assume the widely studied consensus method (see Fig. 10.2) [30, 35, 37, 42], going one step further and analyzing the way that the consensus measures are obtained. Usual resolution methods for GDM problems based on the consensus strategy are composed by two different processes [30] (see Fig. 10.3):

1. *Consensus process:* Clearly, in any decision process, it is preferable that the experts reach a high degree of consensus on the solution set of alternatives. Thus, this process refers to how to obtain the maximum degree of consensus or agreement among the experts on the solution alternatives.
2. *Selection process:* This process consists in how to obtain the solution set (ranking) of alternatives from the opinions on the alternatives given by the experts.

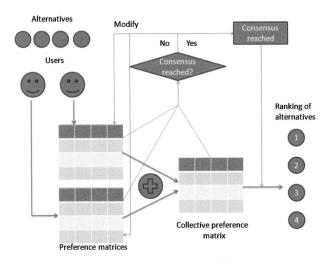

Fig. 10.2 Resolution process of a GDM

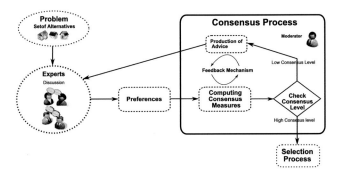

Fig. 10.3 Classical consensus reaching process

A consensus reaching process in a GDM problem is an iterative process composed by several discussion rounds, in which experts are expected to modify their preferences according to the advice given by the moderator.

Usually, to achieve consensus among the experts, it is necessary to provide the whole group of experts with some advice (feedback information) on how far the group is from consensus, what are the most controversial issues (alternatives), whose preferences are in the highest disagreement with the rest of the group, how their change would influence the consensus degree, and so on.

In such a way, the moderator plays a key role in this process. Normally, the moderator is a person who does not participate in the discussion but knows the preferences of each expert and the level of agreement during the consensus process. He is in charge of supervising and driving the consensus process toward success, i.e., to achieve the maximum possible agreement and reduce the number of experts outside of the consensus in each new consensus round.

Usually, the moderator carries out three main tasks (See Fig. 10.3):

1. To compute the consensus measures.
2. To check the level of agreement.
3. To produce some advice for those experts that should change their minds.

10.3   Classical Consensus Models

Currently, we can observe that the soft consensus models have been a hot topic in the last years [34, 53]. In the recent related literature we can find different consensus models according to different criteria as reference domain, concept of coincidence, generation method of recommendations, and guidance measures. According to the reference domain used to compute the soft consensus measures, we find some approaches based on the expert set and others on the alternative set. According to the coincidence concept used to compute the soft consensus measures, we find some consensus models based on strict coincidence among preferences, consensus models based on soft coincidence among preferences, and consensus models based on coincidence among solutions. And according to the generation method of recommendations we find some models in which the moderator is who guides the consensus reaching process and generates the recommendations to the experts to increase the consensus level in the next round of consensus, and on the other hand, we also find other models in which the process is guided automatically, without moderator's participation. Finally, according to the guiding measures we can find some consensus models guided only by soft consensus measures and consensus models guided using various kind of measures, as for example, both soft consensus and consistency measures. In this section, we describe these different approaches.

10.3.1   Reference Domain Based Consensus Models

As we have already mentioned, we can find two different kinds of consensus models according to the reference domain used to obtain the consensus measures. On the one hand, consensus measures focused on the expert set have been proposed in [13, 24, 41–43].

On the other hand, consensus measures focused on the alternative set have been proposed in [8, 9, 30, 31, 33, 37]. In these consensus models, the consensus measures are computed at the three different levels of representation of a preference relation:

1. Level of preference, which indicates the consensus degree existing among all the m preference values attributed by the m experts to a specific preference.
2. Level of alternative, which allows us to measure the consensus existing over all the alternative pairs where a given alternative is present.
3. Level of preference relation, which evaluates the social consensus, that is, the current consensus existing among all the experts about all the preferences.

This measure structure allows us to find out the consensus state of the process at different levels. For instance, it is possible to identify which experts are close to the consensual solutions, or in which alternatives the experts are having more trouble to reach consensus.

10.3.2 Coincidence Method Based Consensus Models

As aforementioned, soft consensus measures are based on the coincidence concept [31], and we can identify three different consensus approaches to compute them: (1) *consensus models based on strict coincidence among preferences*, (2) *consensus models based on soft coincidence among preferences*, and (3) *consensus models based on coincidence among on solutions*. We describe them in more detail in the following subsections [10]:

1. *Consensus models based on Strict coincidence among preferences.* This consensus approach assumes only two possible values: 1 if the opinions are equal and, otherwise, a value of 0. Therefore, the advantage of this approach is that the computation of the consensus degrees is simple and easy. However, the drawback of this approach is that the consensus degrees obtained do not reflect the real consensus situation because it only assigns values of 1 or 0 when comparing the experts' opinions, and, for example, we obtain a consensus value 0 for two different preference situations as (very high, high) and (very high, low), when clearly in the second case the consensus value should be lower than in the first case. Some consensus models using this method can be found in [31, 40].
2. *Consensus models based on Soft coincidence among preferences.* In this approach, similarity criteria among preferences are used to compute the coincidence concept but, in this case, it is assumed that the coincidence concept is a gradual concept which could be assessed with different degrees defined in [0, 1]. The advantage of this approach is that the consensus degrees obtained reflect better the real consensus situation. However, the drawback of this approach is that the computation of the consensus degrees is more difficult because we need to define similarity criteria to compute the consensus measures, and, sometimes it is not possible to define these similarity measures directly. Some examples of consensus models using this method are the following [6, 8, 23, 33, 37, 40–42]. It is worth noting that is the most popular method to compute the soft consensus measures.
3. *Consensus models based on coincidence among solutions.* In this case, similarity criteria among the solutions obtained from the experts' preferences are used to compute the coincidence concept and different degrees assessed in [0, 1] are assumed [4, 35]. Basically, the advantage of this approach is that the consensus degrees are obtained comparing not the opinions but the position of the alternatives in each solution, what allows us to reflect the real consensus situation in each moment of the consensus reaching process. However, the drawback of this approach is that the computation of the consensus degrees is more difficult than

in the above approaches because we need to define similarity criteria and it is necessary to apply a selection process before obtaining the consensus degrees. It is worth noting that the computation of the consensus degrees is more complex.

10.3.3   Generation of Recommendations Method Based Consensus Models

The generation of recommendations to the experts is an important point in order to increase the consensus level in the different consensus rounds. From this point of view, the first consensus models proposed in the literature [6, 30, 31, 40–42, 44] can be considered as basic approaches based on a moderator. In such consensus models the moderator is the person that is monitoring the agreement in each moment of the consensus process and is in charge of supervising and addressing the consensus process towards success, i.e., to achieve the maximum possible agreement and reduce the number of experts outside of the consensus in each new consensus round. However, the moderator can introduce some subjectivity in the process. To overcome this problem, making more effective and efficient the decision making processes, new consensus models have been proposed by substituting the moderator figure or providing moderator with better analysis tools.

On the one hand, consensus models incorporating a feedback mechanism substituting the moderator's actions have been developed [11, 33, 35, 37, 59]. In these approaches, proximity measures are calculated to evaluate the proximity between the individual experts' preferences and the collective one. These proximity measures allow to identify the preference values provided by the experts that are contributing less to reach a high consensus state. In such a way, the feedback mechanism gives advice to those experts to find out the changes they need to make in their opinions to obtain a solution with better consensus degree.

Finally, consensus models have been proposed using a novel data mining tool [46], the so called action rules [55], to stimulate and support the discussion in the group. The purpose of an action rule is to show how a subset of flexible attributes should be changed to obtain an expected change of the decision attribute for a subset of objects characterized by some values of the subset of stable attributes. According to it, these action rules are used to indicate and suggest to the moderator with which experts and with respect to which options it may be expedient to deal.

10.3.4   Guidance Measures Based Consensus Models

Using preference relations, the pairwise comparison helps the experts to provide their preferences by focusing only on two elements once at a time. Thus this allows by reducing uncertainty and hesitation while leading to the higher of consistency. However, the definition of a preference relation does not imply any kind of consistency property, and, due to the complexity of most GDM problems, experts' preferences can be inconsistent [15, 22, 73]. Fortunately, the lack of consistency can be quantified and monitored [18, 31, 36], and it has been used as a parameter to validate the final solution obtained after consensus reaching process [14, 22, 31, 73]. Some consensus models, using both consistency and consensus measures to guide the consensus processes have been proposed in [11, 31, 33]. Usually, in these approaches a consensus/consistency level is calculated as a weighted aggregation of the consistency level and the consensus degree, and it is used as a control parameter to decide if the consensus reaching process must finish and the selection process can be applied. The use of the consistency measures in the consensus models avoids misleading solutions, which cannot be detected by the consensus models using only consensus degrees [6, 30, 37, 40, 41, 44].

10.4   New Challenges on Consensus Models

Once the classical consensus models have been described, it is worth noting that in the most recent literature, we can find some challenges on consensus models which still have to be solved. In such a way, some new consensus models have arisen as a consequence of the new features of the modern real-world applications. In this section we have identify the following ones:

- Social networks and trust based consensus models.
- Agent theory based consensus models.
- Heterogeneous contexts based consensus models.
- Dynamic and changeable consensus models.
- Visualizations tools based consensus models.
- New preference structures based consensus models.
- New guidance measures based consensus models.
- Adaptive consensus models.

In the following subsections, we analyze all of them.

10.4.1   Social Networks and Trust Based Consensus Models

Social networks, [70], present some characteristics that differentiate them to the situations in which the consensus models existing in the literature have usually been applied. For instance, on the one hand, social networks present thousands of users, but it is possible that many of them do not directly participate in the decision process. On the other hand, it is a common issue that some of the users might not be able to collaborate during a whole decision process, but only in a part. In addition, there is a real time communication among its members and it is typical that users exchange opinions through their interaction with other users. This interaction is habitually local in the sense that only neighboring agents in the network exchange information, establishing trust relationships among them [71]. Anyway, social networks have become a dominant force in society and the collective opinions given in a social network can determine the path society takes.

To address these kind of situations, a new consensus model has been proposed in [3]. This approach follows an scheme based on the one presented in Sect. 10.2.2, but it incorporates several important differences in order to deal with a large number of experts. The most important difference is the inclusion of a previous step, prior to the consensus phase, in which the large group of experts is simplified into a "selected experts group" or delegates by using a trust based delegation algorithm, trying to maintain the diversity on the opinions of the whole group. Once this simplification is made, the experts that have not been selected will provide information about the trust that the selected experts inspire to them, thus creating a trust network [68, 69]. After this initial step the consensus process begins, but only the selected experts that are allowed to take part in the process.

10.4.2 Agent Theory Based Consensus Models

In this section, we focuss on [45], where authors present a new architecture of a group decision support system for reaching consensus. Its main features are the use of flexible preference representation, explicit knowledge representation component (ontology) used throughout the whole process, and combination of intelligent textual information and database retrieval support. This approach proposes two extension from previous works:

- A different view of the alternatives that is more suitable for the discussion,
- A more sophisticated representation of the individual preferences.

One the one hand, here, an alternative has a hierarchical structure and may be expressed as a pair $s : (T_s, A_s)$ where T_s is a hierarchy of components and A_s is a set of values of the attributes. This structure of the alternatives is represented as a part of the domain ontology. On the other hand, in previous models, the discussion in the group was treated as a black box. However, in this approach, the arguments used during the discussion are treated as a part of the system. In such a way, the general scheme of the discussion is represented as another ontology. The role of the both ontologies is very broad and comprises the support for the argumentation and expression of the preferences by the individuals; both ontologies combined serve as a source of templates for the generation of arguments during the discussion, and providing an opportunity to analyze consensus on several levels of abstraction; for instance, it may be the case that there is no consensus as to the choice of a particular car, but there is consensus that French cars are preferred. In such a way, this approach provides the members of the group with an information- and knowledge-rich environment, and tools to make an effective and efficient use of such an environment. It helps the agent to gain proper opinions about the issues and opinions and attitudes of other agents, articulates proper testimonies, actively contributes to the discussion, and finally makes sound and informed decisions that can help constructively to run the consensus reaching process.

10.4.3   Heterogeneous Contexts Based Consensus Models

In some GDM situations, it is considered that to each decision maker is assigned an importance degree reflecting his/her importance level or knowledge degree about the problem, and, then, it is defined as a heterogeneous GDM framework [12, 50]. For instance, when several medical experts give their testimonies on the possible illness that a patient presents, there will medical experts with more experience or with more study years than others and, as a consequence, their opinions must not be considered with equal relevance. This heterogeneity has been tackled by assigning a weight value to each decision maker that is used in the aggregation step to model their different importance levels or knowledge degrees. However, it would be desirable to develop consensus models which consider the decision makers' importance weights not only in the aggregation step but also in other steps of the consensus round. A first work following this idea was presented in [56], where fuzzy preference relations were assumed to represent the opinions given by the decision makers. This consensus model was proposed following the idea that the decision makers with lower importance or knowledge level will need more advice than those decision makers that previously have at their disposal a better knowledge about the problem to be solved and, therefore, can make better decisions. As a result, it incorporates a feedback mechanism computing different amount of advice according to the decision makers' importance level. However, it has some drawbacks:

- The solution obtained tries to obey the fuzzy majority principle but there could exists a limit scenario in which the tyranny of the minority is accomplished if the excellence group is very small inside the group of decision makers.
- It is not able to detect when a high importance decision maker is wrong or inconsistent.

10.4.4   Dynamic and Changeable Consensus Models

When the decision process is slow or it takes a long time, the set of feasible alternatives is dynamic because their availability or feasibility could change through the decision making time. For example, in e-commerce decision frameworks, where the alternatives are the items that could be bought, it is possible that the availability

of some of these items changes while experts are discussing and making the decision, even, new good items might become available. However, classical consensus
models are defined within static frameworks, it is assumed that the number of alternatives acting in the GDM problem remains fixed throughout the decision making
process. To solve these situations, new consensus models have been proposed in
[57, 58]. These approaches provide a method which allows to remove and insert new
alternatives into the discussion process. It has two different phases: remove old bad
alternatives, and insert new good alternatives. The first phase manages situations in
which some alternatives of the discussion subset are not available at the moment due
to some dynamic external factors or because the experts have evaluated them poorly.
Therefore, the system checks the availability and the evaluation of each alternative
in the current discussion subset. If some alternative is not available or has an evaluation lower than a threshold, the system looks for a new good alternative in the
new alternatives subset. If this subset is empty, the system uses the supply subset
of alternatives provided by the expert at the beginning of the decision process and
that were not taken into account then because of the impossibility to compare all the
alternatives at the same time. Then, the system asks for the experts' opinions about
the replacement and acts according to them. The second case manages the opposite
situation, when some new alternatives have emerged. The system checks if some new
good alternatives have appeared in the new alternatives subset due to some dynamic
external factors. If this is the case, the system has to identify the worst alternatives of
the current discussion subset. To do this, the system uses the evaluation of all alternatives again to choose the worst alternatives. Then, the system asks for the experts'
opinions about the replacement and acts according to them.

10.4.5   Visualizations Tools Based Consensus Models

The arrival of the new information and communication technologies have allowed
the development of new collaboration and information tools for the decision makers
being able to find solutions to GDM problems in which they cannot meet together
with the others [3, 66]. However, in GDM situations where the decision makers do
not have the possibility of discussing together it is possible that they may not have a
clear idea about the current level of consensus achieved among all of them. In typical
GDM situations, the decision makers gather together to discuss their preferences
about the different alternatives and, therefore, it is to some extent easy to decide
which decision makers have related preferences just by attending to the discussions
among the decision makers. One the one hand, the decision makers may join or form
distinct groups to better debate and reason out about the pros and cons of all the
alternatives. On the other hand, it is more easy for the decision makers to influence
the others and to detect if some of them are trying to bias the consensus process if
they know the consensus state. However, it is very probably that decision makers
need some guidance to establish connections among them and to obtain a clear view
of the consensus state when direct communication is not possible. In order to help

the decision makers in such situations, visual elements could be used as they may allow to the decision makers to have a more profound and clear view about the current consensus process and about which decision makers have alike or different preferences about the alternatives. Furthermore, visual elements can help the decision makers to detect if others are trying to bias the consensus process.

Some initial efforts have been done in this direction [2, 53, 54, 57, 72], but it still is an early stage of development and several future challenges have to be faced.

10.4.6   New Preference Structures Based Consensus Models

Recently, new preference structures for representing the decision makers' opinions have been proposed. On the one hand, in [67], Torra presented the hesitant fuzzy sets, a new extension of fuzzy sets, motivated by the common difficulty that often appears when the membership degree of an element must be established and there are some possible values that make to hesitate about which one would be the right one. Since then, a quick growth and applicability of the hesitant fuzzy sets can be found in the specialized literature [38, 60]. On the other hand, when linguistic information is used to represent the preferences given by the decision makers, different linguistic computational models can be used [29], which rely on the special semantics of the linguistic terms, usually fuzzy numbers in the unit interval, and the linguistic aggregation operators are based on aggregation operators in [0, 1]. However, in [48], a new linguistic computational model based on discrete fuzzy numbers whose support is a subset of consecutive natural numbers was presented ensuring the accuracy and consistency of the model, and in which no underlying membership functions are needed.

10.4.7   New Guidance Measures Based Consensus Models

Soft consensus measures represent the level of agreement among the decision makers and, in consequence, they have usually been modeled mathematically via a similarity function measuring how close decision makers' opinions or preferences are. Similarity functions are defined based on the use of a metric describing the distance between decision makers' opinions or preferences. Different metrics or distance functions have been proposed to implement in consensus models as, for instance: the Manhattan distance, the Euclidean distance, the Dice distance, the Cosine distance, or the Jaccard distance [19]. The distance function used to calculate the similarity among the opinions given by the decision makers affects the convergence of the decision process towards a consensus solution. Therefore, it is very important how to select the distance function according to the characteristics of a particular GDM problem because different distance functions can produce significantly different results. In [16], using fuzzy preference relation to represent the opinions provided by the decision makers, it was proved that the Manhattan and the Euclidean distances increase the global consensus level as the number of decision makers increases. On the other hand, the Cosine and the Dice distances result in a fairly similar consensus levels regardless of the number of decision makers, whereas the Jaccard distance function produces the lowest global consensus levels, being fairly stable in value regardless of the number of decision makers.

10.4.8   Adaptive Consensus Models

The automatic consensus models existing in the literature present the same behavior in all discussion rounds of the consensus reaching process although the conditions of GDM problem change [11, 33, 35, 37]. However, when the level of agreement between the experts is "high", a few number of changes of opinions from some of the experts might lead to consensus in a few discussion rounds. On the contrary, when the level of agreement among the experts is "low", a high number of changes of opinions and many group discussion rounds might be necessary for consensus to be achieved. In other words, the number of changes in different stages of a consensus process is clearly related to the actual level of agreement. Following this idea, an adaptive consensus model has been proposed by Mata et al. in [49]. It is based on a refinement process of the consensus process that allows to increase the agreement and reduce the number of experts' preferences that should be changed after each consensus round, adapting the search for the furthest experts' preferences to the existent agreement in each round of consensus. Therefore, the number of changes of preferences suggested to experts after each consensus round will be smaller according to the favorable evolution of the level of agreement. To do so, three levels of agreement are distinguished: very low, low and medium consensus. Each level of consensus involves to carry out the search for the furthest preferences in a different way: when the consensus degree is very low, it will search for the furthest preferences on all experts, while if the consensus degree is medium, the search will be limited to the furthest experts. This adaptive model increases the convergence toward the consensus and reduces the number of rounds to reach it.

10.5   Concluding Remarks

In this chapter we have identified and analyzed the different existing models to compute soft consensus measures in fuzzy group decision making problems. To do so, we have presented four different kind of models, depending if they are based on the reference domain, on the coincidence method, on the generation of recommendation method or on the guidance measures used to drive the consensus process.

Additionally, we have analyzed and presented some challenges to draw the attention of the researchers because they are not completely solved or have still not been addressed. Among them, we can remark some consensus models based on new technologies as social networks or agent theory. Some others are based on non homogeneous contexts or on dynamic and changeable situations. Finally, other new models try to improve the consensus process by incorporating some visualization tools, allowing the use of some new preference structures and new guidance measures or modeling adaptive processes. We think these challenges will contribute this research topic continue being a hot topic in the future.

Acknowledgments This paper has been developed with the financing of FEDER funds in TIN2013-40658-P and Andalusian Excellence Project TIC-5991.

References

1. S. Alonso, F.J. Cabrerizo, F. Chiclana, F. Herrera, E. Herrera-Viedma, Group decision-making with incomplete fuzzy linguistic preference relations. Int. J. Intell. Syst. **24**(2), 201–222 (2009)
2. S. Alonso, E. Herrera-Viedma, F.J. Cabrerizo, F. Chiclana, F. Herrera, Visualizing consensus in group decision making situations, in *Proceedings of the IEEE International Conference on Fuzzy Systems*, pp. 1823–1828 (2007)
3. S. Alonso, I.J. Pérez, F.J. Cabrerizo, E. Herrera-Viedma, A linguistic consensus model for web 2.0 communities. Appl. Soft Comput. **13**(1), 149–157 (2013)
4. D. Ben-Arieh, Z. Chen, Linguistic-labels aggregation and consensus measure for autocratic decision making using group recommendations. IEEE Trans. Syst. Man Cybern.-Part A: Syst. Hum. **36**(3), 558–568 (2006)
5. J.C. Bezdek, B. Spillman, R. Spillman, A fuzzy relation space for group decision theory. Fuzzy Sets Syst. **1**(4), 255–268 (1978)
6. G. Bordogna, M. Fedrizzi, G. Pasi, A linguistic modeling of consensus in group decision making based on OWA operators. IEEE Trans. Syst. Man Cybern.-Part A: Syst. Hum. **27**(1), 126–133 (1997)
7. C.T. Butler, A. Rothstein, *On Conflict and Consensus: A Handbook on Formal Consensus Decision Making* (Takoma Park, 2006)
8. F.J. Cabrerizo, S. Alonso, E. Herrera-Viedma, A consensus model for group decision making problems with unbalanced fuzzy linguistic information. Int. J. Inf. Technol. Decis. Mak. **8**(1), 109–131 (2009)
9. F.J. Cabrerizo, R. Heradio, I.J. Pérez, E. Herrera-Viedma, A selection process based on additive consistency to deal with incomplete fuzzy linguistic information. J. Univ. Comput. Sci. **16**(1), 62–81 (2010)
10. F.J. Cabrerizo, J.M. Moreno, I.J. Pérez, E. Herrera-Viedma, Analyzing consensus approaches in fuzzy group decision making: advantages and drawbacks. Soft. Comput. **14**(5), 451–463 (2010)
11. F.J. Cabrerizo, I.J. Pérez, E. Herrera-Viedma, Managing the consensus in group decision making in an unbalanced fuzzy linguistic context with incomplete information. Knowl.-Based Syst. **23**(2), 169–181 (2010)
12. F.J. Cabrerizo, E. Herrera-Viedma, W. Pedrycz, A method based on pso and granular computing of linguistic information to solve group decision making problems defined in heterogeneous contexts. Eur. J. Oper. Res. **230**(3), 624–633 (2013)
13. C. Carlsson, D. Ehrenberg, P. Eklund, M. Fedrizzi, P. Gustafsson, P. Lindholm, G. Merkuryeva, T. Riissanen, A.G.S. Ventre, Consensus in distributed soft environments. Eur. J. Oper. Res. **61**(1–2), 165–185 (1992)

14. F. Chiclana, E. Herrera-Viedma, F. Herrera, S. Alonso, Some induced ordered weighted averaging operators and their use for solving group decision making problems based on fuzzy preference relations. Eur. J. Oper. Res. **182**(1), 383–399 (2007)
15. F. Chiclana, F. Mata, L. Martinez, E. Herrera-Viedma, S. Alonso, Integration of a consistency control module within a consensus model. Int. J. Uncertain. Fuzziness Knowl.-Based Syst. **16**(Suppl. 1), 35–53 (2008)
16. F. Chiclana, J.M. Tapia-Garcia, M.J. del Moral, E. Herrera-Viedma, A statistical comparative study of different similarity measures of consensus in group decision making. Inf. Sci. **221**, 110–123 (2013)
17. L. Coch, J.R.P. French, Overcoming resistance to change. Hum. Relat. **1**(4), 512–532 (1948)
18. V. Cutello, J. Montero, Fuzzy rationality measures. Fuzzy Sets Syst. **62**(1), 39–54 (1994)
19. M.M. Deza, E. Deza, *Encyclopedia of Distances* (Springer, New York, 2009)
20. D. Dubois, J.L. Koning, Social choice axioms for fuzzy set aggregation. Fuzzy Sets Syst. **43**(3), 257–274 (1991)
21. D. Dubois, H. Prade, C. Testemale, Weighted fuzzy pattern matching. Fuzzy Sets Syst. **28**(3), 313–331 (1988)
22. M. Fedrizzi, M. Fedrizzi, R.A. Marques-Pereira, On the issue of consistency in dynamical consensual aggregation, in *Technologies for Constructing Intelligent Systems*, ed. by B. Bouchon-Meunier, J. Gutierrez-Rios, L. Magdalena, R.R. Yager (Springer, New York, 2002), pp. 129–137
23. M. Fedrizzi, J. Kacprzyk, H. Nurmi, Consensus degrees under fuzzy majorities and fuzzy preferences using OWA (ordered weighted average) operators. Control Cybern. **22**, 71–80 (1993)
24. M. Fedrizzi, J. Kacprzyk, S. Zadrozny, An interactive multi-user decision support system for consensus reaching processes using fuzzy logic with linguistic quantifiers. Decis. Support Syst. **4**(3), 313–327 (1988)
25. M. Fedrizzi, G. Pasi, Fuzzy logic approaches to consensus modeling in group decision making, in *Intelligent Decision and Policy Making Support Systems*, ed. by D. Ruan, F. Hardeman, K. Van Der Meer (Springer-Verlag, Berlin-Heidelberg, 2008), pp. 19–37
26. J. Fodors, M. Roubens, *Fuzzy Preference Modelling and Multicriteria Decision Support* (Kluwer Academic Publishers, Dordrecht, 1994)
27. J.R.P. French, A formal theory of social power. Psychol. Rev. **63**(3), 181–194 (1956)
28. F. Harary, On the measurement of structural balance. Behav. Sci. **4**(4), 316–323 (1959)
29. F. Herrera, S. Alonso, F. Chiclana, E. Herrera-Viedma, Computing with words in decision making: foundations, trends and prospects. Fuzzy Optim. Decis. Making **8**(4), 337–364 (2009)
30. F. Herrera, E. Herrera-Viedma, J.L. Verdegay, A model of consensus in group decision making under linguistic assessments. Fuzzy Sets Syst. **78**(1), 73–87 (1996)
31. F. Herrera, E. Herrera-Viedma, J.L. Verdegay, A rational consensus model in group decision making using linguistic assessments. Fuzzy Sets Syst. **88**(1), 31–49 (1997)
32. F. Herrera, E. Herrera-Viedma, J.L. Verdegay, Choice processes for non-homogeneous group decision making in linguistic setting. Fuzzy Sets Syst. **94**(3), 287–308 (1998)
33. E. Herrera-Viedma, S. Alonso, F. Chiclana, F. Herrera, A consensus model for group decision making with incomplete fuzzy preference relations. IEEE Trans. Fuzzy Syst. **15**(5), 863–877 (2007)
34. E. Herrera-Viedma, F.J. Cabrerizo, J. Kacprzyk, W. Pedrycz, A review of soft consensus models in a fuzzy environment. Inf. Fusion **17**, 4–13 (2014)
35. E. Herrera-Viedma, F. Herrera, F. Chiclana, A consensus model for multiperson decision making with different preference structures. IEEE Trans. Syst. Man Cybern.-Part A: Syst. Hum. **32**(3), 394–402 (2002)
36. E. Herrera-Viedma, F. Herrera, F. Chiclana, M. Luque, Some issues on consistency of fuzzy preference relations. Eur. J. Oper. Res. **154**(1), 98–109 (2004)
37. E. Herrera-Viedma, L. Martinez, F. Mata, F. Chiclana, A consensus support system model for group decision-making problems with multigranular linguistic preference relations. IEEE Trans. Fuzzy Syst. **13**(5), 644–658 (2005)

38. G. Hesamian, M. Shams, Measuring similarity and ordering based on hesitant fuzzy linguistic term sets. J. Intell. Fuzzy Syst. **28**(2):983–990 (2015)
39. J. Kacprzyk, Group decision making with a fuzzy linguistic majority. Fuzzy Sets Syst. **18**(2), 105–118 (1986)
40. J. Kacprzyk, On some fuzzy cores and 'soft' consensus measures in group decision making, in *The Analysis of Fuzzy Information*, ed. by J.C. Bezdek (CRC Press, Boca Raton, 1987), pp. 119–130
41. J. Kacprzyk, M. Fedrizzi, 'Soft' consensus measures for monitoring real consensus reaching processes under fuzzy preferences. Control Cybern. **15**(3–4), 309–323 (1986)
42. J. Kacprzyk, M. Fedrizzi, A 'soft' measure of consensus in the setting of partial (fuzzy) preferences. Eur. J. Oper. Res. **34**(3), 316–325 (1988)
43. J. Kacprzyk, M. Fedrizzi, A 'human-consistent' degree of consensus based on fuzzy logic with linguistic quantifiers. Math. Soc. Sci. **18**(3), 275–290 (1989)
44. J. Kacprzyk, M. Fedrizzi, H. Nurmi, Group decision making and consensus under fuzzy preferences and fuzzy majority. Fuzzy Sets Syst. **49**(1), 21–31 (1992)
45. J. Kacprzyk, S. Zadrozny, Soft computing and web intelligence for supporting consensus reaching. Soft. Comput. **14**(8), 833–846 (2010)
46. J. Kacprzyk, S. Zadrozny, Z.W. Ras, How to support consensus reaching using action rules: a novel approach. Int. J. Uncertain. Fuzziness Knowl.-Based Syst. **18**(4), 451–470 (2010)
47. J. Lu, G. Zhang, D. Ruan, Intelligent multi-criteria fuzzy group decision making for situation assessments. Soft. Comput. **12**(3), 289–299 (2008)
48. S. Massanet, J.V. Rieranad, J. Torrens, E. Herrera-Viedma, A new linguistic computational model based on discrete fuzzy numbers for computing with words. Inf. Sci. **258**, 277–290 (2014)
49. F. Mata, L. Martinez, E. Herrera-Viedma, An adaptive consensus support model for group decision-making problems in a multigranular fuzzy linguistic context. IEEE Trans. Fuzzy Syst. **17**(2), 279–290 (2009)
50. J. Montero, Aggregation of fuzzy opinion in a non-homogeneous group. Fuzzy Sets Syst. **25**(1), 15–20 (1987)
51. J. Montero, The impact of fuzziness in social choice paradoxes. Soft. Comput. **12**, 177–182 (2008)
52. H. Nurmi, Fuzzy social choice: a selective retrospect. Soft. Comput. **12**, 281–288 (2008)
53. I. Palomares, F.J. Estrella, L. Martinez, F. Herrera, Consensus under a fuzzy context: taxonomy, analysis framework afryca and experimental case of study. Inf. Fusion **20**, 252–271 (2014)
54. I. Palomares, L. Martinez, F. Herrera, Mentor: a graphical monitoring tool of preferences evolution in large-scale group decision making. Knowl.-Based Syst. **58**, 66–74 (2014)
55. Z. Pawlak, Information systems theoretical foundations. Inf. Syst. **6**(3), 205–218 (1981)
56. I.J. Pérez, F.J. Cabrerizo, S. Alonso, E. Herrera-Viedma, A new consensus model for group decision making problems with non homogeneous experts. IEEE Trans. Syst. Man Cybern.: Syst. **44**(4), 494–498 (2014)
57. I.J. Pérez, F.J. Cabrerizo, E. Herrera-Viedma, A mobile decision support system for dynamic group decision-making problems. IEEE Trans. Syst. Man Cybern.-Part A: Syst. Hum. **40**(6), 1244–1256 (2010)
58. I.J. Pérez, F.J. Cabrerizo, E. Herrera-Viedma, Group decision making problems in a linguistic and dynamic context. Expert Syst. Appl. **38**(3), 1675–1688 (2011)
59. I.J. Pérez, R. Wikstrom, J. Mezei, C. Carlsson, E. Herrera-Viedma, A new consensus model for group decision making using fuzzy ontology. Soft. Comput. **17**(9), 1617–1627 (2013)
60. R.M. Rodriguez, L. Martinez, V. Torra, Z.S. Xu, F. Herrera, Hesitant fuzzy sets: state of the art and future directions. Int. J. Intell. Syst. **29**(6), 495–524 (2014)
61. M. Roubens, Fuzzy sets and decision analysis. Fuzzy Sets Syst. **90**(2), 199–206 (1997)
62. T.L. Saaty, *The Analytic Hierarchy Process: Planning, Priority Setting* (McGraw-Hill, Resource Allocation, New York, 1980)
63. S. Saint, J.R. Lawson, *Rules for Reaching Consensus: A Modern Approach to Decision Making* (Jossey-Bass, 1994)

64. B. Spillman, R. Spillman, J.C. Bezdek, A fuzzy analysis of consensus in small groups, in *Fuzzy Automata and Decision Processes*, ed. by P.P. Wang, S.K. Chang (North-Holland, Amsterdam, 1980), pp. 331–356
65. T. Tanino, Fuzzy preference orderings in group decision making. Fuzzy Sets Syst. **12**(2), 117–131 (1984)
66. M. Tavana, D.T. Kennedy, N-site: a distributed consensus building and negotiation support system. Int. J. Inf. Technol. Decis. Mak. **5**(1), 123–154 (2006)
67. V. Torra, Hesitant fuzzy sets. Int. J. Intell. Syst. **25**(6), 529–539 (2010)
68. P. Victor, C. Cornelis, M.D. Cock, P. Pinheiro da Silva, Gradual trust and distrust in recommender systems. Fuzzy Sets Syst. **160**(10), 1367–1382 (2009)
69. P. Victor, C. Cornelis, M.D. Cock, E. Herrera-Viedma, Practical aggregation operators for gradual trust and distrust. Fuzzy Sets Syst. **184**(1), 126–147 (2011)
70. S. Wasserman, K. Faust, *Social Networks Analysis: Methods and Applications* (Cambridge University Press, Cambridge, 2009)
71. J. Wu, F. Chiclana, A social network analysis trust-consensus based approach to group decision-making problems with interval-valued fuzzy reciprocal preference relations. Knowl.-Based Syst. **59**, 97–107 (2014)
72. J. Wu, F. Chiclana, Visual information feedback mechanism and attitudinal prioritisation method for group decision making with triangular fuzzy complementary preference relations. Inf. Sci. **279**, 716–734 (2014)
73. Z.B. Wu, J.P. Xu, A consistency and consensus based decision support model for group decision making with multiplicative preference relations. Decis. Support Syst. **52**(3), 757–767 (2012)
74. L.A. Zadeh, Fuzzy sets. Inf. Control **8**(3), 338–353 (1965)
75. L.A. Zadeh, The concept of a linguistic variable and its applications to approximate reasoning. Inf. Sci. Part I, II, III, **8, 8, 9**, 199–249,301–357,43–80 (1975)

Chapter 11
Relation Between AHP and Operators Based on Different Scales

E. Cables, M.T. Lamata and J.L. Verdegay

Abstract Obtaining the value of the weights in any decision problem is of great importance, because it can change the course of action for the final decision. The value of these weights is approximate due to the vagueness and ambiguity of the data. Our study is based on the Analytic Hierarchy Process and its relation with the Prioritized Aggregation Operators. We propose their obtaining starting from a proportionality relationship, and we study the main properties of the prioritized operator with proportionality ratio and linear scale.

11.1 Introduction

There are many problems in the real life which require making decisions; this implies the election of an alternative from the alternative set, using for it the Multiple Criteria Decision Analysis methods (MCDA). These methods consist of two differentiated phases: the first one is referred to the obtaining of the information, while the second one are related with the aggregation and the exploitation.

One of the steps of MCDA, is the assessment of the importance of each criterion. For that, it is necessary to associate to each criterion a subjective weight given by an expert [8] or known beforehand as consequence of experience. Another option is to apply an operator that will be associated with the criteria; thus, we can obtain different weights vector as we make use of ratio scales, interval scales or simply ordinal scales.

E. Cables
Dpto. Informatica, Universidad de Holguin "Oscar Lucero Moya", Holguin, Cuba
e-mail: elio@decsai.ugr.es

M.T. Lamata · J.L. Verdegay (✉)
Dpto. Ciencias de la Computacion E Inteligencia Artificial,
Universidad de Granada, Granada, Spain
e-mail: verdegay@decsai.ugr.es

M.T. Lamata
e-mail: mtl@decsai.ugr.es

© Springer International Publishing Switzerland 2016   155
T. Calvo Sánchez and J. Torrens Sastre (eds.), *Fuzzy Logic and Information Fusion*,
Studies in Fuzziness and Soft Computing 339, DOI 10.1007/978-3-319-30421-2_11

The importance information is usually provided by human experts who explicitly state which criterion is more important for every pair of criteria. The number of pairwise comparisons required grows very quickly so the process may become too tedious for the experts and the resulting comparison matrix may be inconsistent.

It is well known that depending on the weights, the result of the aggregation step can be different. The proposal of [17, 18], about operators priority as well as a variety of analytic functions [1, 2, 11, 19], are particularly important for the development of this work.

For such reason, the weights vector represents an essential part to obtain the final solution. The following vectors that represent the optimistic, pessimistic, and neutral attitude of the decision-maker, respectively, can be adopted:

1. $W = W^* = (1, 0, 0, \ldots, 0)^T$
2. $W = W_* = (0, 0, 0, \ldots, 1)^T$
3. $W = W_{Ave} = (1/n, 1/n, 1/n, \ldots, 1/n)^T$

Or any vector $W = (w_1, w_2, \ldots, w_n)$, such that satisfies the conditions $\sum_{i=1}^{n} w_i = 1$ and $w_i \in [0, 1]$.

A MCDA model broadly cited in the academic literature is the Analytic Hierarchy Process (AHP), developed by [13]. The main characteristic is that it is based on a ratio scale and in pairwise comparisons between the criteria to assess their relative importance [10], being $n(n-1)/2$ the number of comparisons required. This number can be very high, and therefore a tedious process for the decision-makers, mainly when one has to make a high number of comparisons. Moreover, there exist a risk of obtaining an inconsistent matrix. In order to address these problems, we propose a method where we only need to sort the criteria according to their importance and the preference valuations among those $(n-1)$ adjacent criteria. The result is the weights vector associated with the criteria. In this case, we will say that C_i is preferred to C_j $(C_i \succ C_j)$, if $w_i > w_j$ where w_i and w_j are the weights of those criteria in the weights vector. This relationship can be expressed by means of an intervals scale or a ratio scale. To do this, in the next section we review the AHP model. Next, we describe the Prioritized Aggregation Operators and the operators based on ratio scales. After that, we explain the linear operators and the SMARTER operator. Finally we show the relationship between these four models and draw conclusions.

11.2   The Analytic Hierarchy Process Method (AHP)

The Analytic Hierarchy Process (AHP methodology) [13], has been accepted by the international scientific community as a robust and flexible multi-criteria decision-making tool for dealing with complex decision problems. AHP has been applied to many decision problems such as energy policy [9], project selection [3], measuring business performance, and evaluation of advanced manufacturing technology [5]. Basically, AHP has three underlying concepts:

- structuring the complex decision problem as a hierarchy of goal, criteria and alternatives.
- pairwise comparison of elements at each level of the hierarchy with respect to each criterion on the preceding level.
- vertical synthesis of the judgements over the different levels of the hierarchy.

In this case, we only apply the method in order to obtain the criteria weights. We assume that the quantified judgements provided by the decision-maker on pairs of criteria (C_i, C_j) are represented in an $n \times n$ matrix as in the following (Table 11.1):

The c_{12} value is supposed to be an approximation of the relative importance of C_1 to C_2, i.e., $c_{12} \approx (w_1/w_2)$. This can be generalized and the statements below can be concluded:

1. $c_{ij} \approx (w_i/w_j), \quad i, j = 1, 2, \ldots, n$
2. $c_{ii} = 1, \quad i = 1, 2, \ldots, n$
3. If $c_{ij} = \alpha, \alpha \neq 0$, then $c_{ji} = 1/\alpha, \quad i = 1, 2, \ldots, n$
4. If C_i is more important than C_j then $c_{ij} \simeq (w_i/w_j) > 1$
5. $c_{ij} = c_{ik} \cdot c_{kj}$

The values assigned to c_{ij} according to Saaty scale (see Table 11.2) are usually in the interval of 1–9 or their reciprocals. This implies that matrix C should be a positive and reciprocal matrix (i.e. $c_{ij}^{-1} = c_{ji}$) with 1s in the main diagonal and hence the decision-maker needs only to provide value judgements in the upper triangle of the matrix.

It can be shown that the number of judgments (L) needed in the upper triangle of the matrix is:

$$L = n(n - 1)/2,$$

where n is the size of the matrix C.

If the weights $w_i, \ i = 1, 2, \ldots, n$ were known the matrix of comparisons would be as follows:

$$M = \begin{pmatrix} w_1/w_1 & w_1/w_1 & \ldots & w_1/w_n \\ w_2/w_1 & w_2/w_2 & \ldots & w_2/w_n \\ \ldots & & \ldots & \\ w_n/w_1 & w_n/w_2 & \ldots & w_n/w_n \end{pmatrix}$$

Table 11.1 Comparison decision matrix

	C_1	C_2	...	C_j	...	C_m
C_1	c_{11}	c_{12}	...	c_{1j}	...	c_{1m}
C_2	c_{21}	c_{22}	...	c_{2j}	...	c_{2m}
...	...	...	...	...	...	...
C_i	...	...	...	c_{ij}	...	...
...	...	...	...	...	...	...
C_n	c_{n1}	c_{n2}	...	c_{nj}	...	c_{nm}

Table 11.2 Saatys preferences in the pairwise comparison process

Verbal judgments of preferences between criterion i and criterion j	Labels	Numerical rating
C_i is equally important than C_j	EI	1
C_i is slightly more important than C_j	sMI	3
C_i is strongly more important than C_j	SMI	5
C_i is very strongly more important than C_j	VMI	7
C_i is extremely more important than C_j	EMI	9
Intermediate values		2, 4, 6, 8

If we wanted to get the vector of weights from this matrix the following system of equations must be solved:

$$\begin{pmatrix} w_1/w_1 \ w_1/w_2 \ \dots \ w_1/w_n \\ w_2/w_1 \ w_2/w_2 \ \dots \ w_2/w_n \\ \dots \qquad \dots \\ w_n/w_1 \ w_n/w_2 \ \dots \ w_n/w_n \end{pmatrix} \cdot \begin{pmatrix} w_1 \\ w_2 \\ \dots \\ w_n \end{pmatrix} = \lambda \begin{pmatrix} w_1 \\ w_2 \\ \dots \\ w_n \end{pmatrix}$$

Or equivalently, we would get the eigenvector associated with the λ, where λ is an eigenvalue of M and w is the associated eigenvector. The M matrix is reciprocal and its range is equal to 1 since each row is a constant multiple of the first. Therefore, this matrix has a unique non-zero eigenvalue $\lambda = n$. This happens if the matrix is entirely consistent, which doesn't happen very often in practice. Therefore, it is necessary to define a consistency index for accepting or not the judgements expressed by the decision-maker.

The consistency index (CI) is given by $CI = (\lambda_{max} - n)/(n - 1)$, where λ_{max} is the principal eigenvalue of the matrix. If the expert shows some minor inconsistency, then $(\lambda_{max} > n)$ and Saaty proposes the following measure of the consistency index:

$$CR = CI/RI$$

where RI is the average value of CI obtained in [4] and showns in Table 11.3. The consistency ratio CR is a measure of how a given matrix compares to a purely random matrix in terms of their consistency indices. A value of the consistency ratio $CR \leq 0.1$ is considered acceptable. Larger values of CR require the DM to revise his judgements.

Table 11.3 Random index for different matrix orders

n	1–2	3	4	5	6	7	8	9	10
RI	0.0	0.524	0.881	1.108	1.247	1.341	1.405	1.449	1.485

Example 1 Consider the following matrix provided by an expert and its correspond-ing quantitative decision matrix (see Table 11.2).

$$
\begin{pmatrix}
1 & EI.sMI & sMI & SMI & VMI \\
 & 1 & EI.sMI & sMI.SMI & SMI.VMI \\
 & & 1 & EI.sMI & sMI.SMI \\
 & & & 1 & sMI \\
 & & & & 1
\end{pmatrix}
\equiv
\begin{pmatrix}
1 & 2 & 3 & 5 & 7 \\
1/2 & 1 & 2 & 4 & 6 \\
1/3 & 1/2 & 1 & 2 & 4 \\
1/5 & 1/4 & 1/2 & 1 & 3 \\
1/7 & 1/6 & 1/4 & 1/3 & 1
\end{pmatrix}
$$

Then, $\lambda_{max} = 5.09$, $CI = 0.022$ and the associated weights vector is:

$$W = [0.433, 0.272, 0.155, 0.088, 0.043]$$

Instead of working with the AHP we can define an operator that is based on a ratio scale (as well as AHP). In order to compute the weights vector, we only need to compare $(n - 1)$ criteria. Therefore, the main advantage of this operator lies in the reduction of the number of questions in the questionnaire and the elimination of the inconsistency problem.

11.3 The Prioritized Aggregation Operators

In 2008, Yager [17] introduces the Prioritized Aggregation Operators. In this work, we will start from this conception and we will consider a collection of criteria $C = (C_1, C_2, \ldots, C_n)$, where an order of priority exists among them. This entails that a linear order exists in the following form:

$$C_1 \succ C_2 \succ C_3 \succ \cdots \succ C_n,$$

According to this, the criterion C_i has a higher priority that the criterion C_k, *if* $i < k$. Also, for any alternative x and the criterion C_i, the value $C_i(x) \in [0.1]$ gives the satisfaction grade of the criterion C_i for the alternative x.

1. In order to compute the weights vector, we depart from the aforementioned order relation and satisfaction grades $C_i(x)$ of each criterion, and we proceed as follows: Let $S_k = C_k(x)$.
2. For each criterion let $u_i = T_i$ be its non-normalized weight, which can be obtained through the following recurrent expression:

$$
T_i = \begin{cases}
1 & \text{if } i = 1 \\
S_{i-1} \cdot T_{i-1} & \text{if } i \neq 1
\end{cases}
\tag{11.1}
$$

Finally, the normalized weights are obtained through the expression:

$$w_i = \frac{u_i}{\sum_{j=1}^{n} u_j}$$

It is clear that these weights satisfy:

$$\sum_{i=1}^{n} w_i = 1 \ \ and \ \ w_i \in [0, 1]$$

With them we can define the orness measure

$$\alpha(W) = \frac{1}{n-1} \sum_{i=1}^{n} (n-1)w_i$$

11.4 Operator Based on Ratio Scale

This development is based on the specification of a vector that represents the proportionality ratio between weights [12, 16].

Definition 11.1 If the relationship among the weights satisfies $w_i > w_{i+1}$, $\forall i = 1, 2, \ldots, n-1$ which assumes $\alpha \geq 0.5$, then the following relation is called proportionality ratio of adjacent weights of the weights vector W

$$z_i = \begin{cases} \frac{w_i}{w_{i+1}} & \text{if } w_{i+1} \neq 0 \\ 0 & \text{if } w_{i+1} = 0 \end{cases} \tag{11.2}$$

The above definition amounts to say that, for each weights vector formed by elements, we can obtain a new vector V with dimension $(n-1)$ which contains the proportionality ratio of all the adjacent weights, $V = \{v_1, v_2, \ldots, v_{n-1}\}$ where $v_i = 1/z_i$.

The elements of the weights vector are positive $w_i \geq 0$; then, according to Definition 11.1, it is also satisfied that $v_i \geq 0$, $\forall v_i$, with $i = 1, 2, \ldots, n-1$, and therefore $z_i = 1/v_i$ also are positive. Any vector whose elements are all greater or equal to zero can be considered as the proportionality ratio of the adjacent weights of a given weights vector.

Consider expression (11.2) and the ratio proportionality vector $Z = (z_1, z_2, \ldots, z_{n-1})$ Let W be the weights vector $W = (w_1, w_2, \ldots, w_n)$, then, by Definition 11.1, we can verify that

$$w_2 = z_1 \cdot w_1$$
$$w_3 = z_2 \cdot w_2$$
$$\vdots$$
$$w_{n-1} = z_{n-2} \cdot w_{n-2}$$
$$w_n = z_{n-1} \cdot w_{n-1}$$

When applying an iterative substitution, we obtain:

$$w_n = z_1 \cdot z_2 \cdot z_3 \cdots z_{n-1} \cdot w_1 = \prod_{k=n-1}^{1} z_k \cdot w_1$$

Therefore in the general case, $\forall i = 1, 2, \ldots, n$, we have

$$w_i = \prod_{k=i-1}^{1} z_k \cdot w_1$$

Since $\sum_{i=1}^{n} w_i = 1$ we have

$$\sum_{i=1}^{n-1} \prod_{k=i-1}^{1} z_k \cdot w_1 + w_1 = 1 \iff \left(\sum_{i=1}^{n-1} \prod_{k=i}^{n-1} z_k + 1 \right) \cdot w_1 = 1$$

then

$$w_1 = \frac{1}{\sum_{i=1}^{n-1} \prod_{k=i}^{n-1} z_k + 1}$$

Therefore, the starting point is the sorting of the criteria in descending importance and quantification of the importance given by the vector V where $v_i \geq 1$, $\forall i$ and $z_j = \frac{1}{v_j} \in [0, 1]$ represent the proportion of reverse superiority criteria C_j with respect to C_{j-1}. The weights are obtained recursively using expression (11.1).

$$w_i = \begin{cases} 1/\sum_{i=1}^{n-1} \prod_{k=i}^{n-1} z_k + 1 & \text{if } i = 1 \\ w_1 \cdot \prod_{k=1}^{i-1} z_k & \text{if } i \neq 1 \end{cases} \qquad (11.3)$$

The operator thus defined satisfies the properties

$$\sum_{i=1}^{n} w_i = 1 \ \text{ and } \ w_i \in [0, 1]$$

Theorem 11.1 *Let V be a vector of proportionality ratio. If the elements of vector V verify that $\forall v_i : v_i = 1$, then the resulting weights vector W is the average weights vector $W = (\frac{1}{n}, \frac{1}{n}, \frac{1}{n}, \ldots, \frac{1}{n})$.*

Proof

$$w_1 = \frac{1}{\sum_{j=1}^{n-1} \prod_{k=i}^{j} 1 + 1} \iff w_1 = \frac{1}{\sum_{j=1}^{n-1} 1 + 1} \iff w_1 = \frac{1}{n-1+1} = \frac{1}{n}$$

The elements w_i such that $i > 1$, are given by:

$$w_i = \frac{1}{n} \prod_{k=1}^{i-1} 1 \iff w_i = \frac{1}{n}$$

We obtain the weights vector $W = (\frac{1}{n}, \frac{1}{n}, \frac{1}{n}, \ldots, \frac{1}{n})$.

Example 2 1: Consider an organized set of five criteria, given by an expert in descending order of importance $C_1 \succ C_2 \succ C_3 \succ C_4 \succ C_5$ and the importance between the valuations $I(.,.)$ given by Table 11.1 as follows:

$$I(C_1, C_2) = EI.sMI, \quad I(C_1, C_3) = sMI, \quad I(C_1, C_4) = SMI, \quad I(C_1, C_5) = VMI$$

This means that the valuations of the importance of the consecutive criteria are $C_1 {}^2 C_2 {}^{3/2} C_3 {}^{5/3} C_4 {}^{7/5} C_5$ having applied AHP properties.

According to this, the vector indicating the linguistic assessment is $v_j = [2, 3/2, 5/3, 7/5]$ and therefore the inverse vector is given by $z_j = [1/2, 2/3, 3/5, 5/7]$. Then, using expression (11.3) for calculating the weights, we obtain:

$$W = [0.459, \ 0.2295, \ 0.153, \ 0.0918, \ 0.065]$$

11.5 Operator Based on Linear Scale

In this case we will consider the results of [11].

The general form of the linear functions is given by:

$$f_g(x) = a \cdot x + \frac{1}{n} - a \cdot \left(\frac{1+n}{2}\right) \tag{11.4}$$

which depends on the value of the selected slope $a \in \left]\frac{-2}{n \cdot (n-1)}, 0\right[$. The set of straight lines that provide weights vector with these features are included in the bundle of straight lines limited by $f_{sup}(x)$ and $f_{inf}(x)$ in Fig. 11.1.

Fig. 11.1 Region of the linear functions set with $\alpha \leq 0$

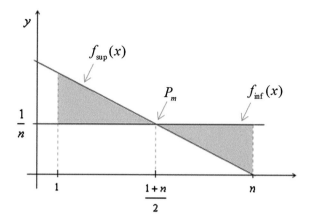

Then when $a = -2/n \cdot (n - 1)$ we obtain the function

$$f_{sup(x)} = \frac{-2}{n \cdot (n - 1)} \cdot x + \frac{2}{n - 1} \tag{11.5}$$

And the corresponding weights vector is [5/15, 4/15, 3/15, 2/15, 1/15] and $\alpha = 2/3$.

In the same form, if a = 0 the weights vector will be [1/5, 1/5, 1/5, 1/5, 1/5] that corresponds to the line $f_{inf}(x)$.

11.6   SMARTER Method

Edwards [6] originally described SMART as the whole process of rating alternatives and weighting attributes. With SMART the weights are obtained in two steps [6, 15]:

1. Rank the importance of the changes in the attributes from the worst attribute levels to the best levels.
2. Make ratio estimates of the relative importance of each attribute relative to the one ranked lowest in importance.

But, [7] also presented a new version, SMARTER, which only uses the ranking of attributes to derive the weights. The idea is to use the centroid method of [14] so that the weight of an attribute ranked to be i-th is

$$v_i = \frac{1}{\sum_{k=i}^{n} \frac{1}{k}}$$

where n is the number of attributes.

Definition 11.2 Let $C = \{C_1, C_2, \ldots, C_n\}$ be a criteria set with an ordering associated to it, given by $C_i \succ C_{i+1}, \forall i = 1, 2, \ldots, n-1$. The weights are obtained by the following recurrent equation:

$$w_i = \begin{cases} w_{i+1} + 1/i \cdot n & \text{if } i \neq n \\ \\ 1/n \cdot n & \text{if } i = n \end{cases} \tag{11.6}$$

$\forall w_i \in [0, 1]$ and $\sum_{i=1}^{n} w_i = 1$.
It is important to verify that $\sum_{i=1}^{n} w_i = 1$

$$\sum_{i=1}^{n} w_i = \frac{1}{n \cdot n} + \left(\frac{1}{n \cdot n} + \frac{1}{n \cdot (n-1)} \right) + \cdots + \left(\frac{1}{n \cdot n} + \frac{1}{n \cdot (n-1)} + \cdots + \frac{1}{2n} \right)$$
$$+ \left(\frac{1}{n \cdot n} + \frac{1}{n \cdot (n-1)} + \cdots + \frac{1}{2n} + \frac{1}{n} \right)$$
$$\sum_{i=1}^{n} w_i = n \frac{1}{n \cdot n} + (n-1) \frac{1}{n \cdot (n-1)} + \cdots + 2 \frac{1}{2n} + \frac{1}{n}$$

then
$$\sum_{i=1}^{n} w_i = 1.$$

Example 2 We consider Example 1 but now we assume the only information about the criteria is ordinal, $C_1 \succ C_2 \succ C_3 \succ C_4 \succ C_5$. For $n = 5$ and applying expression (11.6), these values are:

$v_5 = \frac{1}{5} \left(\frac{1}{5} \right) = 0.0400$

$v_4 = 0.0400 + \frac{1}{5} \left(\frac{1}{4} \right) = 0.1028$

$v_3 = 0.1028 + \frac{1}{5} \left(\frac{1}{3} \right) = 0.1567$

$v_2 = 0.1567 + \frac{1}{5} \left(\frac{1}{2} \right) = 0.2567$

$v_1 = 0.2567 + \frac{1}{5} = 0.4567$

In this way we have obtained the vector of weights as:

$$[0.4567, 0.2567, 0.1567, 0.0900, 0.0400]$$

and $\alpha = 0.75$

11.7 Relation Between the Weights Given by the Four Models

The weights are plotted in Fig. 11.2. This figure reflects small differences between the results using AHP and the operators calculated using the ratio scale or the ordinal scale of the SMARTER, but there are significant differences with the linear operator.

Through the last column of Table 11.4, which reflect the orness coefficient, it can be seen that AHP, ratio scale and ordinal scale yield very similar values. However, the value corresponding to the linear scale is much lower than the others.

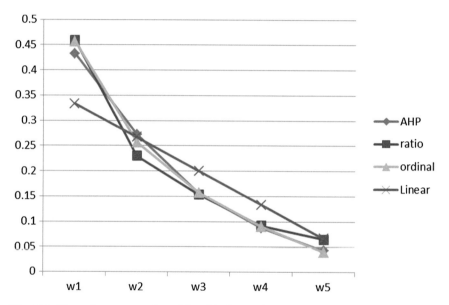

Fig. 11.2 Figure that represent the weight of the four options

Table 11.4 Values of weights obtained by AHP and the three operators

Weights	w_1	w_2	w_3	w_4	w_5	α
AHP	0.4330	0.2720	0.1550	0.0880	0.0430	0.7365
Ratio	0.4590	0.2295	0.1530	0.0918	0.0650	0.7305
Ordinal	0.4567	0.2567	0.1567	0.1028	0.0400	0.7500
Linear	0.3333	0.2666	0.2000	0.1333	0.0666	0.6666

11.8 Conclusions

Taking into account the aspects analyzed previously, the following conclusions can be drawn.

If all the values in the vector of proportionality ratio are equal to 1, then the average weights vector obtained is $W = (\frac{1}{n}, \frac{1}{n}, \frac{1}{n}, \ldots, \frac{1}{n})$. The proofs for obtaining the weights $W = (1, 0, 0, \ldots, 0)$ and $W = (0, 0, 0, \ldots, 1)$ have been omitted.

Moreover, the weights vector associated with AHP is very similar to that obtained by the proportionality ratio, but the latter has the advantage of not requiring the entire array of pairwise comparisons, but only the first row of the matrix. This also avoids the problem of inconsistency.

Although in this particular case the SMARTER method provides a vector that is similar to the other methods, it should be noticed that, if the matrix of Example 1 was different, then the vector obtained with AHP may or may not be similar to that obtained with SMARTER.

However, this will not be the case if we compare the weights vector of AHP and the ratio scale operator that are similar in all the cases. Also, it is easy to see that linear operators are different from the previous ones.

Acknowledgments This work is partially supported by FEDER funds, the DGICYT and Junta de Andaluca under projects TIN2014-55024-P and P11-TIC-8001, respectively.

References

1. B. Ahn, On the properties of owa operator weights functions with constant level of orness. IEEE Trans. Fuzzy Syst. **14**(4), 511–515 (2006)
2. B. Ahn, H. Park, Least-squared ordered weighted averaging operator weights. Int. J. Intell. Syst. **23**(1), 33–49 (2008)
3. K. Al-Harbi, M. Al-Subhi, Application of the AHP in project management. Int. J. Project Manage. **19**(1), 19–27 (2001)
4. J.A. Alonso, M.T. Lamata, Consistency in the analytic hierarchy process: a new approach. Int. J. Uncertainty, Fuzziness Knowl.-Based Syst. **14**(04), 445–459 (2006)
5. F. Chan, M. Chan, N. Tang, Evaluation methodologies for technology selection. J. Mater. Process. Technol. **107**(1), 330–337 (2000)
6. W. Edwards, How to use multiattribute utility measurement for social decisionmaking. IEEE Trans. Syst. Man Cybern. **7**(5), 326–340 (1977)
7. W. Edwards, F. Barron, Smarts and smarter: Improved simple methods for multiattribute utility measurement. Organ. Behav. Hum. Decis. Process. **60**(3), 306–325 (1994)
8. M.S. García-Cascales, M.T. Lamata, Multi-criteria analysis for a maintenance management problem in an engine factory: rational choice. J. Intell. Manuf. **22**(5), 779–788 (2011)
9. M.S. García-Cascales, M.T. Lamata, J.M. Sánchez-Lozano, Evaluation of photovoltaic cells in a multi-criteria decision making process. Ann. Oper. Res. **199**(1), 373–391 (2012)
10. M. Lamata, Ranking of alternatives with ordered weighted averaging operators. Int. J. Intell. Syst. **19**(5), 473–482 (2004)
11. M.T. Lamata, E. Cables Perez, Owa weights determination by means of linear functions. Mathware Soft Comput. **16**(2), 107–122 (2009)

12. M.T. Lamata, E. Cables Pérez, Obtaining owa operators starting from a linear order and preference quantifiers. Int. J. Intell. Syst. **27**(3), 242–258 (2012)
13. T. Saaty, *The Analytic Hierarchy Process* (McGraw-Hill, New York, 1980)
14. T. Solymosi, J. Dombi, A method for determining the weights of criteria: the centralized weights. Eur. J. Oper. Res. **26**(1), 35–41 (1986)
15. D. Von Winterfeldt, W. Edwards, et al., *Decision Analysis and Behavioral Research*, vol. 604 (Cambridge, University Press Cambridge, 1986)
16. Z. Xu, Q. Da, Approaches to obtaining the weights of the ordered weighted aggregation operators. Dongnan Daxue Xuebao (Ziran Kexue Ban)/Journal of Southeast University (Natural Science Edition), **33**(1), 94–96 (2003)
17. R. Yager, Prioritized aggregation operators. Int. J. Approximate Reasoning **48**(1), 263–274 (2008). cited By 85
18. R. Yager, Prioritized owa aggregation. Fuzzy Optim. Decis. Making **8**(3), 245–262 (2009). cited By 62
19. R. Yager, D. Filev, Parameterized and-like and or-like OWA operators. Int. J. General Syst. **22**(3), 297–316 (1994)

Chapter 12
Evolutionary Fuzzy Systems: A Case Study in Imbalanced Classification

A. Fernández and F. Herrera

Abstract The use of evolutionary algorithms for designing fuzzy systems provides them with learning and adaptation capabilities, resulting on what is known as Evolutionary Fuzzy Systems. These types of systems have been successfully applied in several areas of Data Mining, including standard classification, regression problems and frequent pattern mining. This is due to their ability to adapt their working procedure independently of the context we are addressing. Specifically, Evolutionary Fuzzy Systems have been lately applied to a new classification problem showing good and accurate results. We are referring to the problem of classification with imbalanced datasets, which is basically defined by an uneven distribution between the instances of the classes. In this work, we will first introduce some basic concepts on linguistic fuzzy rule based systems. Then, we will present a complete taxonomy for Evolutionary Fuzzy Systems. Then, we will review several significant proposals made in this research area that have been developed for addressing classification with imbalanced datasets. Finally, we will show a case study from which we will highlight the good behavior of Evolutionary Fuzzy Systems in this particular context.

12.1 Introduction

Among all available strategies to be used in real application areas of engineering, those related to Computational Intelligence or Soft Computing have typically shown a good behavior [1]. In addition, we must stress that the collaboration among the components of Computational Intelligence can further improve the results than applying them on isolation. For this reason, hybrid approaches have attracted considerable attention in this community. Among them, the most popular is maybe the synergy

A. Fernández
Department of Computer Science, University of Jaén, 23071 Jaén, Spain
e-mail: alberto.fernandez@ujaen.es

F. Herrera (✉)
Department of Computer Science and Artificial Intelligence,
University of Granada, 18071 Granada, Spain
e-mail: herrera@decsai.ugr.es

© Springer International Publishing Switzerland 2016
T. Calvo Sánchez and J. Torrens Sastre (eds.), *Fuzzy Logic and Information Fusion*,
Studies in Fuzziness and Soft Computing 339, DOI 10.1007/978-3-319-30421-2_12

between Fuzzy Rule Based Systems (FRBSs) [2] and Evolutionary Computation [3, 4] leading to Evolutionary Fuzzy Systems (EFSs) [5, 6].

The automatic definition of an FRBS can be seen as an optimization or search problem. Regarding the former, the capabilities of Evolutionary Algorithms (EAs) [7] makes them an appropriate global search technique. They aim to explore a large search space for suitable solutions, only requiring a performance measure. In addition to their ability to find near optimal solutions in complex search spaces, the generic code structure and independent performance features of EAs allow them to incorporate a priori knowledge. In the case of FRBSs, this a priori knowledge may be in the form of linguistic variables, fuzzy membership function parameters, fuzzy rules, number of rules and so on. Furthermore, this approach has been recently extended by using Multi-Objective Evolutionary Algorithms (MOEAs) [8, 9], which can consider multiple conflicting objectives, instead of a single one. The hybridization between MOEAs and FRBSs is currently known as Multi-Objective Evolutionary Fuzzy Systems (MOEFSs) [10].

As stated previously, the adapting capabilities and goodness of EFSs has made their use to be spread successfully into different Data Mining areas [11]. Among these, possibly the most common application is related to classification problems. When working in this framework, we may find that they frequently present a very different distribution of examples inside their classes, which is known as the problem of imbalanced datasets [12, 13]. In the context of binary problems, the positive or minority class is represented by few examples, since it could represent a "rare case" or because the acquisition of this data is costly. For these reasons, the minority class is usually the main objective from the learning point of view. Therefore, the cost related to a poor classification of one example of this class is usually be greater than on the majority class.

Linguistic FRBSs have shown the achievement of a good performance in the context of classification with imbalanced datasets [14]. Specifically, linguistic fuzzy sets allow the smoothing of the borderline areas in the inference process, which is also a desirable behavior in the scenario of overlapping, which is known to highly degrade the performance in this context [15, 16]. In accordance with the former, and with aims at improving the behaviour and performance of these systems, a wide number of approaches have been proposed in the field of EFS for addressing classification with imbalanced datasets [15, 17, 18].

In this chapter, we will first introduce the existent taxonomy for the different types of EFSs, together with their main properties. Then, we focus on the main aim in this contribution, which is to present the use of EFSs in imbalanced classification, and to provide a list of the most relevant contributions in this area of work. Finally, we will show the goodness of this type of approaches presenting a case study on the topic over highly imbalanced datasets using the GP-COACH-H algorithm [19], a fuzzy rule-based technique based on genetic programming specifically designed to address the imbalance in data.

For achieving these objectives, the remainder of this chapter is organized as follows. In Sect. 12.2, we provide an overview of FRBSs. In Sect. 12.3, we focus our attention to EFSs. Section 12.4 is devoted to the application of EFSs in classification with imbalanced datasets, describing the features of this problem, presenting those EFSs approaches that have been designed for addressing this task, and introducing a brief case study to excel the good behaviour of EFSs in this work area. Finally, in Sect. 12.5, we provide some concluding remarks of this work.

12.2 Fuzzy Rule Based Systems

The basic concepts which underlie fuzzy systems are those of linguistic variable and fuzzy IF-THEN rule. A linguistic variable, as its name suggests, is a variable whose values are words rather than numbers, e.g., small, young, very hot and quite slow. Fuzzy IF-THEN rules are of the general form: IF antecedent(s) THEN consequent(s), where antecedent and consequent are fuzzy propositions that contain linguistic variables. A fuzzy IF-THEN rule is exemplified by "IF the temperature is high THEN the fan-speed should be high". With the objective of modeling complex and dynamic systems, FRBSs handle fuzzy rules by mimicking human reasoning (much of which is approximate rather than exact), reaching a high level of robustness with respect to variations in the system's parameters, disturbances, etc. The set of fuzzy rules of an FRBS can be derived from subject matter experts or extracted from data through a rule induction process.

In this section, we present a brief overview of the foundations of FRBSs, with the aim of illustrate the way they behave. In particular, in Sect. 12.2.1, we introduce the important concepts of fuzzy set and linguistic variable. In Sect. 12.2.2, we deal with the basic elements of FRBSs. Finally, in Sect. 12.2.3 we describe the fuzzy inference system proposed by Mamdani for the output of an FRBS, as in this work we focus on linguistic systems.

12.2.1 Preliminaries: Fuzzy Set and Linguistic Variable

A *fuzzy set* is distinct from a crisp set in that it allows its elements to have a degree of membership. The core of a fuzzy set is its membership function: a surface or line that defines the relationship between a value in the set's domain and its degree of membership. In particular, according to the original ideal of Zadeh [20], membership of an element x to a fuzzy set A, denoted as $\mu_A(x)$ or simply $A(x)$, can vary from 0 (full non-membership) to 1 (full membership), i.e., it can assume all values in the interval [0, 1]. Clearly, a fuzzy set is a generalization of the concept of a set whose membership function can take only two values {0, 1}.

We must point out that this is clearly a generalization and extension of multi-valued logic, in which degrees of truth are introduced in terms of the aforementioned membership functions. These functions can be seen as mapping predicates into FSs (or more formally, into an ordered set of fuzzy pairs, called a fuzzy relation). Fuzzy logic can thus be defined as a logic of approximate reasoning that allows us to work

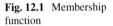

Fig. 12.1 Membership function

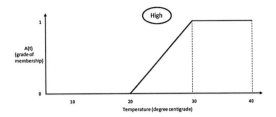

with FSs [21, 22]. In this manner, it allows a simplicity and flexibility which makes them superior with respect to classical logic for some complex problems. This can be achieved as they are able to cope with vague, imprecise or uncertain concepts that human beings use in their usual reasoning [23].

The value of $A(x)$ describes a degree of membership of x in A. For example, consider the concept of *high temperature* in an environmental context with temperatures distributed in the interval [0, 40] defined in degree centigrade. Clearly 0 °C is not understood as a high temperature value, and we may assign a null value to express its degree of compatibility with the high temperature concept. In other words, the membership degree of 0 °C in the class of high temperatures is zero. Likewise, 30 °C and over are certainly high temperatures, and we may assign a value of 1 to express a full degree of compatibility with the concept. Therefore, temperature values in the range [30, 40] have a membership value of 1 in the class of high temperatures. From 20 to 30 °C, the degree of membership in the fuzzy set high temperature gradually increases, as exemplified in Fig. 12.1, which actually is a membership function $A : T \rightarrow [0, 1]$ characterizing the fuzzy set of high temperatures in the universe $T = [0, 40]$. In this case, as temperature values increase they become more and more compatible with the idea of high temperature.

Linguistic variables are variables whose values are not numbers but words or sentences in a natural or artificial language. This concept has clearly been developed as a counterpart to the concept of a numerical variable. In concrete, a linguistic variable L is defined as a quintuple [24–26]: $L = (x, A, X, g, m)$, where x is the base variable, $A = \{A_1, A_2, \ldots, A_N\}$ is the set of *linguistic terms* of L (called *term-set*), X is the domain (universe of discourse) of the base variable, g is a syntactic rule for generating linguistic terms and m is a semantic rule that assigns to each linguistic term its *meaning* (a fuzzy set in X). Figure 12.2 shows an example of a linguistic variable *Temperature* with three linguistic terms "Low, Medium and High". The base variable is the temperature given in appropriate physical units.

Each underlying fuzzy set defines a portion of the variable's domain. But this portion is not uniquely defined. Fuzzy sets overlap as a natural consequence of their elastic boundaries. Such an overlap not only implements a realistic and functional semantic mechanism for defining the nature of a variable when it assumes various data values but also provides a smooth and coherent transition from one state to another.

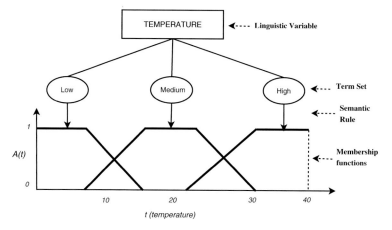

Fig. 12.2 Example of linguistic variable *Temperature* with three linguistic terms

12.2.2 Basic Elements of FRBSs

The essential part of FRBSs is a set of IF-THEN linguistic rules, whose antecedents and consequents are composed of fuzzy statements, related by the dual concepts of fuzzy implication and the compositional rule of inference.

An FRBS is composed of a *knowledge base* (KB), that includes the information in the form of IF-THEN fuzzy rules;

IF *a set of conditions are satisfied*
THEN *a set of consequents can be inferred*

and an inference engine module that includes:

- A *fuzzification interface*, which has the effect of transforming crisp data into fuzzy sets.
- An *inference system*, that uses them together with the KB to make inference by means of a reasoning method.
- A *defuzzification interface*, that translates the fuzzy rule action thus obtained to a real action using a defuzzification method.

Linguistic models are based on collections of IF-THEN rules, whose antecedents are linguistic values, and the system behavior can be described in natural terms. The consequent is an output action or class to be applied. For example, we can denote them as:

$$R_j : \text{ IF } x_{p1} \text{ IS } A_{j1} \text{ AND } \cdots \text{ AND } x_{pn} \text{ IS } A_{jn} \text{ THEN } y \text{ IS } B_j$$

with $j = 1$ to L, and with x_{p1} to x_{pn} and y being the input and output variables, with A_{j1} to A_{jn} and B_j being the involved antecedents and consequent labels, respectively. They are usually called *linguistic* FRBSs or *Mamdani* FRBSs [27].

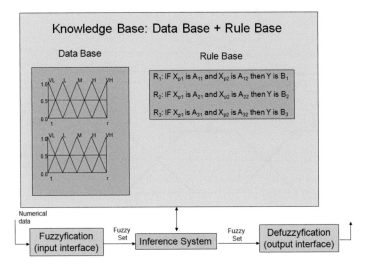

Fig. 12.3 Structure of an FRBS

In linguistic FRBSs, the KB is comprised by two components, a *data base* (DB) and a *rule base* (RB).

- A DB, containing the linguistic term sets considered in the linguistic rules and the membership functions defining the semantics of the linguistic labels.
  Each linguistic variable involved in the problem will have associated a fuzzy partition of its domain representing the fuzzy set associated with each of its linguistic terms. Reader is referred to recall Fig. 12.2 where we showed an example of fuzzy partition with three labels. This can be considered as a discretization approach for continuous domains where we establish a membership degree to the items (labels), we have an overlapping between them, and the inference engine manages the matching between the patterns and the rules providing an output according to the rule consequents with a positive matching. The determination of the fuzzy partitions is crucial in fuzzy modeling [28], and the granularity of the fuzzy partition plays an important role for the FRBS behavior [29].
- A RB, comprised of a collection of linguistic rules that are joined by a rule connective ("also" operator). In other words, multiple rules can be triggered simultaneously for the same input.

The generic structure of an FRBS is shown in Fig. 12.3.

12.2.3 Mamdani Fuzzy Inference Process

The inference engine of FRBSs acts in a different way depending of the kind of problem (classification or regression) and the kind of fuzzy rules. It always includes a fuzzification interface that serves as the input to the fuzzy reasoning process, an inference system that infers from the input to several resulting outputs (fuzzy set, class, etc.) and the defuzzification interface or output interface that converts the fuzzy sets obtained from the inference process into a crisp action that constitutes the global output of the FRBS, in the case of regression problems, or provide the final class associated to the input pattern according to the inference model.

According to Mamdani principles [30], the fuzzy inference process includes five parts, which contain a very simple structure of "max-min" operators, specifically fuzzification of the input variables, application of the fuzzy operator (AND or OR) in the antecedent, implication from the antecedent to the consequent, aggregation of the consequents across the rules and defuzzification. These five operations can be compressed into three basic steps, which are described below:

Step 1. **Computation of the Matching Degree.** The first step is to take the inputs and determine the degree to which they belong to each of the fuzzy sets considering the membership functions. In order to compute the matching degree to which each part of the antecedent is satisfied for each rule, a conjunction operator C is applied. Specifically, Mamdani recommended the use of the minimum t-norm.

$$\mu_{A_j}(x_p) = C(\mu_{A_{j1}}(x_{p1}), \dots, \mu_{A_{jn}}(x_{pn})), \qquad j = 1, \dots, L. \qquad (12.1)$$

Step 2. **Apply an Implication Operator.** In this step, the consequent is reshaped using a function associated with the antecedent (a single number). The input for the implication process is a single number given by the antecedent, and the output is a fuzzy set. Implication is implemented for each rule. Usually, two approaches for the implication operator I are employed, i.e. minimum t-norm, which truncates the output fuzzy set, and product t-norm, which scales the output fuzzy set. Mamdani also recommended the use of the minimum t-norm in this case.

$$\mu_{B'_j}(y) = I(\mu_{A_j}(x_p), \mu_{B_j}(y)) \qquad j = 1, \dots, L. \qquad (12.2)$$

Step 3. **Defuzzification process.** Decisions are based on the testing of all of the rules in a fuzzy inference system, so rules must be combined in order to make a decision. There are two modes of obtaining the output value of a fuzzy system, namely *"aggregation first, defuzzification after"* and *"defuzzification first, aggregation after"*. The defuzzification method suggested by Mamdani considers the first method using the *centre of gravity* of the individual fuzzy sets aggregated with the maximum connective *also*.

$$\mu_{B(y)} = \bigcup_j \mu_{B'_j}(y) \qquad (12.3)$$

$$y_0 = \frac{\int_y y \cdot \mu_B(y) dy}{\int_y \mu_B(y)} \qquad (12.4)$$

12.3 Evolutionary Fuzzy Systems: Taxonomy and Analysis

EFSs are a family of approaches that are built on top of FRBSs, whose components
are improved by means of an evolutionary learning/optimization process as depicted
in Fig. 12.4. This process is designed for acting or tuning the elements of a fuzzy
system in order to improve its behavior in a particular context. Traditionally, this
was carried out by means of Genetic Algorithms, leading to the classical term of
Genetic Fuzzy Systems [5, 31–33]. In this chapter, we consider a generalization of
the former by the use of EAs [7].

The central aspect on the use of EAs for automatic learning of FRBSs is that the
design process can be analyzed as a search problem in the space of models, such as
the space of rule sets, membership functions, and so on. This is carried out by means
of the coding of the model in a chromosome. Therefore, the first step in designing an
EFS is to decide which parts of the fuzzy system are subjected to optimization by the
EA coding scheme. Hence, EFS approaches can be mainly divided into two types
of processes: tuning and learning. Additionally, we must make a decision whether
to just improve the accuracy/precision of the FRBS or to achieve a tradeoff between
accuracy and interpretability (and/or other possible objectives) by means of a MOEA.
Finally, we must stress that new fuzzy set representations have been designed, which
implies a new aspect to be evolved in order to take the highest advantage of this
approach.

This high potential of EFSs implies the development of many different types of
approaches. In accordance with the above, and considering the FRBSs' components
involved in the evolutionary learning process, a taxonomy for EFS was proposed by
Herrera in [33] (please refer to its thematic Website at http://sci2s.ugr.es/gfs/). More
recently, in [6] we extended the former by distinguishing among the learning of the

Fig. 12.4 Integration of an
EFS on top of an FRBS

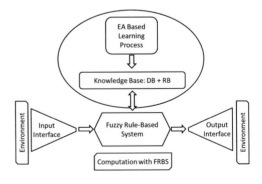

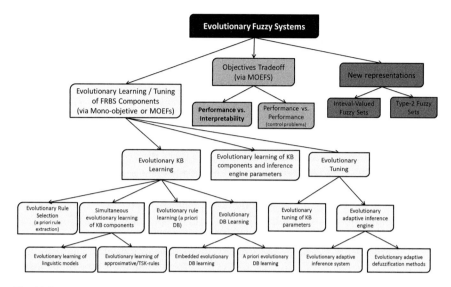

Fig. 12.5 Evolutionary fuzzy systems taxonomy

FRBSs' elements, the EA components and tuning, and the management of the new fuzzy sets representation. This novel EFS taxonomy is depicted in Fig. 12.5.

In order to describe this taxonomy tree of EFSs, this section is arranged as follows. First, we present those models according to the FRBS components involved in the evolutionary learning process (Sect. 12.3.1). Afterwards, we focus on the multi-objective optimization (Sect. 12.3.2). Finally, we provide some brief remarks regarding the parametrized construction for new fuzzy representations (Sect. 12.3.3).

12.3.1 Evolutionary Learning and Tuning of FRBSs' Components

When addressing a given Data Mining problem, the use of any fuzzy sets approach is usually considered when an interpretable system is sought, when the uncertainty involved in the data must be properly managed, or even when a dynamic model is under consideration. Then, we must make the decision on whether a simple FRBS is enough for the given requirements, or if a more sophisticated solution is needed, thus exchanging computational time for accuracy.

As introduced previously, this can be achieved either by designing approaches to learn the KB components, including an adaptive inference engine, or by starting from a given FRBS, developing approaches to tune the aforementioned components. Therefore, we may distinguish among the evolutionary KB learning, the evolutionary learning of KB components and inference engine parameters, and the evolutionary tuning. These approaches are described below, which can be performed via a standard mono-objective approach or a MOEA.

12.3.1.1   Evolutionary KB Learning

The following four KB learning possibilities can be considered:

1. *Evolutionary rule selection.* In order to get rid of irrelevant, redundant, erroneous and/or conflictive rules in the RB, which perturb the FRBS performance, an optimized subset of fuzzy rules can be obtained [34].
2. *Simultaneous evolutionary learning of KB components.* Working in this way, there is possibility of generating better definitions of these components [35]. However, a larger search space is associated with this case, which makes the learning process more difficult and slow.
3. *Evolutionary rule learning.* Most of the approaches proposed to automatically learn the KB from numerical information have focused on the RB learning, using a predefined DB [36].
4. *Evolutionary DB learning.* A DB generation process allows the shape or the membership functions to be learnt, as well as other DB components such as the scaling functions, the granularity of the fuzzy partitions, and so on. Two possibilities can be used: "a priori evolutionary DB learning" and "embedded evolutionary DB learning [37]."

12.3.1.2   Evolutionary Learning of KB Components and Inference Engine Parameters

This area belongs to a hybrid model between adaptive inference engine and KB components learning. These type of approaches try to find high cooperation between the inference engine via parameters adaptation and the learning of KB components, including both in a simultaneous learning process [38].

12.3.1.3   Evolutionary Tuning

With the aim of making the FRBS perform better, some approaches try to improve the preliminary DB definition or the inference engine parameters once the RB has been derived. The following three tuning possibilities can be considered (see the sub-tree under "evolutionary tuning").

1. *Evolutionary tuning of KB parameters.* A tuning process considering the whole KB obtained is used a posteriori to adjust the membership function parameters, i.e. the shapes of the linguistic terms [39].
2. *Evolutionary adaptive inference systems.* This approach uses parameterized expressions in the inference system, sometimes called adaptive inference sys-

tems, for getting higher cooperation among the fuzzy rules without losing the linguistic rule interpretability [40].
3. *Evolutionary adaptive defuzzification methods.* When the defuzzification function is applied by means of a weighted average operator, i.e. parameter based average functions, the use of EAs can allow us to adapt these defuzzification methods [41].

12.3.2 Approaches for Optimizing Several Objectives

Traditionally, the efforts in developing EFSs were aimed at improving the accuracy/precision of the FRBS in a mono-objective way. However, in current applications the interest of researchers in obtaining more interpretable linguistic models has significantly grown [42]. The hitch is that accuracy and interpretability represent contradictory objectives. A compromise solution is to address this problem using MOEAs [8] leading to a set of fuzzy models with different tradeoffs between both objectives instead of a biased one. These hybrid approaches are known as MOEFSs [10] that, in addition to the two aforementioned goals, may include any other kind of objective, such as the complexity of the system, the cost, the computational time, additional performance metrics, and so on.

In this case, the division of this type of techniques is first based on the multi-objective nature of the problem faced and second on the type of FRBS components optimized. Regarding the previous fact, those of the second level present a clear correspondence with the types previously described for EFSs in the previous section.

Here, we will only present a brief description for each category under consideration. For more detailed descriptions or an exhaustive list of contributions see [10] or its associated Webpage (http://sci2s.ugr.es/moefs-review/).

12.3.2.1 Accuracy-Interpretability Trade-Offs

The comprehensibility of fuzzy models began to be integrated into the optimization process in the mid 1990s [43], thanks to the application of MOEAs to fuzzy systems. Nowadays, researchers agree on the need to consider two groups of interpretability measures, complexity-based and semantic-based ones. While the first group is related to the dimensionality of the system (simpler is better) the second one is related to the comprehensibility of the system (improving the semantics of the FRBS components). For a complete survey on interpretability measures for linguistic FRBSs see [42].

The differences between both accuracy and interpretability influence the optimization process, so that researchers usually include particular developments in the proposed MOEA making it able to handle this particular trade-off. An example can be seen in [44] where authors specifically force the search to focus on the most accurate solutions.

12.3.2.2   Performance Versus Performance (Control Problems)

In control system design, there are often multiple objectives to be considered, i.e. time constraints, robustness and stability requirements, comprehensibility, and the compactness of the obtained controller. This fact has led to the application of MOEAs in the design of Fuzzy Logic Controllers.

The design of these systems is defined as the obtaining of a structure for the controller and the corresponding numerical parameters. In a general sense, they fit with the tuning and learning presented for EFSs in the previous section. In most cases, the proposal deals with the postprocessing of Fuzzy Logic Controller parameters, since it is the simplest approach and requires a reduced search space.

12.3.3   Novel Fuzzy Representations

Classical approaches on FRBSs make use of standard fuzzy sets [20], but in the specialized literature we found extensions to this approach with aim to better represent the uncertainty inherent to fuzzy logic. Among them, we stress Type-2 fuzzy sets [45] and Interval-Valued Fuzzy Sets (IVFSs) [46] as two of the main exponents of new fuzzy representations.

Type-2 fuzzy sets reduce the amount of uncertainty in a system because this logic offers better capabilities to handle linguistic uncertainties by modeling vagueness and unreliability of information. In order to obtain a type-2 membership function, we start from the type-1 standard definition, and then we blur it to the left and to the right. In this case, for a specific value, the membership function, takes on different values, which are not all weighted the same. Therefore, we can assign membership grades to all of those points.

For IVFS [46], the membership degree of each element to the set is given by a closed sub-interval of the interval [0,1]. In such a way, this amplitude will represent the lack of knowledge of the expert for giving an exact numerical value for the membership. We must point out that IVFSs are a particular case of type-2 fuzzy sets, having a zero membership out of the ranges of the interval.

In neither case, there is a general design strategy for finding the optimal fuzzy models. In accordance with the former, EAs have been used to find the appropriate parameter values and structure of these fuzzy systems.

In the case of type-2 fuzzy models, EFSs can be classified into two categories [47]: (1) the first category assumes that an "optimal" type-1 fuzzy model has already been designed, and afterwards a type-2 fuzzy model is constructed through some sound augmentation of the existing model [48]; (2) the second class of design methods is concerned with the construction of the type-2 fuzzy model directly from experimental data [49].

Regarding IVFS, current works initialize type-1 fuzzy sets as those defined homogeneously over the input space. Then, the upper and lower bounds of the interval for each fuzzy set are learnt by means of a weak-ignorance function (amplitude tuning) [50], which may also involve a lateral adjustment for the better contextualization of the fuzzy variables [51]. Finally, in [52] IVFS are built ad-hoc, using an interval-valued restricted equivalence functions within a new interval-valued fuzzy reasoning method. The parameters of these equivalence functions per variable are learnt by means of an EA, which is also combined with rule selection in order to decrease the complexity of the system.

12.4 The Use of Evolutionary Fuzzy Systems for Classification with Imbalanced Datasets

As EFSs have improved their performance from its initial models, they have been applied to novel challenges like the problem of classification with imbalanced datasets [12, 13]. This classification scenario has gained recognition in the last years as its importance comes from its presence in numerous real-world problems and the necessity of using specific approaches to address them.

In Sect. 12.4.1, we will briefly introduce the problem of classification with imbalanced datasets, outlining the approaches that are usually applied in the area and the evaluation metrics that are specifically used in this case. Then, in Sect. 12.4.2, we will provide an analysis over the EFS approaches that have been proposed to handle datasets with imbalanced distributions. Finally, one of the EFS approaches that we have developed in the topic, GP-COACH-H [19], is further described in Sect. 12.4.3 together with an experimental analysis to prove its usefulness in the imbalanced scenario.

12.4.1 Introduction to Classification with Imbalanced Datasets

The classification problem with imbalanced datasets arises when the number of examples belonging to one class is negligible with respect to the number of examples that represent the other classes [53–55]. In this problem, it is precisely the underrepresented class, also known as the minority or positive class, the one which needs to be properly identified, as its incorrect identification usually entails high costs [56, 57]. This fact contrasts with the more represented classes, also known as majority or negative classes, which are typically well identified.

The importance of this problem comes from its presence in a high number of realworld problems, becoming one of the top challenges in data mining research [58]. We can find imbalanced distributions in areas like risk management [59], bioactivity of chemical substances [60], fraud detection [61], system failure detection [62] or medical applications [63, 64], just mentioning some of them.

Standard classification algorithms are not usually able to provide a good identification of samples belonging to the minority class as they are guided by global search measures, like a percentage of well-classified examples. Thus, following the standards models are created trying to cover as many samples as possible while maintaining their simplicity. In these cases, the developed models properly identify many examples of the majority class, as they represent the higher number of examples of the whole dataset. However, minority class examples are not usually covered because their representation is small and therefore, it could have no influence in the learning stage and hence no classification rule is created for them.

In this context, the imbalance ratio (IR) [65], is traditionally used to determine how difficult a classification problem with imbalanced datasets is. Specifically, it is defined as the quotient between the number of examples belonging to the majority class and the number of samples belonging to the minority class, $IR = \frac{\#numMaj}{\#numMin}$. Although the imbalanced distribution poses a major challenge to classifiers, there are also some data intrinsic characteristics that difficult the classification with imbalanced datasets, further degrading the performance of methods than when these issues arise separately [12]. These data intrinsic characteristics include the small sample size or lack of density problem [66], the overlapping of the samples belonging to each class [16], the presence of small disjuncts in the data [67], the existence of borderline [68] or noisy samples [69] and the dataset shift [70].

Numerous approaches have been proposed to address the problem of classification with imbalanced datasets. They are commonly divided into approaches at the data-level, at the algorithm-level, and cost-sensitive learning, all of which can be embedded into an ensemble learning scheme:

- Approaches at the data-level [71, 72] modify the original training set to obtain a more or less balanced dataset that can be addressed using standard classification algorithms. These modifications to the dataset can be performed generating additional examples associated to the minority class (oversampling) or removing examples from the minority class (undersampling).
- Algorithm-level approaches [73] modify existing standard classification methods in order to enhance the identification of the minority class examples. These modifications may include the use of imbalanced measures to guide the search, a limitation of procedures designed to generalize the models or even new operations specifically designed to focus on the minority class.
- Cost-sensitive learning solutions combine approaches at the algorithm-level and the data-level for imbalanced classification considering the variable costs of misclassifying an instance as belonging to the other class [74, 75]. In imbalanced datasets, the misclassification costs associated to a minority class instance are higher than the costs associated to the misclassification of a majority class instance $C(min, maj) > C(maj, min)$, as the minority class is the main interest in the learning process.
- Ensembles have also been adapted to imbalance learning [76] obtaining a high performance when applied. In general, these new ensemble approaches introduce

in their way of running some cost-sensitive learning [77] or data preprocessing features [78, 79].

Finally, when considering the evaluation of the performance of classifiers in this context, we must proceed carefully. The most common measure of performance, the overall accuracy in classification, is not appropriate in a dataset with an uneven class distribution, as a high value in the measure can be obtained correctly classifying the instances associated to the majority class, even when all the minority class samples are not properly identified. This situation is completely undesirable as the minority class is the most interesting from the learning point of view. For this reason, more appropriate performance metrics are used in the imbalanced classification scenario.

The geometric mean (GM) [80] of the true rates is one measure that is able to avoid the problems related to the traditional accuracy metrics and is defined as:

$$GM = \sqrt{sensitivity \cdot specificity} \tag{12.5}$$

where $sensitivity = \frac{TP}{TP+FN}$ and $specificity = \frac{TN}{FP+TN}$. The sensitivity and specificity values represent the true rate for each class, computed from TP and TN which are the true rate for the minority and majority instances and the FP and FN which are the rate for the false minority samples and majority examples respectively. This metric tries to balance the performance over the two classes, combining both objectives into one.

12.4.2 EFS Approaches for Imbalanced Classification

EFSs have evolved and addressed new challenges and problems since they were first proposed. There are several proposals of EFS for imbalanced datasets; some of them study the impact and improvement of these systems modifying some of the fuzzy components while others introduce new operations in the methods without changing the basic fuzzy inference process. As the EFS methods applied to imbalanced classification are quite varied, we have organized them in four groups considering how they approach the imbalanced classification problem, namely, with data-level approaches, algorithm-level approaches, cost-sensitive learning and ensembles.

12.4.2.1 EFS and Data-Level Approaches

Data preprocessing techniques have been used together with EFSs because of their versatility as they are independent of the classifier used. FRBSs have demonstrated a good performance [14] in the imbalanced classification scenario, especially for over-sampling techniques. This has encouraged the development of different fuzzy based classifiers together with preprocessing approaches. One of the most popular oversampling techniques, the "Synthetic Minority Oversampling TEchnique" (SMOTE) algorithm [72], has been extensively used in combination with EFSs.

One of the EFSs approximations developed for imbalanced datasets is the one described in [81]. In it, an adaptive inference system with parametric conjunction operators is presented. To deal with the imbalance, the SMOTE algorithm is used to balance the datasets. The idea presented in the paper is based on the suitability of adaptive t-norms, like the Dubois t-norm where a parameter α modifies how the t-norm behaves. Using the CHC evolutionary algorithm [82] with the GM performance measure as evaluation function, the α parameter can be learned for the whole RB or for each specific rule improving the overall performance of the system.

An analysis about the evolutionary tuning of the KB for classification with imbalanced datasets is performed in [83]. In order to avoid the imbalanced problem, the SMOTE algorithm is again used to obtain a balanced distribution of the train set. In this work, a genetic process based on the CHC evolutionary algorithm is used to learn the lateral displacement of the DB using the 2-tuples linguistic representation [84]. This lateral translation can be learned over the full DB for the complete RB or adapting each set of fuzzy labels according to each specific rule in the RB. In this way, the performance of the Chi et al. method [85] and the FH-GBML classifier [86] is improved.

The imbalanced problem can emerge in conjunction with other problems like the availability of low quality data. Therefore, the uncertainty that needs to be managed does not refer only to the difficult identification of samples for each class but also to the values associated to the input values of the samples. In [18], several preprocessing techniques are adapted to the low quality data scenario to obtain a more or less balanced distribution that can be managed more easily. Specifically, low quality data versions of the ENN [87], NCL [88], CNN [89], SMOTE [72] and SMOTE+ENN [71] algorithms are designed to classify low quality imbalanced data using a genetic cooperative-competitive learning algorithm. The performance of these versions is similar to the one obtained with the preprocessing methods for the standard imbalanced problems.

In [90], a genetic procedure for learning the KB in imbalanced datasets, GA-FS+GL, is proposed. In this case, the SMOTE algorithm is again used to balance the training set. The idea presented in this work is the use of a GA to perform a feature selection and a selection of the granularity of the data base. To perform the feature selection, a binary part of the chromosome is used to determine if an attribute is used or not. To select the granularity of the DB, the algorithm searches for the best performing set of labels considering different number of labels between two and seven using equally distributed triangular membership functions. The approach is tested over the Chi et al. method [85] obtaining competitive results.

The data intrinsic characteristics can degrade the performance of imbalanced classifiers in a further extent than when they appear in isolation [12]. In [91], the impact of dataset shift over imbalanced classifiers is studied. Specifically, two partitioning techniques for the validation of classifiers are compared: the stratified cross-validation and a novel cross-validation approach named DOB-SCV [92], an approach that tries to limit the covariate shift that is induced when partitions are created. The FARC-HD classifier [93] is selected to compare how the dataset shift changes the behavior of EFS classifiers. The results obtained show that it is advisable to limit the dataset shift introduced in these processes even when we are using fuzzy systems that can cope with uncertainty and imprecision.

12.4.2.2 EFS and Algorithm-Level Approaches

Modifying operations within the algorithm design to further enhance the correct identification of examples belonging to the minority class is a popular solution to adapt EFS for imbalanced classification. In some cases, these alterations are enough to address the imbalanced distributions, however, in others it is necessary to combine them with preprocessing techniques to further improve the performance of these methods.

One of the approaches that follows a design with specific operations for the imbalance is the one described in [94], renamed in [95] as FLAGID, Fuzzy Logic And Genetic algorithms for Imbalanced Datasets. This approach follow several stages, starting with a first step that is a modified version of the RecBF/DDA algorithm [96]. This first step, creates the membership functions that are going to be used afterwards, creating a smaller number of membership functions for the minority class. The second stage is called ReRecBF and its aim is to simplify the previously created functions so that they cover important regions being able to at least represent a 10 % of the class. Finally, the third stage learns the RB considering the previously generated membership functions using a genetic algorithm procedure that uses the GM as fitness function.

The use of a hierarchical fuzzy rule-based classification system (HFRBCS) for imbalanced classifiers has been considered in [17], using Chi et al.'s method as baseline classifier. Additionally, in [19] authors propose GP-COACH-H, which is also based on a hierarchical system. In both HFRBCSs, the KB is structured following different levels of learning, being the lower levels more general and the higher levels more specific. This type of approaches aim to improve the performance of methods in difficult data areas like the data intrinsic characteristics that further difficult the classification with imbalanced datasets. In a first stage, the SMOTE algorithm is used to balance the data that will be later processed by the hierarchical methods that try to identify the samples in difficult areas. For HFRBCS(Chi) [17], the generation process of the hierarchical rules is based on the Chi et al. method [85]. When the hierarchical rule base has been obtained, a rule selection process to select a subset of rules for the final classifier is used. For GP-COACH-H [19], the hierarchical rule base is obtained modifying the GP-COACH algorithm [97]. In this method, a subsequent step is applied with a genetic tuning of the knowledge base where a combined selection of the rules and a lateral tuning based on the 2-tuples representation is developed. GP-COACH-H will be later described as our selected method for the case of study.

Another algorithm that has been modified for imbalanced classification is the FARC-HD algorithm [93]. Specifically, this method is applied in the case study of intrusion detection systems [98, 99], following two different schemes. In the first case [100], the alterations to the method follow two aspects: the first one is the use of the SMOTE algorithm to preprocess the data for the subsequent operations; whereas the second one is related to the changes introduced in the genetic tuning of the knowledge base phase that is performed in the FARC-HD method. This genetic procedure changes its evaluation function to the GM performance measure. On the other hand, in [101] the FARC-HD EFS baseline algorithm is embedded in a pairwise learning scheme [102]. This is done for being able to improve the recognition of the minority class instances by simplifying the borderline areas in each binary classifier. Finally, Sanz et al. applied an interval-valued fuzzy sets version of FARC-HD in the context of the modeling and prediction of imbalanced financial datasets [103].

Following diverse evolutionary approaches, we can find another proposal for imbalanced classification in [104]. The method is divided in two steps, a first step that applies a feature selection process to reduce the dimensionality of the training set and a second step to generate the fuzzy rules using evolutionary techniques. This second step is divided in several steps as well: first, a differential evolution optimization process is performed to estimate a *radii* value; then, this *radii* value will be used in a subtractive clustering method to generate fuzzy TSK rules; finally, a genetic programming step is used to improve the fuzzy TSK rules and convert them to Mamdani classification rules.

12.4.2.3   EFS and Cost-Sensitive Learning

As in the previous cases, EFSs have also been adapted to classification with imbalanced datasets following cost-sensitive learning, that is, considering the costs within the algorithm to favor the classification of the minority class examples. In [105], the FH-GBML-CS method is proposed, which is a cost-sensitive version of the FH-GBML algorithm [86]. The costs are introduced in the evaluation function of the genetic procedure and in the computation of the rule weight, using the what is called as the cost-sensitive penalized certainty factor, a modified version including costs of the penalized certainty factor [106]. Moreover, the inference process for the fuzzy reasoning method considers the compatibility degree of the example and the fuzzy label together with the cost associated to that example.

In [107], a cost-sensitive MOEFS for imbalanced classification is presented. In this method, the NSGA-II algorithm is used to create an FRBS using the specificity performance measure as an objective, the sensitivity performance measure as another objective and the complexity as a third objective. This complexity measure is computed adding the number of antecedents in all the rules in the RB. Finally, a ROC convex hull technique is used together with the misclassification costs to select the best solution of the pareto of solutions obtained by the multi-objective method.

12.4.2.4   EFS and Ensemble Learning

The existing learning approaches for imbalanced classification have been typically used with weak learners like decision trees instead of using other approaches. In this way, ensembles are rarely combined with EFS in the imbalanced scenario. However, even when this type of systems are not specifically designed for imbalance, we can find proposals that are tested over imbalanced datasets.

For example, in [108], a boosting method based on fuzzy rules is proposed. This method is divided in two stages. In the first one, a fuzzy rule learning method is applied. This method is based on the AdaBoost algorithm [109], which has been modified following an iterative rule learning approach. Rules are created with an evolutionary rule generation method that adapts the rules to the dataset according to the learned weights in each iteration until the performance does not improve or even decreases. When this first stage has finished, a genetic tuning step is performed. There are no specific operations for the imbalanced distributions in this method as it was not specifically designed for imbalanced data, however, the weights associated to each sample in the boosting can favor the correct identification of minority class examples. Furthermore, the method is tested in the land cover classification of remote sensing imagery problem, which can feature imbalanced distributions.

Finally, in [110], three EFS systems for imbalanced classification are compared, which are the GA-FS+GL method described in [90], the GP-COACH-H algorithm presented in [19] and the MOEFS developed in [107]. The authors use 22 datasets to perform this comparison being the MOEFS approach the one with the best performance supported by a Holm test. However, the results extracted from this comparison need to be treated with care, as the AUC performance measure [111] is not computed in the same way for the three methods: for the MOEFS approach, the AUC measure is computed considering all the classifiers obtained in the multi-objective process without applying the ROC convex hull technique that selects one of them, while for the other two methods, the AUC measure is calculated considering the one point formula.

12.4.3   Case Study: Addressing Highly Imbalanced Datasets with GP-COACH-H

Having analyzed how EFS adopt the different strategies for imbalanced datasets, we have selected one of the EFS approaches to demonstrate its effectiveness for classification with imbalanced datasets. In a first step, we will further describe the GP-COACH-H algorithm, a fuzzy rule-based classification system for imbalanced data. Then, we will present the experimental framework associated to the study performed, the result tables and its associated statistical tests.

12.4.3.1   GP-COACH-H: A Hierarchical Genetic Programming Fuzzy Rule-Based Classification System with Rule Selection and Tuning

The GP-COACH-H algorithm [19] is a fuzzy rule-based classification system that has been developed to effectively address imbalanced datasets in arduous imbalanced scenarios, such as highly imbalanced datasets and borderline imbalanced datasets. To do so, the GP-COACH-H algorithm follows an algorithm-level approach as its behavior is modified in order to favor the correct identification of samples belonging to the minority class.

The proposal combines several strategies that are able to obtain a good synergy between them, namely, data preprocessing, a hierarchical linguistic classifier and genetic tuning of KB parameters. As previously mentioned, the data preprocessing modifies the input dataset to ease the learning process of the subsequent classifier.

An HFRBCS [112] extends the standard definition of the KB so it can better model complex search spaces such as imbalanced datasets entangled with some data intrinsic characteristics like borderline examples, overlapping between the classes or even small disjuncts. In these systems, the KB is called hierarchical knowledge base (HKB), as the changes introduced by the hierarchical model affect both the DB and the RB.

An HKB is composed by a set of layers, which represent different granularity levels. Each layer has its own DB and RB which determine the specificity level of description that can be achieved by the model. The RB of a layer can only use the fuzzy linguistic labels defined in the associated DB. The layers are organized in a hierarchical way: a new layer level has a higher number of fuzzy labels than the previous level and the fuzzy labels built in the new level are created preserving the membership function modal points, adding a new linguistic term between each two consecutive terms of the linguistic partition belonging to the previous model. The idea behind the usage of this system is to use a low hierarchical level to describe general areas of data, while using a larger hierarchical level to illustrate more complex areas.

Another strategy used in this method is the genetic tuning of KB parameters. As it was explained in the previous sections, the objective of this component is to enhance the previously learned KB using genetic algorithms so that the final model is able to better characterize the classes of the dataset.

The GP-COACH-H algorithm follows a three stages approach in order to integrate the different strategies proposed. A flowchart of the building of this model can be found in Fig. 12.6. The three stages of the algorithm involve the following operations:

1. *Data preprocessing*: As a first step, the GP-COACH-H algorithm needs to modify the original training set so it displays a more or less balanced distribution. In order to do so, an oversampling scheme is used, following the SMOTE algorithm [72] to generate synthetic samples associated to the minority class.
2. *HKB generation process with an EFS*: Considering the dataset obtained in the previous step, a genetic programming approach is used to generate the HKB. GP-

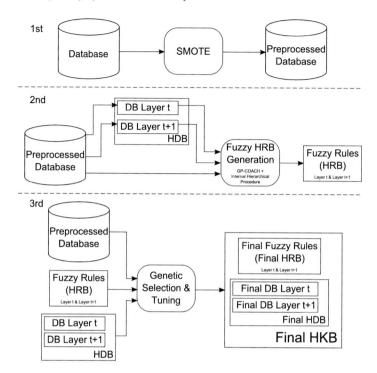

Fig. 12.6 Flowchart of the GP-COACH-H algorithm

COACH-H is based on the GP-COACH algorithm [97], and therefore, it follows
the genetic programming procedure adopted in that method. To obtain a HKB,
some of the GP-COACH steps need to be modified to enable the usage of rules of
different hierarchies in the same population. The first alteration, is the addition of
a new step in each generation of the genetic approach: the identification of *good*
and *bad* rules in the current population. *Bad* rules are discarded and are replaced
by new high granularity rules. Another modification to the genetic process is the
use of a new evaluation function that is able to consider different granularity levels
in the rules. New constraints over the crossover operator need to be performed to
ensure that it is only applied to rules of the same hierarchy.

3. *Genetic tuning of the HKB parameters*: When the building of the HKB has ended,
   a genetic tuning process of the HKB parameters is started to further adapt the
   classifier to the available data. In this step, we try to perform a selection of rules that
   demonstrate a good cooperation [34] while also tuning the existing hierarchical
   DBs following a 2-tuples linguistic representation [84]. These optimizations are
   done together with an unique genetic procedure using the CHC evolutionary
   algorithm to profit from the synergy that these optimizations can achieve. To do
   so, each part of the chromosome codifies the rule selection process or the tuning
   adjustment where genetic operators are modified to consider this situation.

12.4.3.2   Experimental Study

To evaluate the performance of the GP-COACH-H algorithm, we have selected 44 highly imbalanced datasets with an IR higher than 9 from the KEEL dataset repository[1] [113]. Table 12.1 summarizes the datasets used in this study, showing for each one the number of examples (#Ex.), the number of attributes (#Atts.), the class name associated to each class (minority and majority), the attribute class distribution and the IR. This table is sorted by increasing order of IR.

To carry out this study, we follow a five-fold cross-validation approach, that is, performing five random partitions of the data where the aggregation of four of them constitutes the training set, and the other partition is considered as the test set. The results presented in this work display the average results obtained in the five partitions.

For the GP-COACH-H algorithm, we have set the parameters of its different components according to what is usual in these domains. For the SMOTE preprocessing part, we consider only the 1-nearest neighbor (using the euclidean distance) to generate the synthetic examples to obtain a balanced dataset. The parameters associated to the fuzzy rule-based classification system are the use of the minimum t-norm as conjunction operator, the maximum t-norm as disjunction operator, the certainty factor is used to compute the rule weight, the normalized sum is used as fuzzy reasoning method and 5 fuzzy labels are used for low granularity rules while 9 fuzzy labels are used for high granularity rules. Considering the genetic part of the building of the model, the number of evaluations used are 20000, the initial population has a size of 200, the α value for the raw fitness is 0.7, the crossover probability is 0.5, the mutation probability is 0.2, the dropping condition probability is 0.15, the insertion probability is 0.15 as well, the tournament size is 2, while the weights associated to the fitness function are $w_1 = 0.8$, $w_2 = w_3 = 0.05$, $w_4 = 0.1$. Finally, the parameters associated to the hierarchical procedure and the last genetic tuning phase are an α of 0.2 to detect *good* and *bad* rules, the number of evaluations is 10000, the population size is 61 and the bits per gene are 30.

To demonstrate the good performance of the GP-COACH-H algorithm for imbalanced datasets, we have selected the C4.5 algorithm [114], a well-known decision tree that has displayed a good behavior for imbalanced datasets [71]. The parameters associated to this method are the ones recommended by the author, namely, the use of a pruned tree, a confidence of 0.25 and 2 as the minimum number of item-sets per leaf. To deal with the imbalance, we have combined C4.5 with the SMOTE+ENN algorithm [71], where the ENN cleaning method [87] is directly applied after the SMOTE algorithm to generalize the borders between the classes. For generating synthetic samples, the 1-nearest neighbor is used to obtain a balanced dataset. For the ENN part, 3-nearest neighbors are considered. In both cases, the euclidean distance is applied.

Furthermore, we have used statistical tests to detect whether there are significant differences among the results achieved by the different tested methods [115]. We will use non-parametric tests as the conditions that guarantee the reliability of parametric

---

[1]http://www.keel.es/imbalanced.php.

Table 12.1 Summary of highly imbalanced datasets

Datasets	#Ex.	#Atts.	Class (maj;min)	%Class(maj; min)	IR
ecoli034vs5	200	7	(p,imL,imU; om)	(10.00, 90.00)	9.00
yeast2vs4	514	8	(cyt; me2)	(9.92, 90.08)	9.08
ecoli067vs35	222	7	(cp,omL,pp; imL,om)	(9.91, 90.09)	9.09
ecoli0234vs5	202	7	(cp,imS,imL,imU; om)	(9.90, 90.10)	9.10
glass015vs2	172	9	(build-win-non_float-proc,tableware, build-win-float-proc; ve-win-float-proc)	(9.88, 90.12)	9.12
yeast0359vs78	506	8	(mit,me1,me3,erl; vac,pox)	(9.88, 90.12)	9.12
yeast02579vs368	1004	8	(mit,cyt,me3,vac,erl; me1,exc,pox)	(9.86, 90.14)	9.14
yeast0256vs3789	1004	8	(mit,cyt,me3,exc; me1,vac,pox,erl)	(9.86, 90.14)	9.14
ecoli046vs5	203	6	(cp,imU,omL; om)	(9.85, 90.15)	9.15
ecoli01vs235	244	7	(cp,im; imS,imL,om)	(9.83, 90.17)	9.17
ecoli0267vs35	224	7	(cp,imS,omL,pp; imL,om)	(9.82, 90.18)	9.18
glass04vs5	92	9	(build-win-float-proc,containers; tableware)	(9.78, 90.22)	9.22
ecoli0346vs5	205	7	(cp,imL,imU,omL; om)	(9.76, 90.24)	9.25
ecoli0347vs56	257	7	(cp,imL,imU,pp; om,omL)	(9.73, 90.27)	9.28
yeast05679vs4	528	8	(me2; mit,me3,exc,vac,erl)	(9.66, 90.34)	9.35
ecoli067vs5	220	6	(cp,omL,pp; om)	(9.09, 90.91)	10.00
vowel0	988	13	(hid; remainder)	(9.01, 90.99)	10.10
glass016vs2	192	9	(ve-win-float-proc; build-win-float-proc, build-win-non_float-proc,headlamps)	(8.89, 91.11)	10.29
glass2	214	9	(Ve-win-float-proc; remainder)	(8.78, 91.22)	10.39
ecoli0147vs2356	336	7	(cp,im,imU,pp; imS,imL,om,omL)	(8.63, 91.37)	10.59
led7digit02456789vs1	443	7	(0,2,4,5,6,7,8,9; 1)	(8.35, 91.65)	10.97
glass06vs5	108	9	(build-win-float-proc,headlamps; tableware)	(8.33, 91.67)	11.00
ecoli01vs5	240	6	(cp,im; om)	(8.33, 91.67)	11.00
glass0146vs2	205	9	(build-win-float-proc,containers,headlamps, build-win-non_float-proc;ve-win-float-proc)	(8.29, 91.71)	11.06
ecoli0147vs56	332	6	(cp,im,imU,pp; om,omL)	(7.53, 92.47)	12.28

Table 12.1 (continued)

Datasets	#Ex.	#Atts.	Class (maj;min)	%Class(maj; min)	IR
cleveland0vs4	177	13	(0; 4)	(7.34, 92.66)	12.62
ecoli0146vs5	280	6	(cp,im,imU,omL; om)	(7.14, 92.86)	13.00
ecoli4	336	7	(om; remainder)	(6.74, 93.26)	13.84
yeast1vs7	459	8	(nuc; vac)	(6.72, 93.28)	13.87
shuttle0vs4	1829	9	(Rad Flow; Bypass)	(6.72, 93.28)	13.87
glass4	214	9	(containers; remainder)	(6.07, 93.93)	15.47
page-blocks13vs2	472	10	(graphic; horiz.line,picture)	(5.93, 94.07)	15.85
abalone9vs18	731	8	(18; 9)	(5.65, 94.25)	16.68
glass016vs5	184	9	(tableware; build-win-float-proc, build-win-non_float-proc,headlamps)	(4.89, 95.11)	19.44
shuttle2vs4	129	9	(Fpv Open; Bypass)	(4.65, 95.35)	20.5
yeast1458vs7	693	8	(vac; nuc,me2,me3,pox)	(4.33, 95.67)	22.10
glass5	214	9	(tableware; remainder)	(4.20, 95.80)	22.81
yeast2vs8	482	8	(pox; cyt)	(4.15, 95.85)	23.10
yeast4	1484	8	(me2; remainder)	(3.43, 96.57)	28.41
yeast1289vs7	947	8	(vac; nuc,cyt,pox,erl)	(3.17, 96.83)	30.56
yeast5	1484	8	(me1; remainder)	(2.96, 97.04)	32.78
ecoli0137vs26	281	7	(pp,imL; cp,im,imU,imS)	(2.49, 97.51)	39.15
yeast6	1484	8	(exc; remainder)	(2.49, 97.51)	39.15
abalone19	4174	8	(19; remainder)	(0.77, 99.23)	128.87

tests may not be satisfied [116]. As we are performing a pairwise comparison, we use the Wilcoxon test to search for statistical differences between two methods. The objective in this comparison is to obtain an adjusted *p*-value, which represents the lowest level of significance of a hypothesis that results in a rejection, which means the detection of significant differences between the methods.

Table 12.2 shows the average GM values in training and test obtained by the algorithms included in the comparison for the 44 highly imbalanced datasets, namely, the GP-COACH-H algorithm and the C4.5 decision tree together with SMOTE+ENN. The best average values in test per dataset are highlighted in bold.

From these results we can observe that the best performing method is GP-COACH-H, showing the good synergy between the elements integrated into this approach. The GP-COACH-H algorithm is able to obtain better results than the C4.5 decision tree combined with SMOTE+ENN, showing that the goodness in its predictive behavior is obtained through the appropriate integration of the EFS with the hierarchical approach and not by the use of the data preprocessing techniques.

Table 12.2   Detailed table of results for GP-COACH-H and SMOTE+ENN+C4.5

Dataset	GP-COACH-H		SMOTE+ENN+C4.5	
	GM_{tr}	GM_{tst}	GM_{tr}	GM_{tst}
ecoli034vs5	0.9833	0.8660	0.9762	**0.8761**
yeast2vs4	0.9647	**0.9304**	0.9745	0.9029
ecoli067vs35	0.9707	**0.7286**	0.9771	0.7206
ecoli0234vs5	0.9966	0.8473	0.9827	**0.8861**
glass015vs2	0.9503	0.6301	0.9066	**0.7788**
yeast0359vs78	0.8919	**0.7189**	0.9213	0.6894
yeast02579vs368	0.9298	0.9107	0.9572	**0.9125**
yeast0256vs3789	0.8348	**0.7982**	0.9173	0.7707
ecoli046vs5	0.9952	0.8677	0.9834	**0.8776**
ecoli01vs235	0.9845	**0.8471**	0.9649	0.8277
ecoli0267vs35	0.9707	**0.9028**	0.9825	0.8061
glass04vs5	0.9909	0.9429	0.9909	**0.9748**
ecoli0346vs5	0.9993	0.8847	0.9884	**0.8946**
ecoli0347vs56	0.9881	**0.8767**	0.9566	0.8413
yeast05679vs4	0.8961	0.6988	0.9197	**0.7678**
ecoli067vs5	0.9849	**0.8671**	0.9740	0.8376
vowel0	0.9947	**0.9465**	0.9943	0.9417
glass016vs2	0.9415	**0.6467**	0.9365	0.6063
glass2	0.9663	0.5886	0.9261	**0.7377**
ecoli0147vs2356	0.9594	**0.8263**	0.9563	0.8119
led7digit02456789vs1	0.9142	**0.9000**	0.9217	0.8370
glass06vs5	0.9975	0.9120	0.9911	**0.9628**
ecoli01vs5	0.9977	**0.8946**	0.9828	0.8081
glass0146vs2	0.9313	**0.7300**	0.9010	0.6157
ecoli0147vs56	0.9852	**0.8372**	0.9608	0.8250
cleveland0vs4	0.9719	**0.8646**	0.9819	0.7307
ecoli0146vs5	0.9952	**0.9194**	0.9850	0.8880
ecoli4	0.9936	**0.9357**	0.9826	0.8947
yeast1vs7	0.8988	0.6900	0.9093	**0.7222**
shuttle0vs4	1.0000	**1.0000**	0.9999	0.9997
glass4	0.9906	0.7303	0.9665	**0.7639**
page-blocks13vs4	0.9994	0.9482	0.9975	**0.9909**
abalone9-18	0.8595	**0.7500**	0.9273	0.6884
glass016vs5	0.9921	**0.8550**	0.9863	0.7738
shuttle2vs4	1.0000	0.9918	1.0000	**1.0000**
yeast1458vs7	0.8952	**0.6304**	0.8717	0.3345
glass5	0.9957	**0.7877**	0.9698	0.5851
yeast2vs8	0.9937	0.7381	0.8923	**0.8033**

(continued)

Table 12.2 (continued)

Dataset	GP-COACH-H		SMOTE+ENN+C4.5	
	GM_{tr}	GM_{tst}	GM_{tr}	GM_{tst}
yeast4	0.9001	**0.8175**	0.8984	0.6897
yeast1289vs7	0.8843	**0.6939**	0.9408	0.5522
yeast5	0.9724	**0.9428**	0.9819	0.9390
ecoli0137vs26	0.9843	**0.7067**	0.9650	0.7062
yeast6	0.9319	**0.8170**	0.9301	0.8029
abalone19	0.8558	**0.5532**	0.8838	0.1550
Mean	0.9576	**0.8175**	0.9549	0.7848

Table 12.3 Wilcoxon test to compare GP-COACH-H against SMOTE+ENN+C4.5 in highly imbalanced datasets

Comparison	R^+	R^-	p-Value
GP-COACH-H versus SMOTE+ENN+C4.5	667.0	323.0	0.0446

R^+ corresponds to the sum of the ranks for GP-COACH-H and R^- to C4.5

To further support this evidence, we use the Wilcoxon test [116] to develop the statistical study that aims to find statistical differences. Table 12.3 shows the rankings associated to each method and the adjusted p-value that has been calculated. The p-value obtained by the Wilcoxon test, 0.0446, is low enough to reject the null hypothesis, which means that statistical differences are found with a degree of confidence near to the 95 %.

To sum up, we have presented an EFS based approach, GP-COACH-H, to deal with highly imbalanced datasets. The approach is based on a genetic programming approach to build the KB which combines several strategies, including the use of data preprocessing, HKB and genetic tuning of the KB to enhance the performance. The experimental results have demonstrated that GP-COACH-H outperforms the C4.5 decision tree combined with data preprocessing over 44 highly imbalanced datasets, becoming a competitive method in the imbalanced classification scenario.

12.5 Concluding Remarks

In this chapter, we have reviewed the topic of EFSs focusing on the application of this type of systems for classification with imbalanced datasets.

With this aim, we have introduced some preliminaries about linguistic fuzzy rule based systems in order to set the context of this work. Then, we have presented a complete taxonomy for the current types of associated methodologies. Specifically, we have distinguished between three approaches, namely the learning of the FRBS' elements, the different schemes regarding the evolutionary components, and finally the optimization of novel fuzzy representations.

Regarding imbalanced classification, we have paid special interest in providing the design principles for those algorithms that have been used in this work area. Among them, we have divided into those solutions applied in conjunction with data level techniques, algorithm-level approaches, cost-sensitive learning, and the ones embedded into ensemble learning.

Finally, we have analyzed and evaluated the good properties and features of EFSs in this context. In order to do so, we have presented a case study with a recent EFS approach, the GP-COACH-H algorithm. Experimental results stressed the goodness of these types of approaches for addressing the problem of classification with imbalanced data.

Acknowledgments This work have been partially supported by the Spanish Ministry of Science and Technology under projects TIN-2012-33856, TIN2014-57251-P; the Andalusian Research Plans P12-TIC-2958, P11-TIC-7765 and P10-TIC-6858; and both the University of Jaén and Caja Rural Provincial de Jaén under project UJA2014/06/15.

References

1. A. Konar, *Computational Intelligence: Principles, Techniques and Applications* (Springer, Berlin, 2005)
2. R.R. Yager, D.P. Filev, *Essentials of Fuzzy Modeling and Control* (Wiley, 1994)
3. D.E. Goldberg, *Genetic Algorithms in Search, Optimization, and Machine Learning* (Addison-Wesley Professional, Upper Saddle River, 1989)
4. J.H. Holland, *Adaptation in Natural and Artificial Systems* (University of Michigan Press, Ann Arbor, 1975)
5. O. Cordon, F. Herrera, F. Hoffmann, L. Magdalena, Genetic fuzzy systems. *Evolutionary Tuning and Learning of Fuzzy Knowledge Bases* (World Scientific, Singapore, Republic of Singapore, 2001)
6. A. Fernandez, V. Lopez, M.J. del Jesus, F. Herrera, Revisiting evolutionary fuzzy systems: taxonomy, applications, new trends and challenges. Knowl. Based Syst. **80**, 109–121 (2015)
7. A.E. Eiben, J.E. Smith, *Introduction to Evolutionary Computation* (Springer, Berlin, 2003)
8. C.A. Coello-Coello, G. Lamont, D. van Veldhuizen, *Evolutionary Algorithms for Solving Multi-objective Problems*, Genetic and Evolutionary Computation, 2nd edn. (Springer, Berlin, 2007)
9. K. Deb, *Multi-objective Optimization using Evolutionary Algorithms* (Wiley, New York, 2001)
10. M. Fazzolari, R. Alcala, Y. Nojima, H. Ishibuchi, F. Herrera, A review of the application of multi-objective evolutionary systems: current status and further directions. IEEE Trans. Fuzzy Syst. **21**(1), 45–65 (2013)
11. J. Han, M. Kamber, J. Pei, *Data Mining: Concepts and Techniques*, 3rd edn. (Morgan Kaufmann, San Mateo, 2011)
12. V. Lopez, A. Fernandez, S. Garcia, V. Palade, F. Herrera, An insight into classification with imbalanced data: empirical results and current trends on using data intrinsic characteristics. Inf. Sci. **250**(20), 113–141 (2013)
13. R.C. Prati, G.E.A.P.A., Batista, D.F. Silva, Class imbalance revisited: a new experimental setup to assess the performance of treatment methods. Knowl. Inf. Syst. **45**(1), 247–270 (2015)

14. A. Fernandez, S. Garcia, M.J. del Jesus, F. Herrera, A study of the behaviour of linguistic fuzzy rule based classification systems in the framework of imbalanced data-sets. Fuzzy Sets Syst. **159**(18), 2378–2398 (2008)
15. S. Alshomrani, A. Bawakid, S.O. Shim, A. Fernandez, F. Herrera, A proposal for evolutionary fuzzy systems using feature weighting: dealing with overlapping in imbalanced datasets. Knowl.-Based Syst. **73**, 1–17 (2015)
16. V. Garcia, R.A. Mollineda, J.S. Sanchez, On the k-NN performance in a challenging scenario of imbalance and overlapping. Pattern Anal. Appl. **11**(3–4), 269–280 (2008)
17. A. Fernandez, M.J. del Jesus, F. Herrera, Hierarchical fuzzy rule based classification systems with genetic rule selection for imbalanced data-sets. Int. J. Approximate Reasoning **50**, 561–577 (2009)
18. A. Palacios, L. Sanchez, I. Couso, Equalizing imbalanced imprecise datasets for genetic fuzzy classifiers. Int. J. Comput. Intell. Syst. **5**(2), 276–296 (2012)
19. V. Lopez, A. Fernandez, M.J. del Jesus, F. Herrera, A hierarchical genetic fuzzy system based on genetic programming for addressing classification with highly imbalanced and borderline data-sets. Knowl. Based Syst. **38**, 85–104 (2013)
20. L.A. Zadeh, Fuzzy sets. Inf. Control **8**, 338–353 (1965)
21. G. Klir, B. Yuan, *Fuzzy Sets and Fuzzy Logic: Theory and Applications* (Prentice-Hall, 1995)
22. H. Zimmermann, Fuzzy set theory. WIREs Comput. Stat. **2**(3), 317–332 (2010)
23. W. Pedrycz, F. Gomide, *An Introduction to Fuzzy Sets: Analysis and Design* (Prentice-Hall, 1998)
24. L.A. Zadeh, The concept of a linguistic variable and its applications to approximate reasoning. part I. Inf. Sci. **8**, 199–249 (1975)
25. L.A. Zadeh, The concept of a linguistic variable and its applications to approximate reasoning. part II. Inf. Sci. **8**, 301–357 (1975)
26. L.A. Zadeh, The concept of a linguistic variable and its applications to approximate reasoning. part III. Inf. Sci. **9**, 43–80 (1975)
27. E.H. Mamdani, Applications of fuzzy algorithm for control a simple dynamic plant. Proc. Inst. Electr. Eng. **121**(12), 1585–1588 (1974)
28. W.H. Au, K.C.C. Chan, A.K.C. Wong, A fuzzy approach to partitioning continuous attributes for classification. IEEE Trans. Knowl. Data Eng. **18**(5), 715–719 (2006)
29. O. Cordon, F. Herrera, P. Villar, Analysis and guidelines to obtain a good fuzzy partition granularity for fuzzy rule-based systems using simulated annealing. Int. J. Approximate Reasoning **25**(3), 187–215 (2000)
30. E.H. Mamdani, Application of fuzzy logic to approximate reasoning using linguistic synthesis. IEEE Trans. Comput. **26**(12), 1182–1191 (1977)
31. O. Cordon, F. Gomide, F. Herrera, F. Hoffmann, L. Magdalena, Ten years of genetic fuzzy systems: current framework and new trends. Fuzzy Sets Syst. **141**, 5–31 (2004)
32. O. Cordon, A historical review of evolutionary learning methods for mamdani-type fuzzy rule-based systems: designing interpretable genetic fuzzy systems. Int. J. Approximate Reasoning **52**(6), 894–913 (2011)
33. F. Herrera, Genetic fuzzy systems: taxonomy, current research trends and prospects. Evol. Intel. **1**(1), 27–46 (2008)
34. H. Ishibuchi, K. Nozaki, N. Yamamoto, H. Tanaka, Selection of fuzzy IF-THEN rules for classification problems using genetic algorithms. IEEE Trans. Fuzzy Syst. **3**(3), 260–270 (1995)
35. A. Homaifar, E. McCormick, Simultaneous design of membership functions and rule sets for fuzzy controllers using genetic algorithms. IEEE Trans. Fuzzy Syst. **3**(2), 129–139 (1995)
36. P. Thrift, Fuzzy logic synthesis with genetic algorithms, in *Proceedings of the 4th International Conference on Genetic Algorithms (ICGA'91)*. pp. 509–513 (1991)
37. O. Cordon, F. Herrera, P. Villar, Generating the knowledge base of a fuzzy rule-based system by the genetic learning of data base. IEEE Trans. Fuzzy Syst. **9**(4), 667–674 (2001)
38. F. Marquez, A. Peregrín, F. Herrera, Cooperative evolutionary learning of linguistic fuzzy rules and parametric aggregation connectors for mamdani fuzzy systems. IEEE Trans. Fuzzy Syst. **15**(6), 1162–1178 (2008)

39. J. Casillas, O. Cordon, M.J. del Jesus, F. Herrera, Genetic tuning of fuzzy rule deep structures preserving interpretability and its interaction with fuzzy rule set reduction. IEEE Trans. Fuzzy Syst. **13**(1), 13–29 (2005)
40. J. Alcala-Fdez, F. Herrera, F.A. Marquez, A. Peregrin, Increasing fuzzy rules cooperation based on evolutionary adaptive inference systems. Int. J. Intell. Syst. **22**(9), 1035–1064 (2007)
41. D. Kim, Y. Choi, S.Y. Lee, An accurate cog defuzzifier design using lamarckian co-adaptation of learning and evolution. Fuzzy Sets Syst. **130**(2), 207–225 (2002)
42. M.J. Gacto, R. Alcala, F. Herrera, Interpretability of linguistic fuzzy rule-based systems: an overview of interpretability measures. Inf. Sci. **181**(20), 4340–4360 (2011)
43. H. Ishibuchi, T. Murata, I. Turksen, Single-objective and two-objective genetic algorithms for selecting linguistic rules for pattern classification problems. Fuzzy Sets Syst. **8**(2), 135–150 (1997)
44. M.J. Gacto, R. Alcala, F. Herrera, Adaptation and application of multi-objective evolutionary algorithms for rule reduction and parameter tuning of fuzzy rule-based systems. Soft. Comput. **13**(5), 419–436 (2009)
45. N.N. Karnik, J.M. Mendel, Q. Liang, Type-2 fuzzy logic systems. IEEE Trans. Fuzzy Syst. **7**(6), 643–658 (1999)
46. R. Sambuc, Function Φ−Flous, Application a l'aide au Diagnostic en Pathologie Thyroidienne. Ph.D. thesis, University of Marseille (1975)
47. O. Castillo, P. Melin, Optimization of type-2 fuzzy systems based on bio-inspired methods: a concise review. Inf. Sci. **205**, 1–19 (2012)
48. O. Castillo, P. Melin, A.A. Garza, O. Montiel, R. Sepulveda, Optimization of interval type-2 fuzzy logic controllers using evolutionary algorithms. Soft. Comput. **15**(6), 1145–1160 (2011)
49. C. Wagner, H. Hagras, A genetic algorithm based architecture for evolving type-2 fuzzy logic controllers for real world autonomous mobile robots, in *FUZZ-IEEE*. pp. 1–6. IEEE (2007)
50. J.A. Sanz, A. Fernandez, H. Bustince, F. Herrera, Improving the performance of fuzzy rule-based classification systems with interval-valued fuzzy sets and genetic amplitude tuning. Inf. Sci. **180**(19), 3674–3685 (2010)
51. J. Sanz, A. Fernandez, H. Bustince, F. Herrera, A genetic tuning to improve the performance of fuzzy rule-based classification systems with interval-valued fuzzy sets: degree of ignorance and lateral position. Int. J. Approximate Reasoning **52**(6), 751–766 (2011)
52. J.A. Sanz, A. Fernandez, H. Bustince, F. Herrera, IVTURS: a linguistic fuzzy rule-based classification system based on a new interval-valued fuzzy reasoning method with tuning and rule selection. IEEE Trans. Fuzzy Syst. **21**(3), 399–411 (2013)
53. H. He, E.A. Garcia, Learning from imbalanced data. IEEE Trans. Knowl. Data Eng. **21**(9), 1263–1284 (2009)
54. V. Lopez, A. Fernandez, J.G. Moreno-Torres, F. Herrera, Analysis of preprocessing vs. cost-sensitive learning for imbalanced classification. Open problems on intrinsic data characteristics. Expert Syst. Appl. **39**(7), 6585–6608 (2012)
55. Y. Sun, A.K.C. Wong, M.S. Kamel, Classification of imbalanced data: a review. Int. J. Pattern Recognit. Artif. Intell. **23**(4), 687–719 (2009)
56. C. Elkan, The foundations of cost-sensitive learning, in *Proceedings of the 17th IEEE International Joint Conference on Artificial Intelligence (IJCAI'01)*. pp. 973–978 (2001)
57. B. Zadrozny, J. Langford, N. Abe, Cost-sensitive learning by cost-proportionate example weighting, *in Proceedings of the 3rd IEEE International Conference on Data Mining (ICDM'03)*. pp. 435–442 (2003)
58. Q. Yang, X. Wu, 10 challenging problems in data mining research. Int. J. Inform. Technol. Decis. Making **5**(4), 597–604 (2006)
59. S.J. Lin, M.F. Hsu, Enhanced risk management by an emerging multi-agent architecture. Connection Sci. **26**(3), 245–259 (2014)
60. M. Hao, Y. Wang, S.H. Bryant, An efficient algorithm coupled with synthetic minority oversampling technique to classify imbalanced pubchem bioassay data. Anal. Chim. Acta **806**, 117–127 (2014)

61. R. Oentaryo, E.P. Lim, M. Finegold, D. Lo, F. Zhu, C. Phua, E.Y. Cheu, G.E. Yap, K. Sim, M.N. Nguyen, K. Perera, B. Neupane, M. Faisal, Z. Aung, W.L. Woon, W. Chen, D. Patel, D. Berrar, Detecting click fraud in online advertising: a data mining approach. J. Mach. Learn. Res. **15**, 99–140 (2014)
62. W.Y. Hwang, J.S. Lee, Shifting artificial data to detect system failures. Int. Trans. Oper. Res. **22**(2), 363–378 (2015)
63. B. Krawczyk, G. Schaefer, A hybrid classifier committee for analysing asymmetry features in breast thermograms. Appl. Soft Comput. J. **20**, 112–118 (2014)
64. C. Lu, M. Mandal, Toward automatic mitotic cell detection and segmentation in multispectral histopathological images. IEEE J. Biomed. Health Inform. **18**(2), 594–605 (2014)
65. A. Orriols-Puig, E.B. Mansilla, Evolutionary rule-based systems for imbalanced datasets. Soft. Comput. **13**(3), 213–225 (2009)
66. M. Wasikowski, X.W. Chen, Combating the small sample class imbalance problem using feature selection. IEEE Trans. Knowl. Data Eng. **22**(10), 1388–1400 (2010)
67. G.M. Weiss, in *The Impact of Small Disjuncts on Classifier Learning*, eds. by R. Stahlbock, S.F. Crone, S. Lessmann. Data Mining, Annals of Information Systems, vol. 8 (Springer, 2010), pp. 193–226
68. J. Stefanowski, Overlapping, rare examples and class decomposition in learning classifiers from imbalanced data. Smart Innovation Syst. Technol. **13**, 277–306 (2013)
69. C. Seiffert, T.M. Khoshgoftaar, J. Van Hulse, A. Folleco, An empirical study of the classification performance of learners on imbalanced and noisy software quality data. Inf. Sci. **259**, 571–595 (2014)
70. J.G. Moreno-Torres, T. Raeder, R. Alaiz-Rodriguez, N.V. Chawla, F. Herrera, A unifying view on dataset shift in classification. Pattern Recogn. **45**(1), 521–530 (2012)
71. G.E.A.P.A. Batista, R.C. Prati, M.C. Monard, A study of the behaviour of several methods for balancing machine learning training data. SIGKDD Explor. **6**(1), 20–29 (2004)
72. N.V. Chawla, K.W. Bowyer, L.O. Hall, W.P. Kegelmeyer, SMOTE: synthetic minority oversampling technique. J. Artif. Intell. Res. **16**, 321–357 (2002)
73. V. Lopez, I. Triguero, C.J. Carmona, S. Garcia, F. Herrera, Addressing imbalanced classification with instance generation techniques: IPADE-ID. Neurocomputing **126**, 15–28 (2014)
74. P. Domingos, MetaCost: A general method for making classifiers cost-sensitive, in *Proceedings of the 5th International Conference on Knowledge Discovery and Data Mining (KDD'99)*. pp. 155–164 (1999)
75. B. Zadrozny, C. Elkan, Learning and making decisions when costs and probabilities are both unknown, in *Proceedings of the 7th International Conference on Knowledge Discovery and Data Mining (KDD'01)*. pp. 204–213 (2001)
76. M. Galar, A. Fernandez, E. Barrenechea, H. Bustince, F. Herrera, A review on ensembles for class imbalance problem: bagging, boosting and hybrid based approaches. IEEE Trans. Syst. Man Cybern.—Part C: Appl. Rev. **42**(4), 463–484 (2012)
77. W. Fan, S.J. Stolfo, J. Zhang, P.K. Chan, AdaCost: misclassification cost-sensitive boosting, in *Proceedings of the 16th International Conference on Machine Learning (ICML'96)*. pp. 97–105 (1999)
78. C. Seiffert, T.M. Khoshgoftaar, J. Van Hulse, A. Napolitano, RUSBoost: a hybrid approach to alleviating class imbalance. IEEE Trans. Syst. Man Cybern.—Part A **40**(1), 185–197 (2010)
79. S. Wang, X. Yao, Diversity analysis on imbalanced data sets by using ensemble models, in *Proceedings of the 2009 IEEE Symposium on Computational Intelligence and Data Mining (CIDM'09)*. pp. 324–331 (2009)
80. R. Barandela, J.S. Sanchez, V. Garcia, E. Rangel, Strategies for learning in class imbalance problems. Pattern Recogn. **36**(3), 849–851 (2003)
81. A. Fernandez, M.J. del Jesus, F. Herrera, On the influence of an adaptive inference system in fuzzy rule based classification systems for imbalanced data-sets. Expert Syst. Appl. **36**(6), 9805–9812 (2009)
82. L.J. Eshelman, The CHC adaptive search algorithm: How to have safe search when engaging in nontraditional genetic recombination. In: Rawlin, G. (ed.) Foundations of Genetic Algorithms, pp. 265–283. Morgan Kaufman (1991)

83. A. Fernandez, M.J. del Jesus, F. Herrera, On the 2-tuples based genetic tuning performance for fuzzy rule based classification systems in imbalanced data-sets. Inf. Sci. **180**(8), 1268–1291 (2010)
84. R. Alcala, J. Alcala-Fdez, F. Herrera, A proposal for the genetic lateral tuning of linguistic fuzzy systems and its interaction with rule selection. IEEE Trans. Fuzzy Syst. **15**(4), 616–635 (2007)
85. Z. Chi, H. Yan, T. Pham, Fuzzy Algorithms with Applications to Image Processing and Pattern Recognition (World Scientific, 1996)
86. H. Ishibuchi, T. Yamamoto, T. Nakashima, Hybridization of fuzzy GBML approaches for pattern classification problems. IEEE Trans. Syst. Man Cybern.—Part B **35**(2), 359–365 (2005)
87. D.L. Wilson, Asymptotic properties of nearest neighbor rules using edited data. IEEE Trans. Syst. Man Cybern. **2**(3), 408–421 (1972)
88. J. Laurikkala, Improving identification of difficult small classes by balancing class distribution, in *Proceedings of the Artificial Intelligence in Medicine, 8th Conference on AI in Medicine in Europe (AIME 2001)*. pp. 63–66 (2001)
89. P.E. Hart, The condensed nearest neighbor rule. IEEE Trans. Inf. Theory **14**, 515–516 (1968)
90. P. Villar, A. Fernandez, R.A. Carrasco, F. Herrera, Feature selection and granularity learning in genetic fuzzy rule-based classification systems for highly imbalanced data-sets. Int. J. Uncertainty Fuzziness Knowl. Based Syst. **20**(3), 369–397 (2012)
91. V. Lopez, A. Fernandez, F. Herrera, Addressing covariate shift for genetic fuzzy systems classifiers: a case of study with FARC-HD for imbalanced datasets, in *Proceedings of the 2013 IEEE International Conference on Fuzzy Systems (FUZZ-IEEE 2013)*. pp. 1–8 (2013)
92. J.G. Moreno-Torres, J.A. Saez, F. Herrera, Study on the impact of partition-induced dataset shift on k-fold cross-validation. IEEE Trans. Neural Netw. Learn. Syst. **23**(8), 1304–1313 (2012)
93. J. Alcala-Fdez, R. Alcala, F. Herrera, A fuzzy association rule-based classification model for high-dimensional problems with genetic rule selection and lateral tuning. IEEE Trans. Fuzzy Syst. **19**(5), 857–872 (2011)
94. V. Soler, J. Cerquides, J. Sabria, J. Roig, M. Prim, Imbalanced datasets classification by fuzzy rule extraction and genetic algorithms, in *Proceedings of the 2006 IEEE International Conference on Data Mining (ICDM 2006)*. pp. 330–334 (2006)
95. V. Soler, M. Prim, in *Extracting a Fuzzy System by Using Genetic Algorithms for Imbalanced Datasets Classification: Application on Down's Syndrome Detection*, eds. by D.A. Zighed, S. Tsumoto, Z.W. Ras, H. Hacid. Mining Complex Data, Studies in Computational Intelligence, vol. 165 (Springer, 2009), pp. 23–39
96. M.R. Berthold, K.P. Huber, Constructing fuzzy graphs from examples. Intell. Data Anal. **3**(1), 37–53 (1999)
97. F.J. Berlanga, A.J. Rivera, M.J. del Jesus, F. Herrera, GP-COACH: genetic Programming-based learning of COmpact and ACcurate fuzzy rule-based classification systems for High-dimensional problems. Inf. Sci. **180**(8), 1183–1200 (2010)
98. S. Axelsson, Research in intrusion-detection systems: a survey. Technical Report 98–17, Department of Computer Engineering, Chalmers University of Technology, Goteborg, Sweden (1998)
99. W. Lee, S. Stolfo, A framework for constructing features and models for intrusion detection systems. ACM Trans. Inform. Syst. Secur. **3**(4), 227–261 (2000)
100. S.M. Gaffer, M.E. Yahia, K. Ragab, Genetic fuzzy system for intrusion detection: analysis of improving of multiclass classification accuracy using KDDCup-99 imbalance dataset, in *Proceedings of the 2012 12th International Conference on Hybrid Intelligent Systems (HIS 2012)*. pp. 318–323 (2012)
101. S. Elhag, A. Fernandez, A. Bawakid, S. Alshomrani, F. Herrera, On the combination of genetic fuzzy systems and pairwise learning for improving detection rates on intrusion detection systems. Expert Syst. Appl. **42**(1), 193–202 (2015)
102. T. Hastie, R. Tibshirani, Classification by pairwise coupling. Ann. Stat. **26**(2), 451–471 (1998)

103. J. Sanz, D. Bernardo, F. Herrera, H. Bustince, H. Hagras, A compact evolutionary interval-valued fuzzy rule-based classification system for the modeling and prediction of real-world financial applications with imbalanced data. IEEE Trans. Fuzzy Syst. **23**(4), 973–990 (2015)
104. M. Mahdizadeh, M. Eftekhari, Designing fuzzy imbalanced classifier based on the subtractive clustering and genetic programming, in *Proceedings of the 13th Iranian Conference on Fuzzy Systems (IFSC 2013)*. pp. 318–323 (2013)
105. V. Lopez, A. Fernandez, F. Herrera, A first approach for cost-sensitive classification with linguistic genetic fuzzy systems in imbalanced data-sets, in *Proceedings of the 10th International Conference on Intelligent Systems Design and Applications (ISDA'10)*. pp. 676–681 (2010)
106. H. Ishibuchi, T. Yamamoto, Rule weight specification in fuzzy rule-based classification systems. IEEE Trans. Fuzzy Syst. **13**, 428–435 (2005)
107. P. Ducange, B. Lazzerini, F. Marcelloni, Multi-objective genetic fuzzy classifiers for imbalanced and cost-sensitive datasets. Soft. Comput. **14**(7), 713–728 (2010)
108. D.G. Stavrakoudis, J.B. Theocharis, G.C. Zalidis, A boosted genetic fuzzy classifier for land cover classification of remote sensing imagery. ISPRS J. Photogrammetry Remote Sens. **66**(4), 529–544 (2011)
109. R.E. Schapire, A brief introduction to boosting. In: Proceedings of the 16th International Joint Conference on Artificial Intelligence, Vol. 2 (IJCAI'99). pp. 1401–1406 (1999)
110. M. Antonelli, P. Ducange, F. Marcelloni, A. Segatori, Evolutionary fuzzy classifiers for imbalanced datasets: an experimental comparison, in *Proceedings of the 2013 Joint IFSA World Congress and NAFIPS Annual Meeting (IFSA/NAFIPS 2013)*. pp. 13–18 (2013)
111. J. Huang, C.X. Ling, Using AUC and accuracy in evaluating learning algorithms. IEEE Trans. Knowl. Data Eng. **17**(3), 299–310 (2005)
112. O. Cordon, F. Herrera, I. Zwir, Linguistic modeling by hierarchical systems of linguistic rules. IEEE Trans. Fuzzy Syst. **10**(1), 2–20 (2002)
113. J. Alcala-Fdez, A. Fernandez, J. Luengo, J. Derrac, S. Garcia, L. Sanchez, F. Herrera, KEEL data-mining software tool: data set repository, integration of algorithms and experimental analysis framework. J. Multi-Valued Logic Soft Comput. **17**(2–3), 255–287 (2011)
114. J.R. Quinlan, *C4.5: Programs for Machine Learning* (Morgan Kaufmann Publishers, San Mateo-California, 1993)
115. S. Garcia, A. Fernandez, J. Luengo, F. Herrera, A study of statistical techniques and performance measures for genetics-based machine learning: accuracy and interpretability. Soft. Comput. **13**(10), 959–977 (2009)
116. D. Sheskin, *Handbook of Parametric and Nonparametric Statistical Procedures*, 2nd edn. (Chapman & Hall/CRC, 2006)

Chapter 13
Mayor-Torrens t-norms in the Fuzzy Mathematical Morphology and Their Applications

P. Bibiloni, M. González-Hidalgo, S. Massanet, A. Mir
and D. Ruiz-Aguilera

Abstract Fuzzy mathematical morphology has been extensively used in many different applications such as edge detection, noise reduction and shape and pattern recognition. The fundamentals of this morphology are based on an appropriate selection of the operators involved, namely the conjunction and implication. In this work we investigate the use of the Mayor-Torrens family of t-norms, from both theoretical and practical point of view. The results suggest that competitive results can be obtained by using the t-norms of this family.

13.1 Introduction

The class of t-norms is one of the most important and studied types of fuzzy conjunctions. Originally developed to generalise the classical triangle inequality in the context of probabilistic metric spaces, they play a major role in many fields. These operations, which are closely related to the associativity equation, are essential in the theory of probabilistic metric spaces, in many-valued logics, in fuzzy sets theory to model the intersection of fuzzy sets and in the theory of fuzzy measures and integrals, among others. All these applications have led to the proposal of several families of t-norms, parametrised by one or more parameters, in order to be able to pick out the t-norm which best suits the requirements of the researcher (see [21] for more details).

P. Bibiloni · M. González-Hidalgo (✉) · S. Massanet · A. Mir · D. Ruiz-Aguilera
Department of Mathematics and Computer Science, University of the Balearic Islands,
Palma, Spain
e-mail: manuel.gonzalez@uib.es

P. Bibiloni
e-mail: p.bibiloni@uib.es

S. Massanet
e-mail: s.massanet@uib.es

A. Mir
e-mail: arnau.mir@uib.es

D. Ruiz-Aguilera
e-mail: daniel.ruiz@uib.es

© Springer International Publishing Switzerland 2016
T. Calvo Sánchez and J. Torrens Sastre (eds.), *Fuzzy Logic and Information Fusion*,
Studies in Fuzziness and Soft Computing 339, DOI 10.1007/978-3-319-30421-2_13

The existence of this large bunch of t-norms has enabled researchers to apply them in image processing techniques with notable success. In particular, they play a key role in the fuzzy mathematical morphology based on fuzzy conjunctions and fuzzy implications. This theory introduced by De Baets in [8] and developed in further works [20] is a generalization of the binary morphology using concepts and techniques from the fuzzy set theory [3]. It has proved to be useful to extract features, describe shapes and recognize patterns in digital images and consequently, it is widely acknowledged as a key technique in edge detection, noise reduction, pattern recognition and object segmentation.

Focusing in the fuzzy conjunction, different types of these operators have been used in the literature. Indeed, t-norms in [0, 1] (see [8, 20]), uninorms [12] and discrete t-norms [14] have been considered to generate specific fuzzy mathematical operators with competitive results in the previously mentioned applications. Restricting ourselves to the case of t-norms, traditionally only the class of nilpotent Archimedean t-norms had been considered since this class is the only one that is able to generate fuzzy mathematical operators satisfying all the desirable algebraical properties [20]. However, if the purpose is restricted to apply the fuzzy mathematical operators in some specific application, only some properties are mandatory and consequently, many more t-norms are suitable to generate these operators (see [17]) which could improve the performance of the algorithms. This was the goal pursued in [17] where the performance of the edge detector based on the fuzzy morphological edge detector for several t-norms was analysed improving drastically the traditional edge detector. Therefore, there is much room for improvement by taking into account different t-norms, specially the well-known families.

One of these families is the Mayor-Torrens family of t-norms, which was introduced in [26] as a solution to a functional equation relating all the values of the t-norm with respect to those of the diagonal. This family depends on a parameter $\lambda \in [0, 1]$ and generates continuous t-norms which are ordinal sums of one Lukasiewicz t-norm summand. Indeed, these t-norms form a parametrised family from the minimum t-norm to the Lukasiewicz t-norm, two of the most important t-norms. This is a major reason to investigate the application of these t-norms in the fuzzy mathematical morphology. Thus, this work is focused on one hand, on the study of which algebraical properties hold for the fuzzy morphological operators arising from the family of Mayor-Torrens t-norms and, on the other hand, to analyse the performance of their fuzzy morphological operators in edge detection, noise reduction, pattern recognition and vessel segmentation in angiography.

This work is organised as follows. After recalling the basic definitions and results needed throughout this work in Sect. 13.2, we will study in Sect. 13.3 which algebraical properties are satisfied by the morphological operators when we consider Mayor-Torrens t-norms as fuzzy conjunctions. Then, in Sect. 13.4, we will analyse the performance of these morphological operators in edge detection, noise reduction, pattern recognition and vessel segmentation in angiography. Finally, Sect. 13.5 contains some conclusions derived from the study carried out in this work.

13.2 Preliminaries

In this chapter we will recall definitions and properties related to fuzzy operations such us t-norms, t-conorms, fuzzy negations and fuzzy implications, which will be used throughout the work. First we begin with the definition of fuzzy conjunction.

Definition 13.1 A binary operation $C : [0, 1]^2 \to [0, 1]$ is called a *fuzzy conjunction* if it is increasing with respect to each variable and it satisfies $C(1, 1) = 1$ and $C(0, 1) = C(1, 0) = 0$.

Some of the most studied fuzzy operations are t-norms and t-conorms, used as generalisations of fuzzy conjunctions and disjunctions respectively, which are defined as follows.

Definition 13.2 A *t-norm* (*t-conorm*) is a commutative, associative and increasing binary function $T : [0, 1]^2 \to [0, 1]$ ($S : [0, 1]^2 \to [0, 1]$) with neutral element 1 (0), i.e., $T(1, x) = T(x, 1) = x$ ($S(0, x) = S(x, 0) = x$) for all $x \in [0, 1]$.

The most studied t-norms, that will be used in this work, are the minimum t-norm $T_M(x, y) = \min(x, y)$, the product t-norm $T_P(x, y) = x \cdot y$, the Łukasiewicz t-norm $T_{LK}(x, y) = \max(x + y - 1, 0)$, the nilpotent minimum t-norm $T_{nM}(x, y) = \begin{cases} 0, & \text{if } x + y \leq 1, \\ \min(x, y), & \text{otherwise,} \end{cases}$ and the drastic t-norm $T_D(x, y) = \begin{cases} \min(x, y), & \text{if } \max(x, y) = 1, \\ 0, & \text{otherwise.} \end{cases}$

These operators can also be defined in other frameworks such as the discrete chain $L_n = \{0, 1, \dots, n\}$ where the definition is analogous, by having n as neutral element. In this case they are called *discrete t-norms*. For more details, see [14].

Another important operation considered in the fuzzy framework is the fuzzy negation.

Definition 13.3 A *fuzzy negation* is a decreasing function $N : [0, 1] \to [0, 1]$ such that $N(0) = 1$ and $N(1) = 0$. If a continuous fuzzy negation is strictly decreasing it will be called *strict*. If a continuous negation is involutive, $N(N(x)) = x$ for all $x \in [0, 1]$, then it will be called *strong*.

The most used negation is the standard negation, given by $N_C(x) = 1 - x$ for all $x \in [0, 1]$. T-conorms are dual operators of t-norms by means of the standard negation, as it is stated in the next result.

Proposition 13.1 *A function $S : [0, 1]^2 \to [0, 1]$ is a t-conorm if and only if there exists a t-norm T such that for all $(x, y) \in [0, 1]^2$, $S(x, y) = 1 - T(1 - x, 1 - y)$.*

By means of this duality, the maximum t-conorm $S_M(x, y) = \max(x, y)$, the probabilistic sum $S_P(x, y) = x + y - x \cdot y$, and the Łukasiewicz t-conorm $S_{LK}(x, y) = \min(x + y, 1)$ can be obtained from T_M, T_P and T_{LK}, respectively.

Next, we give the definition of fuzzy implications.

Definition 13.4 ([11]) A binary operator $I : [0, 1]^2 \rightarrow [0, 1]$ is called a *fuzzy implication* if it is decreasing in the first variable, increasing in the second one and it satisfies $I(0, 0) = I(1, 1) = 1$ and $I(1, 0) = 0$.

The implications that will be used in the work are the Kleene-Dienes implication $I_{KD}(x, y) = \max(1-x, y)$, and the Łukasiewicz implication $I_{LK}(x, y) = \min(1, 1 - x + y)$, and the Gödel implication $I_{GD}(x, y) = \begin{cases} 1, & \text{if } x \leq y, \\ y, & \text{if } x > y. \end{cases}$

An important property that can be satisfied by fuzzy implications is the *left neutrality principle*, that is,

$$I(1, y) = y, \quad y \in [0, 1]. \tag{NP}$$

There are many classes of fuzzy implications. For a complete review on this topic, we recommend Chaps. 2, 3 and 5 in [1], a whole book devoted to these operators. More details on fuzzy implications can be also found in the surveys [2, 25]. Two well-known ways of constructing fuzzy implications from other operators are the following:

- *Residual implications.* Given a conjunction C such that $C(1, x) > 0$ for all $x > 0$, the binary operator

$$I_C(x, y) = \sup\{z \in [0, 1] \mid C(x, z) \leq y\}, \quad \text{for all } (x, y) \in [0, 1]^2,$$

  is a fuzzy implication called the residual implication of C.
- *(S, N)-implications.* Consider N a strong negation and S a t-conorm. The operator

$$I_{S,N}(x, y) = S(N(x), y), \quad \text{for all } (x, y) \in [0, 1]^2,$$

  is a fuzzy implication called the (S, N)-implication generated from S and N.

We have the following particular cases: $I_{T_{LK}} = I_{S_{LK}, N_c} = I_{LK}$, $I_{T_M} = I_{GD}$ and $I_{S_M, N_c} = I_{KD}$.

At this point, we can define the basic fuzzy morphological operators based on t-norms such as dilation and erosion. From now on, we will use the following notation: T denotes a t-norm, I a fuzzy implication, A a grey-level image, and B a grey-level structuring element.

Definition 13.5 ([28]) The *fuzzy dilation* $D_T(A, B)$ and the *fuzzy erosion* $E_I(A, B)$ of A by B are the grey-level images defined by

$$D_T(A, B)(y) = \sup_{x \in d_A \cap T_y(d_B)} T(B(x - y), A(x)),$$

$$E_I(A, B)(y) = \inf_{x \in d_A \cap T_y(d_B)} I(B(x - y), A(x)),$$

where d_A denotes the set of points where A is defined, $T_v(A)$ is the translation of a fuzzy set A by $v \in \mathbb{R}^n$ defined by $T_v(A)(x) = A(x - v)$ and "sup" and "inf" denote the usual supremum and infimum operations, respectively.

From the fuzzy erosion and the fuzzy dilation, the fuzzy opening and the fuzzy closing of a grey-level image A by a structuring element B can be defined as follows.

Definition 13.6 ([8]) The *fuzzy closing* $C_{T,I}(A, B)$ and the *fuzzy opening* $O_{T,I}(A, B)$ of A by B are the grey-level images defined by

$$O_{T,I}(A, B) = D_T(E_I(A, B), \overline{B}),$$
$$C_{T,I}(A, B) = E_I(D_T(A, B), \overline{B}).$$

where $\overline{B}(x) = B(-x)$.

13.3 Theoretical Properties of the FMM Based on Mayor-Torrens t-norms

One of the first theoretical problems posed on t-norms was to study in which way the diagonal of a t-norm determines all its values. Dealing with this problem, Mayor and Torrens [26] proposed and solved the following functional equation for continuous t-norms T:

$$T(x, y) + |x - y| = T(\max(x, y), \max(x, y)), \quad \text{whenever } T(x, y) > 0.$$

This equation has as solutions the family of continuous t-norms expressed by

$$T_\lambda^{MT}(x, y) = \begin{cases} \max(0, x + y - \lambda), & \text{if } x, y \in [0, \lambda], \\ \min(x, y), & \text{otherwise.} \end{cases}$$

This family depends on the parameter $\lambda \in [0, 1]$, and it is known as the Mayor-Torrens family of t-norms. As particular cases, we have that $T_0^{MT} = T_M$ and $T_1^{MT} = T_{LK}$. In Fig. 13.1, the structure of these t-norms can be observed for several values of λ. As it can be noted, the Mayor-Torrens t-norms are ordinal sums of T_{LK} in $[0, \lambda]^2$. Using the notation of [21], it holds that $T_\lambda^{MT} = (\langle 0, \lambda, T_{LK}\rangle)$.

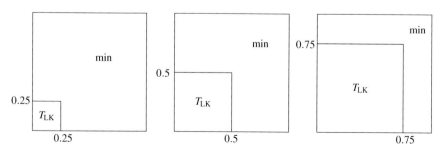

Fig. 13.1 Structure of Mayor-Torrens t-norms for $\lambda = 0.25$ (*left*), $\lambda = 0.5$ (*middle*), and $\lambda = 0.75$ (*right*)

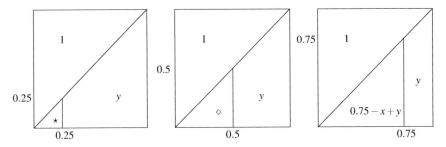

Fig. 13.2 Structure of residual implications generated from Mayor-Torrens t-norms for $\lambda = 0.25$ (*left*), $\lambda = 0.5$ (*middle*), and $\lambda = 0.75$ (*right*), where values in region $\star$ are given by $0.25 - x + y$, and in region $\diamond$ by $0.5 - x + y$

The corresponding residual implications obtained from Mayor-Torrens t-norms are the following:

$$
I_\lambda^{\mathrm{MT}}(x, y) = \begin{cases} 1, & \text{if } y \geq x, \\ \lambda - x + y, & \text{if } y < x \leq \lambda, \\ y, & \text{otherwise.} \end{cases}
$$

Taking particular cases, $I_0^{\mathrm{MT}} = I_{\mathrm{GD}}$ and $I_1^{\mathrm{MT}} = I_{\mathrm{LK}}$. The structure of the residual implications obtained from Mayor-Torrens t-norms can be observed in Fig. 13.2.

If we want to use the t-norms of Mayor-Torrens in the fuzzy mathematical morphology, and its applications, first of all we need to see if fuzzy mathematical operators based on Mayor-Torrens t-norms verify some desirable properties. These properties are duality, monotonicity, interaction with union and intersection, invariance under translation and scaling, extensivity and idempotence, inclusion, commutativity and associativity of the fuzzy dilation, combinations of dilation and erosion, local knowledge and adjunction.

These properties were studied in detail in Sect. 5.5 of [28] for the fuzzy mathematical operators based on a general fuzzy conjunction and a fuzzy implication. Now we will check if the fuzzy mathematical operators based on Mayor-Torrens t-norms verify the conditions presented in the results of the mentioned work, referring to them with the corresponding proposition number.

1. All Mayor-Torrens t-norms are continuous in both variables and the first partial mappings of their residual implications I_λ^{MT} are left-continuous. Also, the second partial mappings of I_λ^{MT} are right-continuous. From all these facts, Proposition 36 in [28] is satisfied, obtaining the relation of the dilation and erosion with respect to the union and intersection of grey-level images.
2. For any $\lambda \in [0, 1]$, we have that for all $(x, y) \in [0, 1]^2$, $y \leq I_\lambda^{\mathrm{MT}}(x, T_\lambda^{\mathrm{MT}}(x, y))$, a sufficient condition for the extensivity of the dilation and the anti-extensivity of the erosion (Proposition 41 (ii)).

3. The inequality $T_\lambda^{MT}(x, I_\lambda^{MT}(x, y)) \leq y$ is satisfied for all $(x, y) \in [0, 1]^2$, which assures Proposition 41 (iii). Besides, this condition and the one presented in Condition 2 are required to fulfill Proposition 42 about the idempotence of the opening and closing and Proposition 45 about the inclusion properties of the dilation, erosion, opening and closing operators.
4. All T_λ^{MT} are semi-norms on $[0, 1]$. This fact assures Proposition 43 about the inclusion of the erosion into the dilation.
5. As all T_λ^{MT} are t-norms, they are associative and commutative. These conditions and Condition 1 assure Proposition 46 about the associativity and commutativity of the dilation and Proposition 47 about the combination of two dilations.
6. All the residual implications from Mayor-Torrens t-norms I_λ^{MT} verify the Exchange Principle, that is, for all $(x, y, z) \in [0, 1]^3$:

$$I_\lambda^{MT}(x, I_\lambda^{MT}(y, z)) = I_\lambda^{MT}(y, I_\lambda^{MT}(x, z)).$$

This condition assures Proposition 48 about the combination of two erosions.
7. The following condition (which is called the Law of Importation) holds for all Mayor-Torrens t-norms and their residual implications:

$$I_\lambda^{MT}(T_\lambda^{MT}(x, y), z) = I_\lambda^{MT}(x, I_\lambda^{MT}(y, z)).$$

Then this condition and Condition 1 of this list assure Proposition 49 about the combination of one dilation and one erosion.
8. The following condition holds for all $(x, y, z) \in [0, 1]^3$:

$$T_\lambda^{MT}(x, z) \leq y \Leftrightarrow I_\lambda^{MT}(x, y) \geq z.$$

This condition assures Proposition 52 about the adjunction property.

We conclude that the Fuzzy Mathematical Morphology based on Mayor-Torrens t-norms satisfy Propositions from 33 to 52 except Proposition 50, which envolves combinations of dilation and erosion, because left-continuity of the second partial mappings of I_λ^{MT} is required.

The next proposition studies the relation of the dilation and the erosion of an image using the t-norms of Mayor-Torrens and their residual implications with respect to the ones obtained with T_M and I_{GD}.

Proposition 13.2 *Let A and B be grey-level images, and T_λ^{MT} be a t-norm in the family of Mayor-Torrens, where $\lambda \in [0, 1]$. If $B(z) \geq \lambda$, for all $z \in d_B$ then we have the following equalities:*

$$E_{I_\lambda^{MT}}(A, B) = E_{I_{GD}}(A, B),$$
$$D_{T_\lambda^{MT}}(A, B) = D_{T_M}(A, B).$$

Proof Let A be an image and B be a structuring element which verifies $B(z) \geq \lambda$, for all $z \in d_B$. The image dilation of A by B using the t-norms family of Mayor-Torrens is:

$$D_{T_\lambda^{MT}}(A, B)(y) = \sup_x T_\lambda^{MT}(B(x-y), A(x)) = \sup_x \min(B(x-y), A(x)) = D_{T_M}(A, B)(y),$$

because if $B(z) \geq \lambda$, $(B(x-y), A(x)) \notin [0, \lambda]^2$. Then $\sup_x T_\lambda^{MT}(B(x-y), A(x)) = \sup_x \min(B(x-y), A(x))$ using the definition of T_λ^{MT}.

The erosion of A by B using the family of Mayor-Torrens is given by:

$$E_{I_\lambda^{MT}}(A, B)(y) = \inf_x I_\lambda^{MT}(B(x-y), A(x)) = E_{I_{GD}}(A, B)(y),$$

because if $B(z) \geq \lambda$,

$$I_\lambda^{MT}(B(x-y), A(x)) = \begin{cases} 1, & \text{if } A(x) \geq B(x-y), \\ A(x), & \text{if } A(x) < B(x-y), \end{cases}$$

which coincides with the erosion obtained from the Gödel implication. ∎

13.4 Applications of the FMM Based on Mayor-Torrens t-norms

Fuzzy Mathematical Morphology has proved to be a useful tool in a wide range of applications such as edge detection, noise removal, shape and pattern recognition and blood vessel segmentation, among many others. This framework is specially suitable to adequately handle the ambiguity and uncertainty inherent to digital images. This feature highlights fuzzy mathematical morphology, not only as a generalization of the binary and gray-scale mathematical morphology, but as one of the leading techniques in image processing improving the results of the other mathematical morphologies.

As it has been already commented, the original fuzzy mathematical morphology is the one based on t-norms as conjunctions and fuzzy implications. This framework has been deeply studied for several applications. In this section, we want to revisit some of these applications and to study the performance of Mayor-Torrens t-norms when taken as particular t-norms in this fuzzy mathematical morphology.

13.4.1 Edge Detection

Edge detection has become a main line of research of the scientific community in recent decades. Some high-level operations in image processing such as segmentation, computer vision or pattern recognition use an edge detection preliminary step whose performance is crucial for the final results of the image processing technique. A plethora of edge detection algorithms have been proposed, but since edges are in fact an intuitive concept which can hardly be described in mathematical terms, the soft computing community (mainly the fuzzy logic one) has specially devoted their efforts to this problem.

Among these fuzzy approaches, the fuzzy mathematical morphology based on t-norms as conjunctions and their residual implications as fuzzy implications have been studied in-depth and some edge detectors based on this framework have been developed. In a first attempt, De Baets in [8, 9] proposed an edge detector based on the fuzzy morphological operators generated by the couple formed by a nilpotent t-norm and its residual implication since this couple of operators is the only one which satisfies all the desirable algebraical properties to lead to a "good" fuzzy mathematical morphology. Since then, nilpotent t-norms which are conjugated with the Łukasiewicz t-norm T_{LK}, and their residual implications, which are conjugated with the Łukasiewicz implication I_{LK}, have been considered to define the fuzzy morphological operators and the edge detector of this theory. However, the recent article [17] has become a landmark in this statement showing that the key property to define an edge detector based on this framework is the extensivity of the fuzzy dilation and the antiextensivity of the fuzzy erosion which are satisfied by many more couples of t-norms and fuzzy implications under some properties of the structuring element, as the following result shows.

Proposition 13.3 *([17]) Let C be a fuzzy conjunction with neutral element 1, I a fuzzy implication that satisfies (NP) and B a grey-level structuring element such that $B(0) = 1$. Then the following inclusions hold:*

$$E_I(A, B) \subset A \subseteq D_C(A, B).$$

Consequently, since t-norms are fuzzy conjunctions with neutral element 1, as in the case of classical morphology, the difference between the fuzzy dilation and the fuzzy erosion of a grey-level image, $D_T(A, B) \setminus E_I(A, B)$, known as the *fuzzy morphological gradient operator*, can be used in edge detection, when a fuzzy implication satisfying (NP) is considered.

In the article, a comparison of the morphological gradients generated from different configurations of t-norms and fuzzy implication was performed concluding that the pair formed by the nilpotent minimum t-norm T_{nM} and the Kleene-Dienes implication I_{KD} was the best configuration according to a performance measure and an image dataset on edge detection. However, the family of Mayor-Torrens t-norms was not included in the study. This is going to be the main goal of this section. Specifically, we are going to compare the results obtained by, on the one hand, the couple formed by the Mayor-Torrens family of t-norms and their (S, N) or R-implication, and on the other hand, the results obtained by the couple (T_{nM}, I_{KD}), leading us to a deeper knowledge in this topic.

13.4.1.1 Edge Detector and Objective Performance Comparison Method

Fuzzy methods of edge detection, the framework which morphological gradients belong to, generate an image where the value of a pixel determines the membership degree of that pixel to the set of edges. This idea contradicts the restrictions given by Canny in [7]. There, a representation of the edge image as a binary image with edges of one pixel width is recommended. Hence, the fuzzy edge image must be thinned and binarised. Indeed, the fuzzy edge image will contain large values where there is a strong image gradient, but to identify edges the broad regions present in areas where the slope is large must be thinned so that only the magnitudes at those points which are local maxima remain. Non Maxima Suppression (NMS), an algorithm proposed by Canny in [7], performs this by suppressing all values along the line of the gradient that are not peak values. NMS has been performed using P. Kovesis' implementation in MATLAB [22]. Specifically, *nonmaxsup.m* function with a radius value of 1.5 and the orientation provided by *featureorient.m* function has been used.

The final step is to binarize the image. We have implemented an automatic non-supervised hysteresis based on the determination of the instability zone of the histogram to find the threshold values [27]. Hysteresis allows to choose which pixels are relevant in order to be selected as edges, using their membership values. Two threshold values T_1, T_2 with $T_1 \leq T_2$ are used. All the pixels with a membership value greater than T_2 are considered as edges, while those which are lower to T_1 are discarded. Those pixels whose membership value is between the two values are selected if, and only if, they are connected with other pixels above T_2. The method needs some initial set of candidates for the threshold values. In this case, the set $\{0.01, \ldots, 0.25\}$ has been introduced, the same one which is used in [27]. In Fig. 13.3, we display the block diagram of the edge detector algorithm proposed in this section and in Fig. 13.4, the intermediate images which are obtained in each step.

Fig. 13.3 Block diagram of the proposed edge detector

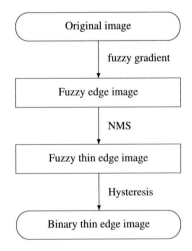

(a)   **(b)**   **(c)**   **(d)**

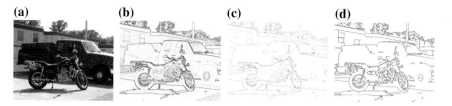

Fig. 13.4 Sequence of the proposed edge detector. **a** Original. **b** Fuzzy edge image. **c** NMS. **d** Binary edge image

However, an objective performance comparison method must be established. The performance comparison can not rely only on the visual comparison of the edge images which is always affected by subjective views. Thus, to compare the results, we need some objective performance measures on edge detection. These measures require, in addition to the binary edge image with edges of one pixel width (DE) obtained by the edge detector we want to evaluate, a reference edge image or *ground truth* edge image (GT) which is a binary edge image with edges of one pixel width containing the real edges of the original image.

There are several measures of performance for edge detection in the literature, see [24, 30]. In this work, we will use the following objective measures to evaluate the similarity between DE and GT:

1. The measure proposed by Pratt [31], *Pratt's figure of merit*, defined as

$$FoM = \frac{1}{\max\{|DE|, |GT|\}} \cdot \sum_{x \in DE} \frac{1}{1 + ad^2},$$

   where $|X|$ is the number of edge pixels of the image X, a is a scaling constant and d is the separation distance between an obtained edge pixel with respect to an ideal one (see [31] for further details). In this work, we will consider $a = 1$ and the Euclidean distance d.

2. The F-measure [32] which is given by the weighted harmonic mean of the precision PR and recall RE, i.e.,

$$F = \frac{2 \cdot PR \cdot RE}{PR + RE},$$

   where

$$PR = \frac{|E|}{|E| + |E_{FP}|}, \text{ and } RE = \frac{|E|}{|E| + |E_{FN}|},$$

   where E is the set of well detected edge pixels, E_{FN} is the set of edges of the GT which have not been detected by the considered edge detector and E_{FP} is the set of edge pixels which have been detected but without any correspondence in the GT.

Fig. 13.5 Some original images (*top*) with their corresponding ground truth edge image (*bottom*)

Larger values of FoM and F ($0 \leq FoM, F \leq 1$) are indicators of a better capability to detect edges.

Consequently, this comparison method needs an image database containing, in addition of the original images, their corresponding ground truth edge images in order to compare the outputs obtained by the different configurations. Thus, we have used the original images and their ground truth edge images of the public image database of the University of South Florida[1] [5]. This image dataset includes 60 natural images with their ground truth edge images. Some of them are displayed in Fig. 13.5. The first fifty images representing the domain of generic object recognition contain a single object approximately centred in the image, appearing unoccluded and set against a natural background for the object. On the other hand, the last ten images are aerial images collected from the DARPA-IU Fort Hood aerial image data set. Specifically, in addition to the 10 aerial images, the image dataset contains 39 indoor and 11 outdoor scenes, and 8 natural and 42 man-made objects. Some images have highlights, reflections or low resolution. In [5], the details about the ground truth edge images and their use for the comparison of edge detectors are specified.

13.4.1.2 Experimental Results

We have compared the edge detector previously presented with three different configurations. Namely, we have considered:

- the nilpotent minimum t-norm T_{nM} and the Kleene-Dienes implication I_{KD},
- the Mayor Torrens t-norms T_λ^{MT} with their residual implications I_λ^{MT} for $\lambda \in \{0.1, 0.2, \ldots, 0.8\}$,

---

[1]It can be downloaded from ftp://figment.csee.usf.edu/pub/ROC/edge_comparison_dataset.tar.gz.

- the Mayor Torrens t-norms T_λ^{MT} with the (S, N)-implications generated by the Mayor-Torrens t-conorms S_λ^{MT} for $\lambda \in \{0.1, 0.2, \ldots, 0.8\}$ and $N = N_C$, the standard negation.

These different edge detectors have been applied to the 60 images of the aforementioned dataset obtaining the values of FoM and F for each image and each detector. For all the configurations, we have considered the same structuring element given by

$$B(i, j) = 1 - \frac{\sqrt{i^2 + j^2}}{\max_{i', j' \in D}\{\sqrt{i'^2 + j'^2}\}}, \quad i, j \in D = \{-2, -1, 0, 1, 2\},$$

which is an Euclidean square structuring element of size 5.

First of all, we have analysed the behaviour of the two configurations based on Mayor-Torrens t-norms according to their λ value. Figure 13.6 shows the evolution of the mean of the 60 obtained values of FoM and F with respect to the λ value, $\lambda \in \{0.1, 0.2, \ldots, 0.8\}$.

Several conclusions can be derived. First, note that the mean value slightly increases from $\lambda = 0.1$ to $\lambda = 0.6$ and then drops abruptly for higher values of λ. The best performance values are obtained by the configuration with the residual implication when we consider FoM and $\lambda \le 0.6$, and by the configuration with the (S, N)-implication when we consider $\lambda \ge 0.7$ for both measures. More specifically, the best configuration is $(T_{0.6}^{\mathrm{MT}}, I_{T_{0.6}^{\mathrm{MT}}})$ according to FoM and $(T_{0.5}^{\mathrm{MT}}, I_{T_{0.5}^{\mathrm{MT}}})$ according to F. In Table 13.1 we can observe that these configurations improve drastically the results obtained by $(T_{\mathrm{nM}}, I_{\mathrm{KD}})$, the best pair according to [17], regarding to the performance measures. In fact, the optimal Mayor-Torrens based configurations obtain better results than $(T_{\mathrm{nM}}, I_{\mathrm{KD}})$ in up to 52 images of the 60 which constitute the dataset (see Fig. 13.7 for more details). This statement is reinforced with the visual observation of the results in Fig. 13.8.

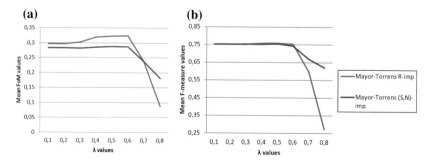

Fig. 13.6 Evolution of the means of the values of the considered measures obtained by the configurations with T_λ^{MT} depending on the λ values. **a** FoM. **b** F-measure

Table 13.1 Mean of the 60 values of FoM and F obtained using different configurations

Configuration (T, I)	FoM	F
$(T_{0.5}^{MT}, I_{0.5}^{MT})$	0.3225	0.7587
$(T_{0.6}^{MT}, I_{0.6}^{MT})$	0.3233	0.7521
(T_{nM}, I_{KD})	0.2772	0.7247

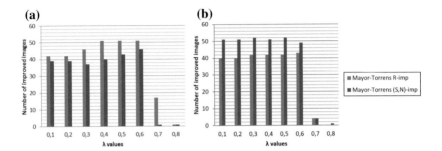

Fig. 13.7 Evolution, depending on the λ values, of the number of images where the configurations with T_{λ}^{MT} obtain better results than (T_{nM}, I_{KD}) according to the performance measures. **a** FoM. **b** F-measure

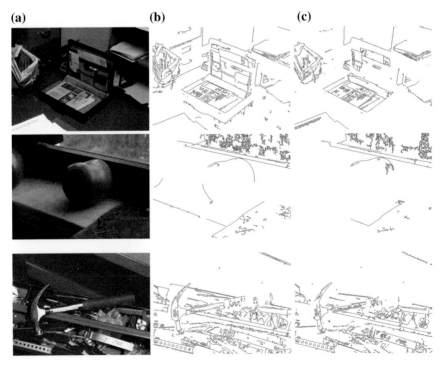

Fig. 13.8 Comparison of the edge images obtained by $(T_{0.5}^{MT}, I_{0.5}^{MT})$ and (T_{nM}, I_{KD}). **a** Original. **b** $(T_{0.5}^{MT}, I_{0.5}^{MT})$. **c** (T_{nM}, I_{KD})

13.4.2 Noise Removal

As we have stated in the introduction, a preprocessing step to remove undesired noise in digital images is mandatory to avoid bad results in later image processing techniques. One of the main topics in noise removal is related to impulse noise reduction. Images are often corrupted with this type of noise due to errors by noisy sensors or transmissions channels leading to an image where only a few pixels are corrupted, but they are very noisy. The effect is similar to sprinkling light and dark dots on the image. One example is the "salt and pepper" noise which arises when transmitting images over noisy digital channels. In this section, we will deal with the salt and pepper noise which we define as follows. Let $x_{i,j}$ be the grey level image of the original image $\mathbf{x}$ at the pixel location (i, j). Denote by $\mathbf{y}$ a noisy image, then the observed grey level at the pixel location (i, j) is given by

$$y_{i,j} = \begin{cases} x_{i,j}, & \text{with probability } 1 - p, \\ v_{i,j}, & \text{with probability } p, \end{cases}$$

where $(i, j) \in \{1, \ldots, M\} \times \{1, \ldots, N\}$, and $v_{i,j}$ is an identically distributed, independent random process with an arbitrary underlying probability density function [33], that is the intensity value of the noisy pixel. If we restrict ourselves to salt and pepper noise, the observed grey level is given by

$$y_{i,j} = \begin{cases} 0, & \text{with probability } p, \\ 255, & \text{with probability } q, \\ x_{i,j}, & \text{with probability } 1 - (p + q), \end{cases}$$

where $r = p + q$ defines the noise level.

Several approaches have been proposed to remove this noise type. In particular, in this section we will consider the method presented in [15], which is an extension of the methods presented in [16, 40], by using fuzzy mathematical morphology with respect to the first one, and a new detection function of the noisy pixels and an improved Block Smart Erase with an adaptive window size with respect to the second one. This version of the algorithm is able to remove the noise from images corrupted up to 90 % of noise preserving the edges and details. Thus, the goal of this subsection will be to study the performance of this method when Mayor-Torrens t-norms are used into the operators involved in this method.

13.4.2.1 Filtering Algorithm and Objective Performance Comparison Method

For the sake of completeness, we will recall the main steps of the filtering algorithm we are dealing with. For more details, we refer the reader to [15]. The algorithm is based on two main steps: the detection step, where the noisy pixels are detected and a filtering step, where only those pixels detected as noisy ones in the first step are denoised. More specifically, the algorithm follows the following steps for each pixel of the image, where A is a corrupted image; B, B_1 and B_2 are structuring elements of sizes 5, 3 and 7, respectively; T is a t-norm and I_T its corresponding residual implication:

1. Detection phase:

   a. Find out the maximum (S_{max}) and minimum (S_{min}) values of a 7×7 window centred at the current pixel.
   b. Compute the following function

   $$d(i, j) = \left| \frac{C_{T,I_T}(O_{T,I_T}(A, B), B)(i, j) + O_{T,I_T}(C_{T,I_T}(A, B), B)(i, j)}{2} - A(i, j) \right|$$

   c. Finally, the detection function is given by

   $$b(i, j) = \begin{cases} 1, & \text{if } (A(i, j) = S_{max} \text{ or } A(i, j) = S_{min}) \text{ and } d(i, j) \geq t, \\ 0, & \text{otherwise,} \end{cases}$$

   where t is a predefined threshold and we conclude that $A(i, j)$ is a corrupted pixel if, and only if, $b(i, j) = 1$.

2. Filtering phase (only applied to the noisy pixels, i.e., those $A(i, j)$ with $b(i, j) = 1$): The filter is defined as

$$A(i, j) = FMMOCS_{T,I_T}(A, (B_1, B_2))(i, j)$$
$$= \frac{BSE(FOCF_{T,I_T}(A, (B_1, B_2)))(i, j)}{2} + \frac{BSE(FCOF_{T,I_T}(A, (B_1, B_2)))(i, j)}{2},$$

where $FOCF$ and $FCOF$ are the so-called alternate filters given by

$$FOCF_{T,I_T}(A, (B_1, B_2)) = C_{T,I_T}(O_{T,I_T}(A, B_1), B_2), \text{ and}$$

$$FCOF_{T,I_T}(A, (B_1, B_2)) = O_{T,I_T}(C_{T,I_T}(A, B_1), B_2)$$

and BSE is the so-called Block Smart Erase algorithm given by the following steps:

   a. Consider an $N \times N$ window centered at the test pixel, starting by $N = 5$.
   b. If $A(i, j) \in \{0, 255\}$ then we have an absolute extreme value and step 3 must be applied. Otherwise, the pixel is not altered.
   c. If an extreme value is detected, assign the median value of the window as its grey-level value. If the median value is again an extreme value, go to step 1 and consider a larger window size N' with $N' = N + 2$.

Since we are going to compare several filtering methods, an objective performance comparison method must be established. The performance comparison can not rely only on the visual comparison of the denoised images which is always affected by subjective views. Thus, a performance measure on noise reduction will be considered, namely *SSIM*. The structural similarity index measure (*SSIM*) was introduced in [37] under the assumption that human visual perception is highly adapted for extracting structural information from a scene. Let I_1 and I_2 be two images of dimensions $M \times N$. In the following, we suppose that I_1 is the original noise-free image and I_2 is the restored image for which some filter has been applied. The measure is defined as follows:

$$SSIM(I_1, I_2) = \frac{(2\mu_1\mu_2 + C_1)}{(\mu_1^2 + \mu_2^2 + C_1)} \cdot \frac{(2\sigma_{12} + C_2)}{(\sigma_1^2 + \sigma_2^2 + C_2)},$$

where $\mu_k, k = 1, 2$ is the mean of the image I_1 and I_2 respectively, σ_k^2 is the variance of each image, σ_{12} is the covariance between the two images, $C_1 = (0.01 \cdot 255)^2$ and $C_2 = (0.03 \cdot 255)^2$.

Larger values of SSIM are indicators of better capabilities for noise reduction and image recovery.

13.4.2.2 Experimental Results

We have compared the filtering method presented previously with two different configurations. Namely, we have considered:

- the Mayor Torrens t-norm $T_{0.8}^{\mathrm{MT}}$ with its residual implication $I_{0.8}^{\mathrm{MT}}$ and as structuring elements B, B_1 and B_2 the square Euclidean structuring elements of sizes 5, 3 and 7 whose grey level values in (i, j) are given by

$$1 - \frac{\sqrt{i^2 + j^2}}{\max_{i', j' \in D}\{\sqrt{i'^2 + j'^2}\}}, \quad i, j \in D = \{-k, -k + 1, \ldots, -1, 0, 1, \ldots, k - 1, k\},$$

  with $k = 2, 1$ and 3 respectively,
- the Łukasiewicz t-norm with its residual implication I_{LK} and a flat sequence of squares of sizes 5, 3 and 7.

Both configurations have used the value $t = 210$. These two configurations have been applied to 15 images of the miscellaneous volume of the USC-SIPI image database of the University of Southern Carolina,[2] which have been previously corrupted with salt and pepper noise (where 255 represents salt and 0 represents pepper) with equal probability, with levels varying from 5 to 95 % with a step of 5 %, and finally a 98 % level. Once the filtered images have been obtained, the *SSIM* values have been computed for each image and each filter.

In Fig. 13.9, the evolution of the mean *SSIM* values obtained by the two different filtering methods from the 15 considered images with different noise levels is shown.

---

[2]This image database can be downloaded from http://sipi.usc.edu/database/misc.tar.gz.

Fig. 13.9 Evolution of the mean *SSIM* values obtained by the two considered configurations of the filter for the considered 15 images when the salt and pepper noise varies between 5 and 98 %

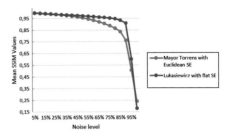

Fig. 13.10 Evolution, depending on the noise levels, of the number of images where the configuration with $T_{0.8}^{MT}$ obtains better results than the configuration with T_{LK} according to SSIM values

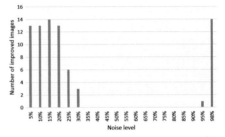

Note that the configuration where Mayor-Torrens t-norms are used obtains similar (or even better results) than the other configuration for low noise levels but its performance is worse for high noise levels, except for 98 % where both results are poor. This behaviour can be also deduced from Fig. 13.10 where for each noise level, the number of images where the Mayor-Torrens t-norm based configuration improves the *SSIM* values of the other one is displayed. It must be highlighted that for noise levels smaller than 25 %, it obtains better results in almost all the images.

In addition, it is necessary to highlight that when flat structuring elements are used, any pair of t-norm and residual implication yields the same results due to their border properties. Thus, Mayor-Torrens t-norms could be also applied with flat structuring elements to obtain notable results also for high noise levels.

This statement is reinforced with the visual observation of the results in Fig. 13.11.

13.4.3 *Shape and Pattern Recognition*

In addition to the fields of edge detection and noise removal, the fuzzy mathematical morphology has shown its potential in shape and pattern recognition. The two main operators of this theory devoted to this goal are the Top-Hat transforms and the Hit-or-Miss operator. Thus, this section will investigate the behaviour of these operators when Mayor-Torrens t-norms are considered into their expressions.

Top-Hat Transform

Firstly, we will define the top-hat transform, which is an operator derived from the morphological opening and the morphological closing, the most elementary morphological filters. These morphological operators, also called basic filters, are defined

(a)   **(b)**   **(c)**   **(d)**

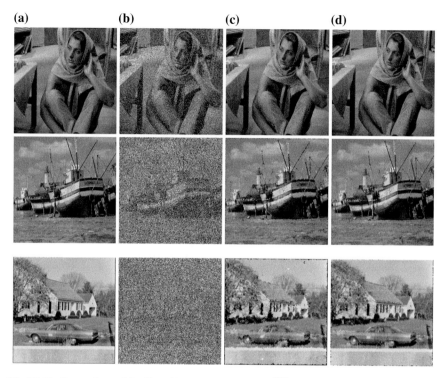

Fig. 13.11 Comparison of the filtered images obtained by $(T_{0.8}^{\mathrm{MT}}, I_{T_{0.8}^{\mathrm{MT}}})$ with Euclidean structuring elements and $(T_{\mathrm{LK}}, I_{LK})$ with flat structuring elements. From *top* to *bottom*, the noise levels are 20, 60 and 85 %. **a** Original image. **b** Corrupted image. **c** $(T_{0.8}^{\mathrm{MT}}, I_{T_{0.8}^{\mathrm{MT}}})$. **d** $(T_{\mathrm{LK}}, I_{LK})$

as the dilation of the erosion and the erosion of the dilation, respectively. They have been already formalised in Definition 13.6.

Under certain situations, the opening is an antiextensive morphological filter and the closing is an extensive morphological filter. So, in a first step, they can be used to remove specific undesired objects in images. This is formalised in Proposition 13.4, proved in [20, Proposition 41].

Proposition 13.4 *Let A and B be grey-level images, T a left-continuous t-norm and I its residual implication. Let us assume that $B(0) = 1$. Then,*

$$O_{T,I}(A, B) \subseteq A \subseteq C_{T,I}(A, B)$$

The size and shape of the structuring element used in these operations determine which structures will be affected. We observe that the opening of a grey level image by a structuring element removes the bright zones smaller than such element by darkening them. On the other hand, the morphological closing helps to remove dark structures that are smaller than the structuring element, brightening thus the dark objects. Larger structuring elements will remove small undesired artifacts in an image, but they will also remove large shapes and they may affect the rest of the structures. Reduced structuring elements may thus be preferred when the images contain small details. Thus, the size and shape of the structuring element are key factors to remove undesired objects of images, but they should also depend on the structures that are not supposed to be affected by these filters.

The top-hat transforms are defined as residual operations, and are in fact two different operations. The *Top-Hat* or *White Top-Hat* is the difference between the original image and its opening, whereas the *Dual Top-Hat* or *Black Top-Hat* is the subtraction between the closing and the original image.

Definition 13.7 ([28]) Let A be a grey-level image, B a structuring element, T a t-norm and I an implication. Then, the *Top-Hat* $\rho_{T,I}(A, B)$ and the *Dual Top-Hat* $\rho^d_{T,I}(A, B)$ of A by B are the grey-level images defined as

$$\rho_{T,I}(A, B) = A - O_{T,I}(A, B),$$
$$\rho^d_{T,I}(A, B) = C_{T,I}(A, B) - A.$$

The top-hat transforms, originally presented by F. Meyer in 1977 (see [34]), find the structures that have been removed by either the opening or the closing. The subtraction operation between the original image and the filtered one increases the contrast of the regions of interest. The top-hat transform enhances the light objects that have been completely removed by the opening, further contrasting them from the background. The dual top-hat, on the other hand, extracts the dark components which have been removed by the closing. Besides, it never takes negative values due to the fact that we will always use (a) a left-continuous t-norm, (b) its residual implication, and (c) a structuring element B such that $B(0) = 1$.

Beyond the mathematical expression, we give a deep insight of its behaviour before presenting its applications. To do so, a template image has been specifically created to visually observe how it is affected by a top-hat operation. It presents a range of shapes and patterns besides a homogeneous intensity gradient. The dual top hat transform is applied to the template with a Mayor-Torrens t-norm with $\lambda = 0.5$, its residual implication and a Gaussian-shaped structuring element. This setting is depicted in Fig. 13.12.

The template is shown in Fig. 13.13a, along with its dual top hat transform, in Fig. 13.13d, and the steps performed to compute it, Fig. 13.13b, c.

It is clear that Fig. 13.13b is the dilation of the original image: the bright areas— which in mathematical morphology are usually considered to be the foreground object—have been enlarged. Figure 13.13c shows the closing of the template, defined as the erosion of its dilation. The thin dark regions that disappeared during the dilation can not be "recovered", in contrast to the large ones. This effect can be observed in the horizontal lines. Besides, the bright objects recover their initial shape, as can be observed in the small and big lighter squares. Therefore, the dual Top Hat—the closing minus the original template—highlights thin areas which are darker than

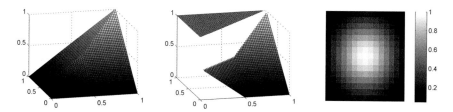

Fig. 13.12 Settings used to perform the dual top-hat transform. From *left* to *right*, the Mayor-Torrens t-norm with $\lambda = 0.5$, its corresponding residual implication and the structuring element, which is a Gaussian shape with $\sigma = 4$ pixels

Fig. 13.13 Dual top-hat transform of a specifically designed template that shows its behaviour in different situations. **a** Template image. **b** Dilation. **c** Closing. **d** Black Top Hat

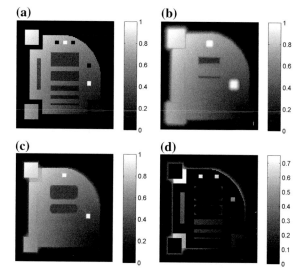

its surrounding neighbourhood. Besides, Fig. 13.13d also shows some artifacts: the edges are slightly highlighted, specially the rounded ones.

Using a Mayor-Torrens t-norm has noticeable effects in the transform. The t-norm (as well as its residual implication) is computed in a pixel-wise fashion: it uses as input one pixel of the image and the corresponding pixel of the shifted structuring element. The Mayor-Torrens t-norm behaves like the minimum operator when $(x, y) \notin [0, \lambda]^2$, being more penalising than the minimum otherwise. Therefore, when the values of the pixels are small (that is, $x \in [0, \lambda]$) and the distance to the pixel evaluated is high (so the structuring element's values are small, $y \in [0, \lambda]$), the dilation is less powerful and the dual top hat provides smaller values. In other words, thin areas surrounded by *really* brighter objects are much more highlighted than thin areas surrounded by *slightly* brighter objects. This effect can be observed by comparing the lower right corner and the upper left one in Fig. 13.13d.

There are several applications of the top-hat transform due to its specific behaviour. In general, the detection of light objects is improved by means of an opening top-hat,

222 P. Bibiloni et al.

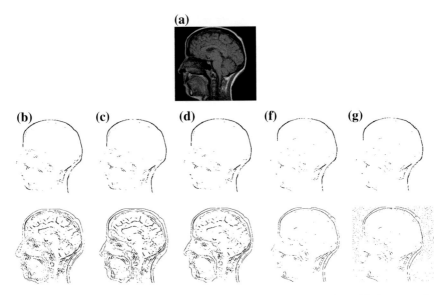

Fig. 13.14 Top-hat (*top*) and dual top-hat (*bottom*) using several approaches with the structuring element used in [14]. The behaviour of the top-hat transforms are here visually observed: it extracts the areas in which the structuring element does not fit and are brighter (top-hat) or darker (dual top-hat) than its neighbouring pixels. **a** Original. **b** $(T_{0.9}^{MT}, I_{0.9}^{MT})$. **c** $(T_{nM}, I_{T_{nM}})$. **d** $(T_{nM25}, I_{T_{nM25}})$. **e** $(T_{LK}, I_{T_{LK}})$. **f** Umbra

in the same way in which the dual top-hat detects dark shapes. In [18] we show applications of the top-hat transform using uninorms as conjunctions in the general framework [20]. In [13, 14], top-hat transforms with discrete t-norms are used to leverage the discretized nature of digital images. In Sect. 13.4.4, this transform is used to segment vessels in retinal images.

In Fig. 13.14, a comparison of the top-hat and the dual top-hat is displayed. We compare the results obtained using the Mayor-Torrens t-norm $(T_{0.9}^{MT}, I_{0.9}^{MT})$ where $I_{0.9}^{MT}$ is the residual implication, with those obtained by the umbra approach (classical grey-level morphology), the Łukasiewicz continuous t-norm $(T_{LK}, I_{T_{LK}})$, the nilpotent minimum t-norm $(T_{nM}, I_{T_{nM}})$, and finally with the nilpotent minimum discrete t-norm $(T_{nM25}, I_{T_{nM25}})$. In the results, the well-known Otsu's threshold method ([29]) has been applied in order to ease their visualization. As we can observe, the Mayor-Torrens pair $(T_{0.9}^{MT}, I_{0.9}^{MT})$, provides competitive results and outperform at naked eye the results obtained by some of the other procedures.

In Fig. 13.15 we can see a comparison of performing the dual top-hat transform with different Mayor-Torrens t-norms. The parameter λ varies across the set $\{0.01, 0.3, 0.9, 0.99\}$. As we can see in Fig. 13.15, the parameter of the Mayor-Torrens t-norm allows to control the similarity of the top-hat transform to the results obtained by the Łukasiewicz continuous t-norm $(T_{LK}, I_{T_{LK}})$. This fact can be visually observed by comparing Fig. 13.14e with Fig. 13.15d. Although the operators based on the Łukasiewicz t-norm are powerful, they may fail to detect some important

(a)   **(b)**   **(c)**   **(d)**

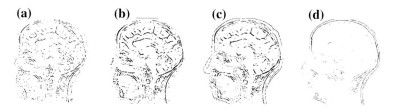

Fig. 13.15 Dual top-hat using several approaches with the structuring element used in [14]. **a** $(T_{0.01}^{MT}, I_{T_{0.01}^{MT}})$. **b** $(T_{0.3}^{MT}, I_{T_{0.3}^{MT}})$. **c** $(T_{0.9}^{MT}, I_{T_{0.9}^{MT}})$. **d** $(T_{0.99}^{MT}, I_{T_{0.99}^{MT}})$

edges. Since the parameter λ provides the aforementioned control, one can find the best trade-off between the "Łukasiewicz behaviour" (for $\lambda = 1$) and the "Minimum behaviour" (for $\lambda = 0$). Therefore, the results obtained can be optimised according to each particular application.

Fuzzy Morphological Hit-or-Miss Transform

Besides the Top-Hat transforms presented previously, in [19], a novel Hit-or-Miss operator derived from the fuzzy mathematical morphology based on t-norms was presented. Some preliminary experimental results provided evidences of the potential of this tool to be feasible to design algorithms for detecting patterns. Specifically, the fuzzy morphological hit-or-miss based on t-norms is defined as follows.

Definition 13.8 ([19]) Let T be a t-norm, I be a fuzzy implication and N be a strong fuzzy negation. The *fuzzy morphological hit-or-miss transform* (FMHMT) of the grey-level image A with respect to the grey-level structuring element $B = (B_1, B_2)$ is defined, for any $y \in d_A$, by

$$FMHMT_{T,I,N}(A, B)(y) = T\left(E_I(A, B_1)(y), E_I(N(A), B_2)(y)\right), \tag{13.1}$$

where $N(A)(x) = N(A(x))$ for all $x \in d_A$.

Next proposition studies the expression of the operator when Mayor-Torrens t-norms and their residual implications are considered in a particular case.

Proposition 13.5 *Let N be a strong negation, A be a grey-level image and $B = (B_1, N(B_1))$ be a grey-level structuring element such that $B_1(x - y) = m$ for all $x \in d_{T_y(B_1)}$ and $y \in d_A$. Suppose that $A(x) = k$ for all $x \in d_A \cap d_{T_y(B_1)}$. Then it holds:*

(i) If $m < k$, then

$$FMHMT_{T_\lambda^{MT}, I_\lambda^{MT}, N}(A, B)(y) = \begin{cases} \lambda - N(m) + N(k), & \text{if } N(\lambda) \leq m, \\ N(k), & \text{if } m < \min(k, N(\lambda)). \end{cases}$$

(ii) *If $m \geq k$, then*

$$FMHMT_{T_\lambda^{MT}, I_\lambda^{MT}, N}(A, B)(y) = \begin{cases} \lambda - m + k, & \text{if } \lambda \geq m, \\ k, & \text{if } m > \max(k, \lambda). \end{cases}$$

Proof First of all, let us suppose that $m < k$. Using Proposition 5.2 in [19], we have:

$$FMHMT_{T_\lambda^{MT}, I_\lambda^{MT}, N}(A, B)(y) = I_\lambda^{MT}(N(m), N(k)).$$

If $m < k$, it verifies $N(m) > N(k)$, because N is a non-increasing function. Using the definition of I_λ^{MT}, we get:

$$I_\lambda^{MT}(N(m), N(k)) = \begin{cases} \lambda - N(m) + N(k), & \text{if } \lambda \geq N(m) > N(k), \\ N(k), & \text{if } N(m) > N(k) \text{ and } \lambda < N(m). \end{cases}$$

The previous expression can be written as:

$$I_\lambda^{MT}(N(m), N(k)) = \begin{cases} \lambda - N(m) + N(k), & \text{if } N(\lambda) \leq m < k, \\ N(k), & \text{if } m < \min(k, N(\lambda)). \end{cases}$$

Next, we suppose that $m \geq k$. In this case, using the same proposition as in the previous case, we have:

$$FMHMT_{T_\lambda^{MT}, I_\lambda^{MT}, N}(A, B)(y) = I_\lambda(m, k).$$

Using the definition of I_λ^{MT}, we get:

$$I_\lambda^{MT}(m, k) = \begin{cases} \lambda - m + k, & \text{if } \lambda \geq m > k, \\ k, & \text{if } m > k \text{ and } \lambda < m. \end{cases}$$

The previous expression can be written as:

$$I_\lambda^{MT}(m, k) = \begin{cases} \lambda - m + k, & \text{if } \lambda \geq m > k, \\ k, & \text{if } m > \max(k, \lambda), \end{cases}$$

and the result follows.   ∎

Once the expression of the operator has been obtained when both the image and the structuring element are constant in some region, we are going to analyse the behaviour of Mayor-Torrens t-norms into this operator and compare it with other configurations. Namely, we will compare the results obtained by the configurations with I_λ^{MT} as fuzzy implication, N_C as fuzzy negation and the t-norms T_λ^{MT}, T_{nM}, T_P, T_{LK}, T_D and T_M.

In the following examples we are going to deal with a synthetic image containing several geometric figures (see Fig. 13.16a) with different grey level values and shapes.

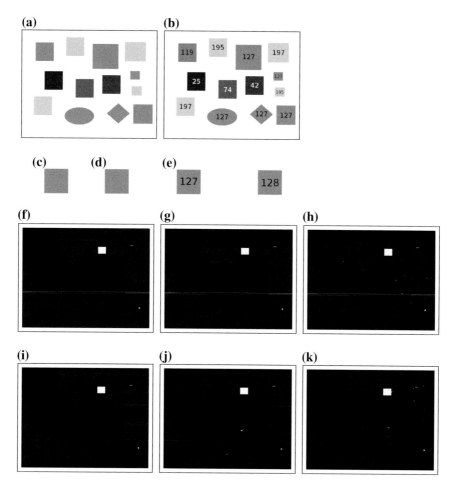

Fig. 13.16 Fuzzy morphological hit-or-miss transform using different t-norms, with the Mayor-Torrens residual implication $I_{0.55}^{MT}$ of the original image displayed in (**a**), using $B = (B_1, B_2)$ as a structuring element. **a** Original image. **b** Grey-level values original image. **c** B_1. **d** B_2. **e** Grey-level values of B_1 and B_2. **f** FMHMT$_{T_D, I_{0.55}^{MT}, N_C}$. **g** FMHMT$_{T_{L.K}, I_{0.55}^{MT}, N_C}$. **h** FMHMT$_{T_{0.55}^{MT}, I_{0.55}^{MT}, N_C}$. **i** FMHMT$_{T_M, I_{0.55}^{MT}, N_C}$. **j** FMHMT$_{T_{nM}, I_{0.55}^{MT}, N_C}$. **k** FMHMT$_{T_P, I_{0.55}^{MT}, N_C}$

The aim is to detect a specific combination of grey-level and square size and test the potentiality of the FMHMT transform using the Mayor-Torrens family of t-norms in shape and pattern recognition. In the figure we can see the structuring element that we have used to compute the FMHMT. Note that, the geometric figures of the image do not have a picture frame around it and consequently, we choose a plain square as foreground structuring element. We have chosen an inner value of 127 since there are several figures with this inner value in the image. We will set $\lambda = 0.55$ in the first experiment and $\lambda = 0.9$ in the second one.

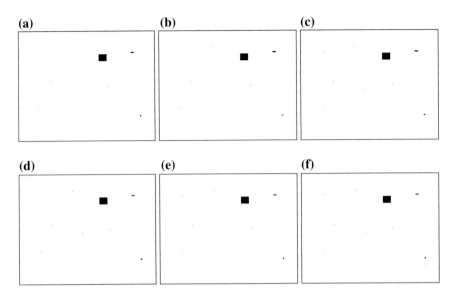

Fig. 13.17 Inverse thresholded image of the FMHMT$_{T,I_{0.55}^{MT},N_C}$ that are displayed in Fig. 13.16 highlighting non-zero pixels. **a** FMHMT$_{T_D,I_{0.55}^{MT},N_C}$. **b** FMHMT$_{T_{L,K},I_{0.55}^{MT},N_C}$. **c** FMHMT$_{T_{0.55}^{MT},I_{0.55}^{MT},N_C}$. **d** FMHMT$_{T_M,I_{0.55}^{MT},N_C}$. **e** FMHMT$_{T_{nM},I_{0.55}^{MT},N_C}$. **f** FMHMT$_{T_P,I_{0.55}^{MT},N_C}$

In the first experiment, in Fig. 13.16f–k, all the squares with a size equal to or greater than the size of the structuring element have been detected by the t-norms $\{T_{0.55}^{MT}, T_{nM}, T_P\}$, but the square with grey-level value of 119 is not detected by the t-norms $\{T_D, T_{LK}, T_M\}$. The grey-level of the pixels belonging to the square that exactly matches the structuring element (in shape and grey-level) is 1, the maximum value, which is predicted by Proposition 5.1 in [19]. In addition, we can observe that white regions larger than one pixel appear into the FMHMT image. This is because the structuring element is included in a geometric figure of the image when it is centred in all the pixels of this region. In addition, the FMHMT value decreases when the difference between the grey-level of the desired shape into the image with respect to the grey-level of the structuring element increases. See the work [19] for a detailed account about the fuzzy hit-or-miss transform.

In addition, we can see the behaviour of the operator when in the image, there are figures with different shapes to the one of the structuring element. The diamond and the ellipse have not been detected by any configuration, and none of them differentiate between these figures and the black background.

In Fig. 13.17 we show the thresholded versions of the FMHMT images that are displayed in Fig. 13.16 to see better how the squares with a size greater than or equal to the structuring element have been found for each of the $\{T, I_\lambda^{MT}, N_C\}$ combinations. Obviously one could argue if an ellipse or another figure has a size larger than the square with the same grey-level, it would be detected as a square in the FMHMT.

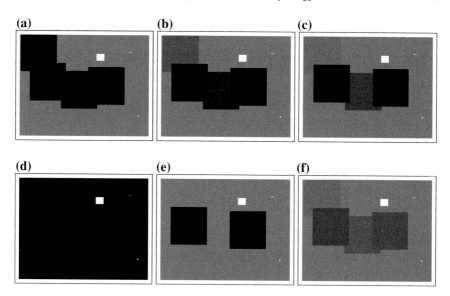

Fig. 13.18 Fuzzy morphological hit-or-miss transform using different t-norms with the Mayor-Torrens residual implication $I_{0.9}^{\mathrm{MT}}$, of the original image displayed in **a**, using as the structuring element $B = (B_1, B_2)$ displayed in **c**, **d**. **a** $\mathrm{FMHMT}_{T_{\mathrm{D}}, I_{0.9}^{\mathrm{MT}}, N_C}$. **b** $\mathrm{FMHMT}_{T_{\mathrm{L,K}}, I_{0.9}^{\mathrm{MT}}, N_C}$. **c** $\mathrm{FMHMT}_{T_{0.9}^{\mathrm{MT}}, I_{0.9}^{\mathrm{MT}}, N_C}$. **d** $\mathrm{FMHMT}_{T_M, I_{0.9}^{\mathrm{MT}}, N_C}$. **e** $\mathrm{FMHMT}_{T_{nM}, I_{0.9}^{\mathrm{MT}}, N_C}$. **f** $\mathrm{FMHMT}_{T_P, I_{0.9}^{\mathrm{MT}}, N_C}$

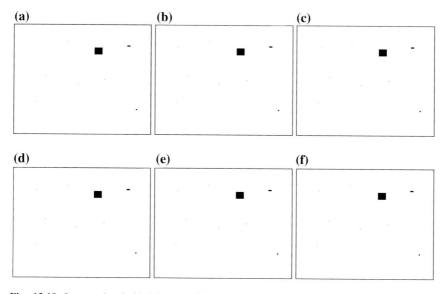

Fig. 13.19 Inverse thresholded image of the $\mathrm{FMHMT}_{T, I_{0.9}^{\mathrm{MT}}, N_C}$ displayed in Fig. 13.18 highlighting non-zero pixels. **a** $\mathrm{FMHMT}_{T_{\mathrm{D}}, I_{0.9}^{\mathrm{MT}}, N_C}$. **b** $\mathrm{FMHMT}_{T_{\mathrm{L,K}}, I_{0.9}^{\mathrm{MT}}, N_C}$. **c** $\mathrm{FMHMT}_{T_{0.9}^{\mathrm{MT}}, I_{0.9}^{\mathrm{MT}}, N_C}$. **d** $\mathrm{FMHMT}_{T_M, I_{0.9}^{\mathrm{MT}}, N_C}$. **e** $\mathrm{FMHMT}_{T_{nM}, I_{0.9}^{\mathrm{MT}}, N_C}$. **f** $\mathrm{FMHMT}_{T_P, I_{0.9}^{\mathrm{MT}}, N_C}$

This is true, but this fact can be easily avoided by surrounding the structuring element with a white picture frame.

A similar experiment is shown in Figs. 13.18 and 13.19, but in this case using the Mayor-Torrens t-norm with parameter value $\lambda = 0.9$. As we can see in Fig. 13.18, no configuration has detected the diamond and the ellipse, and also none of them differentiate between these figures and the black background. Note that, we can see better also the differences between the results obtained by each configuration. Moreover, in Fig. 13.19 we see that now the only configuration that does not detect the square with 119 grey level is $\{T_D, I_{0.9}^{MT}, N_C\}$. In addition, note the similarity between the results obtained by $T_{0.9}^{MT}$ and T_{LK}. This is due to the fact that $T_{0.9}^{MT}$ and $I_{0.9}^{MT}$ are very close to T_{LK} and I_{LK} respectively, and consequently, the obtained results must be close.

13.4.4 Blood Vessel Segmentation in Retinal Images

Eye-fundus images contain information extremely valuable from a medical point of view. For instance, diabetic retinopathy and glaucoma are conditions that can be detected by inspecting them. In this last subsection, we will describe the appearance of blood vessels in eye-fundus photographs and leverage it to design an automatic vessel segmentation algorithm based on Fuzzy Mathematical Morphology and Mayor-Torrens t-norms.

Introduction to Retinal Vessel Segmentation

Eye-fundus photographs are images of the interior surface of the eye taken through the cornea with a special device—a fundus camera. The images typically capture the retina (that is, the photosensitive tissue), the optic disc (a bright circle which corresponds to the termination of the optic nerve) and the macula (a darker area with high-density of cone cells that is responsible for high-resolution vision). Some examples of eye-fundus images are shown in Fig. 13.20.

Computer-based detection of medical conditions from eye-fundus images is a hard task that can be eased by performing a solid first step: detecting the tree-like vessel structure. For instance, diabetic retinopathy is a condition with an increasing

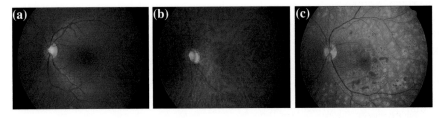

Fig. 13.20 Eye-fundus samples taken from the HRF database [6]. **a** No affection. **b** Glaucoma. **c** Diabetic retinopathy

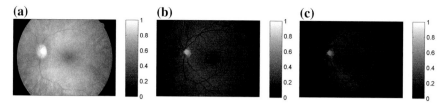

Fig. 13.21 Three channels of the RGB eye-fundus image depicted in Fig. 13.20a—12th image of the HRF database [6]. **a** Red channel. **b** Green channel. **c** Blue channel

impact worldwide, characterized by damaging the retina due to diabetes. Appropriately detecting such damage, besides indicating the presence of diabetes, is essential towards applying an effective treatment, which could greatly reduce the risk of severe vision loss [10]. This condition is characterized by a wide range of indicators depending on its severity. Instances of them are microaneurysms, intraretinal and preretinal hemorrhages, venous beading (indicated by the irregular grow and shrink of their width) and neovascularization (formation of a new microvasculature), among others [38]. To automatically detect these symptoms, the blood vessel segmentation helps to differentiate background noise and artifacts from veins and arteries. Glaucoma is another affection whose detection can leverage a previous vessel segmentation. Besides typically rising the intraocular pressure, the optic disc region suffers changes visually noticeable. Some vessels are located in this area and, since their location and diameter are minimally affected by glaucoma, segmenting them out leaves the area of interest prepared to extract the desired information [4].

The vessel structure has a uniform appearance across different images and pathologies, as can be observed in Fig. 13.20. In this study, we focus on color photography eye-fundus images—with no contrast agent. Figure 13.21 shows the three channels of an image which are represented as greyscale intensities, the setting in which we can apply the aforementioned morphological operators. The red channel is oversaturated and the blue channel is undersaturated, leaving only the green channel with a high contrast, so that it conveys more information. In it, vessels can be geometrically characterized as curvilinear objects—relatively thin and relatively long—whose width varies smoothly and which are connected in a tree-like structure, with bifurcations and eventual crossings. As a photometric feature, the green-channel intensity of vessels is darker than its surrounding background, although their absolute values vary greatly since eye-fundus images present a remarkable non-uniform illumination. The ridge of some vessels, specially arteries, might appear slightly brighter than its limits, an effect called central vessel reflexion. Besides non-uniform illumination, they present background noise—which has been modelized as Gaussian, Uniform, Rayleigh or Laplacian noise [23]. Lastly, they may present a wide variety of objects in addition to the vessel structure, the optic disc and the macula: microaneurysms (small red points), small and big hemorrhages (irregular and connected red splatters) or hard exudates (yellow flecks) are a few examples of them.

Our proposed approach leverages the features that characterize a vessel as such. We use the green channel with a preprocessing procedure to further enhance contrast

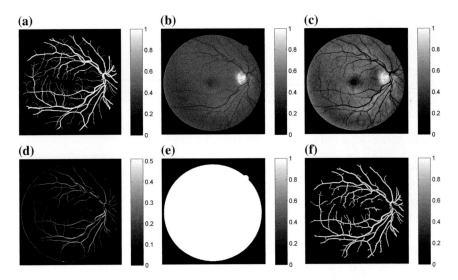

Fig. 13.22 b–f represent the step-by-step behaviour of our algorithm, being the last one the final result. **a** Shows the ground truth, which we aim to mimic. This depiction uses the 2nd image of the DRIVE database [36]. **a** Ground truth. **b** Green channel. **c** Preprocessing. **d** Top Hat transform. **e** Border estimation. **f** Hysteresis

and reduce background noise, followed by the dual top-hat transform with a Mayor-Torrens t-norm, which retains thin and locally dark objects. After it, some post-processing extracts the desired regions. The step-by-step application of the method is depicted in Fig. 13.22.

Since the dual top-hat is the central operation towards distinguishing vessels, we refer the reader to its description and behaviour, presented in Sect. 13.4.3, to provide a deeper understanding beyond its mathematical definition.

Vessel Segmentation Method

All the steps of our automatic segmentation method are shown in Fig. 13.22b–f, whereas Fig. 13.22a contains a vessel segmentation provided by a human expert, which we aim to imitate. Although the central idea is the enhancement by means of the dual top-hat transform, other steps are needed in order to homogenize eye-fundus images and refine the results.

The steps are detailed as follows:

1. In Fig. 13.22b, the green channel is selected out of the RGB eye-fundus image, since it is the one with better contrast, as shown in Fig. 13.21.
2. In Fig. 13.22c, the preprocessing provides robustness and a higher contrast. It consists on (i) a contrast-limited adaptive histogram equalization to enhance local differences in the image without excessive noise amplification and (ii) a Gaussian-shaped smoothing filtering to further remove background noise.

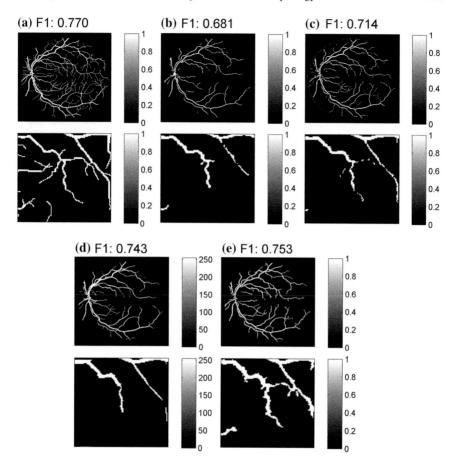

Fig. 13.23 Comparison of different algorithms with the 9th image of the DRIVE database [35], along with their F_1-score with respect to a provided ground truth. The lower row shows a detail of the first image containing subtle vessels, corresponding to $(x, y) \in [200, 300]^2$. **a** shows the segmentation of a second human expert, whereas the other segmentations are computer generated. **a** 2nd GT. **b** Zena. **c** Staal. **d** Soares. **e** Our method

3. The dual top-hat is then computed in Fig. 13.22d. It will extract the areas such that (i) the structuring element does not fit in, and (ii) are darker than its neighbours. This description models, precisely, the vessels contained in the previous step, besides some noise and small artifacts.

4. The black outer background, in which the top-hat transformation creates false positives, is removed in Fig. 13.22e. Firstly, the luminance is computed from the original RGB eye-fundus image. The background is then estimated as a connected component with very low luminance values.

5. In Fig. 13.22f, A hysteresis procedure is applied to extract connected structures. Two thresholds are set, $0 \leq t_1 \leq t_2 \leq 1$. The hysteresis selects the 4-connected components formed by pixels greater than t_1 which, in addition, have at least one pixel greater than t_2.

Results

A number of algorithms have been proposed to perform the same task, either using the same techniques or unrelated ones. Also using mathematical morphology, Zana and Klein [39] presented a method in which a sum of greyscale-morphology top-hats enhance vessels, to later label each pixel as either "positive" or "negative" using the Laplacian operator. Staal et al. [36] presented a method that detected ridges with a gradient-like vector field, followed by a region growing algorithm. Soares et al. [35] extract several pixel-wise features with different Gabor wavelets with which they train a classifier. In Fig. 13.23 we find a comparison between these methods and our own one.

The implementation details used to execute our method are the following: the histogram equalization is computed according to [41], and the subsequently filtering uses a 3×3 Gaussian structuring element; the dual top-hat uses a Mayor-Torrens t-norm with $\lambda = 0.5$ and a 15×15 Gaussian-shaped structuring element, which can be found in Fig. 13.12; lastly, the hysteresis uses $t_1 = 0.125$ and $t_2 = 0.32$, selected by experimentation.

The results provided by our automatic segmentation method successfully mimic those given by experts as manual annotations. They are mainly curvilinear and connected structures which are located where vessels can be appreciated in their corresponding RGB eye-fundus image. Large vessels are almost always captured, whereas the smaller, subtle ones have a lower detection rate. We remark that this behaviour is also observed between different expert segmentations. As we have showed, Fuzzy Mathematical Morphology is a paradigm that can be leveraged to design and easily implement state-of-the-art methodologies, such as the presented vessel segmentation procedure.

13.5 Conclusions

The possibility of considering more t-norms than the continuous nilpotent Archimedean ones in the fuzzy mathematical operators derived from t-norms and fuzzy implications opens an interesting line of research, which could improve substantially the results obtained by this theory in all the fields where it is applied. In this work we have considered the Mayor-Torrens family of t-norms and first, we have studied which desirable algebraical properties are satisfied by the corresponding fuzzy morphological operators. We have proved that they almost satisfy all the properties and more importantly, they satisfy the required properties to be applied to the main applications. Namely, these operators can be applied to edge detection, noise removal, pattern and shape recognition and retinal vessel segmentation.

After that, we have analysed the behaviour of these operators for each of these applications and we have compared the results with the configurations which were used in the original papers. In all cases, Mayor-Torrens t-norms obtain notable and competitive results, specially in edge detection where it outperforms heavily the up to now best configuration.

To sum up, many of the original papers dealing with applications of the fuzzy mathematical morphology based on t-norms should revisit their results by considering more t-norms, and specially, families of t-norms, into their algorithms.

Acknowledgments This project was partially supported by the Spanish project TIN 2013-42795-P. P. Bibiloni also benefited from a fellowship of the *Conselleria d'Educaci, Cultura i Universitats* of the *Govern de les Illes Balears* under an operational program co-financed by the European Social Fund.

References

1. M. Baczyński, B. Jayaram, *Fuzzy Implications*, vol. 231 (Berlin Heidelberg, Studies in Fuzziness and Soft Computing (Springer, 2008)
2. M. Baczyński, B. Jayaram, S. Massanet, J. Torrens, Fuzzy implications: past, present, and future, in *Springer Handbook of Computational Intelligence*, ed. by J. Kacprzyk, W. Pedrycz (Springer, Berlin Heidelberg, 2015), pp. 183–202
3. I. Bloch, H. Maître, Fuzzy mathematical morphologies: a comparative study. Pattern Recognit. **28**, 1341–1387 (1995)
4. R. Bock, J. Meier, L.G. Nyúl, J. Hornegger, G. Michelson, Glaucoma risk index: automated glaucoma detection from color fundus images. Med. Image Anal. **14**(3), 471–481 (2010)
5. K. Bowyer, C. Kranenburg, S. Dougherty, Edge detector evaluation using empirical ROC curves, in *IEEE Conference on Computer Vision and Pattern Recognition (CVPR '99)*, vol. 1, pp. 354–359 (1999)
6. A. Budai, J. Odstricilik, R. Kollar, J. Jan, T. Kubena, G. Michelson, A public database for the evaluation of fundus image segmentation algorithms, in *Proceedings of The Association of Research in Vision and Ophthalmology (ARVO) Annual Meeting, Vancouver, Canada*, pp. 1345–1345 (2011)
7. J. Canny, A computational approach to edge detection. IEEE Trans. Pattern Anal. Mach. Intell. **8**(6), 679–698 (1986)
8. B. De Baets, Fuzzy morphology: a logical approach, in *Uncertainty Analysis in Engineering and Science: Fuzzy Logic, Statistics, and Neural Network Approach*, ed. by B.M. Ayyub, M.M. Gupta (Kluwer Academic Publishers, Norwell, 1997), pp. 53–68
9. B. De Baets, Generalized idempotence in fuzzy mathematical morphology, in Fuzzy Techniques in Image Processing, vol. 52. Studies in Fuzziness and Soft Computing, E.E. Kerre, M. Nachtegael Chap. 2. (Physica, New York, 2000), pp. 58–75
10. Diabetic Retinopathy Study Research Group and others, Photocoagulation treatment of proliferative diabetic retinopathy: clinical application of Diabetic Retinopathy Study (DRS) findings, DRS Report Number 8. Ophthalmology **88**(7), 583–600 (1981)
11. J. Fodor, M. Roubens, *Fuzzy Preference Modelling and Multicriteria Decision Support*. Knowledge Engineering and Problem Solving (Kluwer Academic Publishers, Dordrecht, 1994) (System Theory)
12. M. González, D. Ruiz-Aguilera, J. Torrens, Algebraic properties of fuzzy morphological operators based on uninorms, in *Artificial Intelligence Research and Development*. Frontiers in Artificial Intelligence and Applications, vol. 100 (IOS Press, Amsterdam, 2003), pp. 27–38

13. M. González-Hidalgo, S. Massanet, Closing and opening based on discrete t-norms. Applications to natural image analysis, in *Proceedings of the 7th Conference of the European Society for Fuzzy Logic and Technology, EUSFLAT 2011, Aix-Les-Bains, France*, 18–22 July 2011, pp 358–365 (2011)
14. M. González-Hidalgo, S. Massanet, A fuzzy mathematical morphology based on discrete t-norms: fundamentals and applications to image processing. Soft Comput. **18**(11), 2297–2311 (2014)
15. M. González-Hidalgo, S. Massanet, A. Mir, D. Ruiz-Aguilera, A fuzzy filter for high-density salt and pepper noise removal, in *Advances in Artificial Intelligence, vol. 8109*, Lecture Notes in Computer Science, ed. by C. Bielza, et al. (Springer, Berlin, 2013), pp. 70–79
16. M. González-Hidalgo, S. Massanet, A. Mir, D. Ruiz-Aguilera, High-density impulse noise removal using fuzzy mathematical morphology, in *Proceedings of the 8th Conference of the European Society of Fuzzy Logic and Technology Conference (EUSFLAT 2013)*, ed. by G. Pasi, J. Montero, D. Ciucci (Atlantis Press, Milano, Italy, 2013), pp. 728–735
17. M. González-Hidalgo, S. Massanet, A. Mir, D. Ruiz-Aguilera, On the choice of the pair conjunction-implication into the fuzzy morphological edge detector. IEEE Trans. Fuzzy Syst. **23**(4), 872–884 (2015)
18. M. González-Hidalgo, A. Mir-Torres, D. Ruiz-Aguilera, J. Torrens, Image analysis applications of morphological operators based on uninorms, in *Proceedings of the IFSA-EUSFLAT 2009 Conference, Lisbon, Portugal*, pp. 630–635 (2009)
19. M. González-Hidalgo, S. Massanet, A. Mir, D. Ruiz-Aguilera, A fuzzy morphological hit-or-miss transform for grey-level images: a new approach. Fuzzy Sets Syst. 286, 30-65 (2016)
20. E. Kerre, M. Nachtegael, *Fuzzy Techniques in Image Processing*, vol. 52 (Studies in Fuzziness and Soft Computing (Springer, New York, 2000)
21. E. Klement, R. Mesiar, E. Pap, *Triangular Norms* (Kluwer Academic Publishers, London, 2000)
22. P.D. Kovesi, MATLAB and Octave functions for computer vision and image processing. Centre for Exploration Targeting, School of Earth and Environment, The University of Western Australia. Retrieved from: http://www.csse.uwa.edu.au/_pk/research/matlabfns/ in 1994
23. D. Lesage, E.D. Angelini, I. Bloch, G. Funka-Lea, A review of 3D vessel lumen segmentation techniques: models, features and extraction schemes. Med. Image Anal. **13**(6), 819–845 (2009)
24. C. Lopez-Molina, B. De Baets, H. Bustince, Quantitative error measures for edge detection. Pattern Recognit. **46**(4), 1125–1139 (2013)
25. M. Mas, M. Monserrat, J. Torrens, E. Trillas, A survey on fuzzy implication functions. IEEE Trans. Fuzzy Syst. **15**(6), 1107–1121 (2007)
26. G. Mayor, J. Torrens, On a family of t-norms. Fuzzy Sets Syst. **41**, 161–166 (1991)
27. R. Medina-Carnicer, R. Muoz-Salinas, E. Yeguas-Bolivar, L. Diaz-Mas, A novel method to look for the hysteresis thresholds for the Canny edge detector. Pattern Recognit. **44**(6), 1201–1211 (2011)
28. M. Nachtegael, E. Kerre, Classical and fuzzy approaches towards mathematical morphology, in *Fuzzy Techniques in Image Processing*, vol. 52. E.E. Kerre, M. Nachtegael. Studies in Fuzziness and Soft Computing, Chap. 1 (Physica, New York, 2000), pp. 3–57
29. N. Otsu, A threshold selection method from gray-level histograms. IEEE Trans. Syst. Man Cybern. **9**, 62–66 (1979)
30. G. Papari, N. Petkov, Edge and line oriented contour detection: state of the art. Image Vis. Comput. **29**(2–3), 79–103 (2011)
31. W.K. Pratt, *Digital Image Processing*, 4th edn. (Wiley-Interscience, 2007)
32. C. Rijsbergen, *Information Retrieval* (Butterworths, 1979)
33. S. Schulte, V. De Witte, M. Nachtegael, D. Van der Weken, E. Kerre, Fuzzy random impulse noise reduction method. Fuzzy Sets Syst. **158**(3), 270–283 (2007)
34. J. Serra, *Image Analysis and Mathematical Morphology*, vols. 1, 2 (Academic Press, London, 1982)
35. J.V. Soares, J.J. Leandro, R.M. Cesar Jr., H.F. Jelinek, M.J. Cree, Retinal vessel segmentation using the 2-D Gabor wavelet and supervised classification. IEEE Trans. Med. Imaging **25**(9), 1214–1222 (2006)

36. J. Staal, M.D. Abràmoff, M. Niemeijer, M. Viergever, B. Van Ginneken et al., Ridge-based vessel segmentation in color images of the retina. IEEE Trans. Med. Imaging **23**(4), 501–509 (2004)
37. Z. Wang, A.C. Bovik, H.R. Sheikh, E.P. Simoncelli, Image quality assessment: from error visibility to structural similarity. IEEE Trans. Image Process. **13**(4), 600–612 (2004)
38. C. Wilkinson, F.L. Ferris, R.E. Klein, P.P. Lee, C.D. Agardh, M. Davis, D. Dills, A. Kampik, R. Pararajasegaram, J.T. Verdaguer et al., Proposed international clinical diabetic retinopathy and diabetic macular edema disease severity scales. Ophthalmology **110**(9), 1677–1682 (2003)
39. F. Zana, J.-C. Klein, Segmentation of vessel-like patterns using mathematical morphology and curvature evaluation. IEEE Trans. Image Process. **10**(7), 1010–1019 (2001)
40. D. Ze-Feng, Y. Zhou-Ping, X. You-Lun, High probability impulse noise-removing algorithm based on mathematical morphology. IEEE Signal Process. Lett. **14**(1), 31–34 (2007)
41. K. Zuiderveld, Contrast limited adaptive histogram equalization, in *Graphics Gems IV* (Academic Press Professional Inc, 1994), pp. 474–485

Chapter 14
A Short Dialogue Concerning "What Is" and "What Is Not" with Imprecise Words

Enric Trillas

Abstract What follows is a virtual conversation between two imagined characters, Karl and Carla. They try to debate on how, in fuzzy set algebras, what is not covered under a linguistic label should be represented; that is, and mainly, on both the negation and the opposites of a predicate, and on which fuzzy and crisp expressions of not covered by can be obtained.

14.1 Dialogue

Karl. With crisp sets, and perhaps coming from the strong simplification of natural language mainly forced when only precise terms are tried to be used, only those objects that are in a subset, and those that are not, are taken into account. But in language things are somewhat different since, for instance, negation and antonym are not coincidental. The negation of a word is not a word; for instance, the word "young" is in the dictionary, but "not young" is not, and what it is in it is the word "old", the opposite of young. Why are fuzzy sets algebras built up by only considering the negation?

  Carla. Well, antonyms are also considered and are very important in fuzzy logic, since linguistic variables are constructed by departing from the predicate (giving raise and naming the concept they represent), and its antonym. But it is true that the algebras of fuzzy sets are always defined by only taking into account ways of representing the linguistics and, or, and not, and that modifiers, quantifiers and opposites are left aside. Why? Let me risk at advancing a, perhaps naïve, hypothesis for it. Although Zadeh [1] did introduce fuzzy sets for mathematically representing imprecise, or gradable predicates, non admitting to be specified by a (crisp) subset of the universe in which they are applied to, what was back to his developments, and also to those researchers initially following him, was the naïve theory of sets where intersection, union, and complement is all that is needed. In addition, they were looking at multiple–valued logic as the logic on which fuzzy sets can be based on, instead of the classical logic

E. Trillas (✉)
European Centre for Soft Computing Mieres (Asturias), Mieres, Spain
e-mail: enric.trillas@softcomputing.es

behind crisp, or classical, sets. And it should be remembered that multiple-valued logics are just defined through tables for just the "and", the "or", and the "not", and that the differences among them just lie in these tables. It took some time to realize that not only imprecise words but natural language is the goal and scope of fuzzy logic; that the ground for fuzzy sets is natural language, like matter and energy are the ground of physics, and life the ground of biology. This is what marks a separation between fuzzy logic and multiple-valued logic; the second is a mathematical and logical theory, and the first is closer to an empirical science. By joking a bit, it can be said "Ill teach you differences, by following William Shakespeare's King Lear".

Karl. And also, and in Hillary Putnam's words, that "Logic is as empirical as Geometry". This, even if your words remember me that Russell wrote, in his early book with A.N. Whitehead, "Principia mathematica", that mathematics and logic are merely concerned with "the correct use of a certain small number of words" [2]. Hence, since fuzzy logic is concerned with much more words, and in more complex contexts, my question cannot be seen out of focus.

Carla. No, indeed, it can't. Even more, it is an interesting question that could conduct to challenge what is currently intended to be an algebra of fuzzy sets.

Karl. Possibly, but my current interest just lies in debating on the idea of "complement", that is, on how to represent all that is not covered by a predicate [3]. Namely, given a use of $B = $ big in the universe $X = [0, 10]$, for instance, and once it is specified in X by a fuzzy set m_B, how can the numbers that do not lie under the qualification by B be specified? Let me firstly say that I accept without any doubt that the numbers that are $B' = $ not big can be specified by a membership function $m_{B'} = N_B \circ m_B$, and those that can be called $B^a = $ small, by another $m_{B^a} = m_B \circ s_B$, with N_B and s_B, respectively, a suitable negation function in $[0, 1]$, and s_B a suitable symmetry in X. These two functions should verify, of course, the consistency inequality $m_{B^a} \leq m'_B$, coming from examples like "If the bottle is full, it is not empty" where the predicate is empty and its antonym is "full" [4]. Contrarily to the non-uniqueness of the opposites, there is just one linguistic negation whose membership function can be seen as an upper-limit of those of the antonyms. The opposite affirms, but the negation denies.

Carla. Then your worries lie, if I understand you, on finding crisp sets to which the idea of not covered could be applied. Is not it?

Karl. Yes, it is, but without forgetting the possibility of finding new fuzzy sets for it. If what can be qualified as fuzzy is seen as a matter of degree, the practical use of fuzzy sets is actually a matter of design depending upon the context where the linguistic labels currently apply to, the purpose for its use, as well as the mathematical and computational armamentarium available for its representation. The genetic roots of fuzzy sets lie in the linguistic phenomenon of generating collectives named by a precise or imprecise predicate, like the origin of thermodynamics lies in heat transmission; fuzzy sets are situation-dependent. For instance, the predicate "young" generates in language a collective like it is that of "young Londoners", that we can know by just specifying a membership function among the universe of Londoners; fuzzy sets, or their membership functions, are but states of the, well anchored in language, linguistic collectives.

Carla. Ok. Well, there are some open possibilities to attack what you asks for, like it is, for instance, that offered by the set of those x in X showing more not P than P, that is, the set $A = \{x \in X; m_P(x) \leq m_{P'}(x)\}$. Another possibility is the set whose x show more an antonym of P than P itself, that is, the set $B = \{x \in X; m_P(x) \leq m_{P^a}(x)\}$. Since it should be always $m_{P^a} \leq m_{P'}$, it follows $B \subseteq A$, and B can be seen as the (crisp) kernel of antonymy, and A as the (crisp) kernel of negation [5].

Karl. An example could be in order with the predicate $B = $ big, in the universe of discourse $X = [0, 1]$, supposing $m_B(x) = x$ is the corresponding membership function. Then, if $B' = $ not big, and $B^a = $ small, provided $m_{B^a}(x) = m_B(s(x)) = s(x) \leq m_{B'}(x) = N(m_B(x)) = N(x)$, it can be taken $N = 1 - id$, and $s(x) = \frac{1-x}{1+x}$. Hence,

- $A = \{x \in [0, 1]; x \leq 1 - x\} = [0, 1/2]$, is the kernel of negation,
- $B = \{x \in [0, 1]; x \leq \frac{1-x}{1+x}\} = [0, \sqrt{2} - 1]$, is the kernel of antonymy,

with, of course, $B \subseteq A$. The elements in the kernel of antonymy, $[0, \sqrt{2} - 1]$, are more small than big, and, for what concerns its character as big, its degree varies from 0 to $\sqrt{2} - 1$. Relatively to small, the degrees vary between $\sqrt{2} - 1$ and 1, and the same with respect to not big. The elements in the kernel of negation are those that are more not big than big, and the smallest separation point is, for the taken negation and the symmetry, $\sqrt{2} - 1$: The points that can be truly considered big are those greater than $\sqrt{2} - 1$ and, consequently, in the hypothesis of being necessary to specify which numbers can be properly called big, no doubt that under these N and s, they are those in the interval $(\sqrt{2} - 1, 1]$. This facilitates a way for the specification of imprecise terms; something that actually changes their meaning and that, by this single reason, should be carefully done.

Carla. Since the antonym is not unique, and from the inequality $m_{P^a} \leq m_{P'}$, it follows that the function $\sup\{m_{P^a}; \text{ for all } P^a\} = k_P \leq m_{P'}$, can be seen as a fuzzy kernel of antonymy that can, or cannot, coincide with $m_{P'}$. For instance, with the former example, and since the symmetries s in $[0, 1]$ can be identified with the strong negations N whose supremum is the discontinuous negation $N^*(x) = 1$ if $0 \leq x < 1$, and $N^*(1) = 0$, it follows $k_{\text{big}} = N^*$, specifying the subset $[0, 1)$ of the unit interval. Hence, in this case, the kernel of antonymy is $[0, 1)$, reflecting that 1 is the only element that always will be qualified as big.

Karl. Can it be analogously studied the function $\sup\{m_{P'} = N \circ m_P\}$, for all the negations N consistent with a given symmetrys? If possible, it can be called the fuzzy kernel of negation. For instance, with the former $P = $ big, this fuzzy set is sup N, with $s \leq N$, and hence, is equal to N^*, with which it also specifies the same set $[0, 1)$.

Carla. What happens when P specifies a crisp set **P** in X, that is, the use of P is precise, and $m_P \in \{0, 1\}^X$? The kernel of negation comes from the inequality $m_P(x) \leq 1m_P(x) \Leftrightarrow m_P(x) \leq 1/2$, and, hence, is the set of those x in X such that $m_P(x) = 0$ specifying the complement $\mathbf{P}^c$. Hence, since in the crisp case the kernel of negation reduces to the complement, it seems to be a good candidate for assigning a crisp complement to any fuzzy set.

Karl. When P is precise, and respect to the kernel of antonymy, the situation can be different since, in this case what is fixed is the negation function, but not the symmetry.

Carla. Even more different, since there are many precise predicates for which not a single antonym is in language and for which, sometimes, the antonym is identified with the negation as we did comment. Anyway, there is no reason for adopting this identification as a general rule, since through symmetries it is possible to find antonyms even if to assign them a name can be something, up to some extent, fortuitous. If P is precise and with a membership function m_P, the problem lies in finding symmetries s such that $m_P \circ s \leq 1 - m_P$. Thus, for those x in **P** it should be $m_P(s(x)) = 0$, or $s(x) \in \mathbf{P}^c$, and for those in $\mathbf{P}^c$ it is $m_P(s(x)) \leq 1$ that, were the antonym crisp, could be either $m_P(s(x)) = 0$, or $m_P(s(x)) = 1$. Thus, and in principle, the antonym of a precise predicate can be either precise or imprecise.

Karl. Let me consider a simple example in $X = [0, 10]$, with $P = $ smaller than five , whose membership function is $m_P(x) = 1$ if $0 \leq x < 5$, and $m_P(x) = 0$ if $5 \leq x \leq 10$. Obviously, it is $\mathbf{P} = [0, 5)$, and $\mathbf{P}^c = [5, 10]$. For the symmetries s giving antonyms of P, it should be $m_P(s(x)) \leq 1 - m_P(x)$, implying $m_P(s(x)) = 0$ if x is in **P**, and $m_P(s(x)) \leq 1$ if x is in $\mathbf{P}^c$. Looking in the reverse way, if $s(x) \in [0, 5)$, then $m_P(s(x)) = 1$, and if $s(x) \in [5, 10]$, it follows $m_P(s(x)) = 0$: Hence, the antonym P^a should be, also, a precise predicate whose specification just depends on the chosen symmetry $s : [0, 10] \to [0, 10]$, verifying $s(0) = 10$, and $s(10) = 0$. For instance, with $s(x) = 10 - x$, it is $x \in [0, 5) \Leftrightarrow s(x) \in (5, 10]$, and $x \in [5, 10] \Leftrightarrow s(x) \in [0, 5]$, with $s(5) = 5$, with which it is $\mathbf{P}^a = \mathbf{P}^c$, but with $s(x) = 10\frac{10-x}{10+x}$, with which it is $s(5) = 10/3$, and the antonym is specified by:

- $m_{P^a}(x) = 1 \Leftrightarrow 0 \leq 10\frac{10-x}{10+x} < 5 \Leftrightarrow 10/3 < x \leq 10,$
- $m_{P^a}(x) = 0$, otherwise,

with which is $P^a = (10/3, 10]$ the set specified by the predicate $P^a = $ greater than $10/3$, that is different from $P' = $ not smaller than five.

Carla. Let me observe that, in mathematics, it is not felt the need for considering antonyms; for instance, in reference to the predicate defining the semi-closed interval $[0, 5)$, there are in mathematical language the intervals $[5, 10]$ and $(10/3, 10]$ that are perfectly describable without the need of adding more words. All that is more than a curiosity; it shows that fuzzy logic offers a methodology for obtaining, at each context, antonyms words not existing in language. For instance, there are dictionaries of antonyms where the antonym of "probable" appears as "not probable" instead of the "improbable" appearing in other dictionaries that, nevertheless and like the Webster's, just describes it as "not probable". If not probable, with respect to a probability p, can be represented by $1 - p$, improbable will be notwithstanding represented by a symmetry s such that $p \circ s \leq 1 - p$.

Karl. Yes, but, and for instance, in the case of computing in the real line, the probabilities of "x is P" and "x is not P", by prob (**P**) and prob $(\mathbf{P}^c) = 1- $ prob (**P**), as it happens with the Borel's sets under a Gaussian Normal Law, it also appears how to compute the probability of "x is antonym of P", thanks to prob $(\mathbf{P}^a)$.

Carla. Well, our aim in this short debate just started by looking at what does not properly lie under a linguistic label, and for what in language there is not only the negation, but also the opposites. It is something limiting the usefulness of logic for modeling commonsense reasoning expressed in natural language. In fact, it is really difficult to recognize that, for instance, John is rich without being able to simultaneously recognize that Peter is poor; and, perhaps, to recognize that Laura is not rich without knowing those categories placed between rich and not rich like it is poor. Negation means an exclusion that comprises many un-concrete things, and it does not designate a single concrete thing, it is not a word but two. By its side, the antonym is a word designating something affirmatively more concrete.

Karl. What can be seen by "not covered by P" still deserves more study. This brief dialogue is just a trial for opening the eyes of theoreticians of fuzzy logic towards the true grounds on which it is anchored, and to take into account that fuzzy logic cannot be an isolated mathematical, or logical, subject. Its evolution should be towards an empirical science that, supported on computation and grounded in natural language and commonsense reasoning, can mainly deal with the imprecision and non-random uncertainty of which language and reasoning are, let me say, "infected".

Carla. Yes, but without forgetting that there are still subjects permeating language and of which no mathematical models are known like, for instance, the linguistic phenomenon of ambiguity. Actually, I think that at least without suitable mathematical and computing models for linguistic imprecision, uncertainty, and ambiguity, it will be impossible to undo the so-called knot of Artificial Intelligence. Along the progress of Computing with Words and Perceptions [1], sooner or later it will arrive the time for considering large phrases in themselves. Something that will call for a deep knowledge on the linguistic separation between what is and what is not under a linguistic label, and concerning both how to name it, and which elements are under such un-coverage. Yet, and possibly, the meaning of large phrases is what will allow to determine the meaning of its parts, and not in the reverse, atomic-like form, typical of logic. In this sense, it could be interesting to add that to syntactically capture, with a syntax of fuzzy sets provided by an algebra of them, the semantics of these parts, it can be necessary to consider either the symmetry, or the negation function, at different parts of the universe of discourse since it could be that both the negation and the antonym can change along the discourse. I would say that if, for instance, the universe X can be decomposed in the partition $X = X_1 \bigcup X_2$, then one can consider point-dependent [6] negation functions N_x, and point-dependent symmetries s_x, in such form that for $x \in X_1$ it is taken $N_x = 1 - x$, and for $x \in X^2$, $N_x = \frac{1-x}{1+x}$, and something analogous with the symmetries.

Additionally, let me say how difficult is to imagine an experimental science without including systems of measuring what is involved in its subject; a sentence attributed to Galileo Galilei, "Measure what is measurable, and make measurable what is not so", reflects a deep characteristic for any scientific enterprise that, in the case of imprecision, cannot be necessarily linked to the typical additive law of measures of length, surface, volume, and also of random uncertainty, since there are usually exchanges of information among the different parts that are being considered.

E. Trillas

Karl. For what relates to the word "complement" it is not to be forgot that, in plain language, it refers to adding what can complete something. In fact, language makes two different cuts between "what is under P and "what is not under P', the first by the particle not, and the second by an antonym. Think, for instance with $P = \text{old}$, $P^a = \text{young}$, and $P' = \text{not old}$, where with P^a and P is created the intermediate term "middle aged" as "neither old, nor young", whereas P and P' seem to express a more strong cut. Even if language always tends to find a flexible use of its terms, there are occasions where strong cuts are in order. Nevertheless, there are other situations where a superposition of meanings similar, up to some extent, to the quantum phenomenon of state's superposition, is more convenient for expressing some thoughts. At this respect, may be it should be remembered that fuzzy sets are but the states collectives associated to words can show at each context.

Carla. It seems to me obvious that our dialogue, where very few things are truly new, cannot be conclusive but just tentative. What, perhaps, can be of interest in it, is that some previously hidden hints are leveled up to a first explicit form. Anyway, perhaps and merely, it is nothing else than sugar for the ants.

Acknowledgments This paper is partially supported by the Foundation for the Advancement of Soft Computing (Asturias, Spain).

References

1. L.A. Zadeh, *Computing with Words*. Principal Concepts and Ideas (Springer, New York, 2012)
2. K. Green, *Bertrand Russell, Language and Linguistic Theory* (Continuum Pubs, Easton, 2007)
3. E. Trillas, R. Seising, Turning Around the Ideas of Meaning and Complement (2015) (forthcoming)
4. E. Trillas, C. Moraga, S. Guadarrama, S. Cubillo, E. Castieira, in *Computing with Antonyms*, ed. by M. Nikravesh et altri. Forging New Frontiers. Fuzzy Pioneers I, in the series: Studies in Fuzziness and Soft Computing, vol. 217 (Springer, New York, 2007), pp. 133–154
5. E. Trillas, A.R. de Soto, A reflection on fuzzy complements, in *Proceedings of IFSA-Eusflat Conference-2015*, Gijón, pp. 823–827
6. E. Trillas, A. Pradera, Non-functional fuzzy connectives: the case of negations, in *Proceedings of ESTYLF-2002*, León, pp. 527–532

Printed in the United States
By Bookmasters